D0141639

Martin Aigner
Günter M. Ziegler

Proofs from THE BOOK

Third Edition

Springer
Berlin
Heidelberg
New York
Hong Kong
London
Milan
Paris
Tokyo

Martin Aigner
Günter M. Ziegler

Proofs from THE BOOK

Third Edition

With 250 Figures
Including Illustrations
by Karl H. Hofmann

Martin Aigner
Freie Universität Berlin
Institut für Mathematik II (WE2)
Arnimallee 3
14195 Berlin, Germany
email: aigner@math.fu-berlin.de

Günter M. Ziegler
Technische Universität Berlin
Institut für Mathematik, MA 6-2
Straße des 17. Juni 136
10623 Berlin, Germany
email: ziegler@math.tu-berlin.de

Cataloging-in-Publication Data applied for

A catalog record for this book is available from the Library of Congress

Bibliographic information published by Die Deutsche Bibliothek

Die Deutsche Bibliothek lists this publication in the Deutsche Nationalbibliografie;
detailed bibliographic data is available in the Internet at http://dnb.ddb.de.

Mathematics Subject Classification (2000): 00-01 (General)

ISBN 3-540-40460-0 Springer-Verlag Berlin Heidelberg New York

ISBN 3-540-67865-4 Second edition Springer-Verlag Berlin Heidelberg New York
ISBN 3-540-63698-6 First edition Springer-Verlag Berlin Heidelberg New York

Springer-Verlag Berlin Heidelberg New York
a member of BertelsmannSpringer Science+Business Media GmbH

Printed in Germany

Typeset in LaTeX by the authors
Cover design: de'blik, Berlin

Printed on acid-free paper 46/3142db – 5 4 3 2 1 0

Preface

Paul Erdős

"The Book"

Paul Erdős liked to talk about The Book, in which God maintains the perfect proofs for mathematical theorems, following the dictum of G. H. Hardy that there is no permanent place for ugly mathematics. Erdős also said that you need not believe in God but, as a mathematician, you should believe in The Book. A few years ago, we suggested to him to write up a first (and very modest) approximation to The Book. He was enthusiastic about the idea and, characteristically, went to work immediately, filling page after page with his suggestions. Our book was supposed to appear in March 1998 as a present to Erdős' 85th birthday. With Paul's unfortunate death in the summer of 1996, he is not listed as a co-author. Instead this book is dedicated to his memory.

We have no definition or characterization of what constitutes a proof from The Book: all we offer here is the examples that we have selected, hoping that our readers will share our enthusiasm about brilliant ideas, clever insights and wonderful observations. We also hope that our readers will enjoy this despite the imperfections of our exposition. The selection is to a great extent influenced by Paul Erdős himself. A large number of the topics were suggested by him, and many of the proofs trace directly back to him, or were initiated by his supreme insight in asking the right question or in making the right conjecture. So to a large extent this book reflects the views of Paul Erdős as to what should be considered a proof from The Book.

A limiting factor for our selection of topics was that everything in this book is supposed to be accessible to readers whose backgrounds include only a modest amount of technique from undergraduate mathematics. A little linear algebra, some basic analysis and number theory, and a healthy dollop of elementary concepts and reasonings from discrete mathematics should be sufficient to understand and enjoy everything in this book.

We are extremely grateful to the many people who helped and supported us with this project — among them the students of a seminar where we discussed a preliminary version, to Benno Artmann, Stephan Brandt, Stefan Felsner, Eli Goodman, Torsten Heldmann, and Hans Mielke. We thank Margrit Barrett, Christian Bressler, Ewgenij Gawrilow, Michael Joswig, Elke Pose, and Jörg Rambau for their technical help in composing this book. We are in great debt to Tom Trotter who read the manuscript from first to last page, to Karl H. Hofmann for his wonderful drawings, and most of all to the late great Paul Erdős himself.

Berlin, March 1998 *Martin Aigner · Günter M. Ziegler*

Preface to the Second Edition

The first edition of this book got a wonderful reception. Moreover, we received an unusual number of letters containing comments and corrections, some shortcuts, as well as interesting suggestions for alternative proofs and new topics to treat. (While we are trying to record *perfect* proofs, our exposition isn't.)

The second edition gives us the opportunity to present this new version of our book: It contains three additional chapters, substantial revisions and new proofs in several others, as well as minor amendments and improvements, many of them based on the suggestions we received. It also misses one of the old chapters, about the "problem of the thirteen spheres," whose proof turned out to need details that we couldn't complete in a way that would make it brief and elegant.

Thanks to all the readers who wrote and thus helped us — among them Stephan Brandt, Christian Elsholtz, Jürgen Elstrodt, Daniel Grieser, Roger Heath-Brown, Lee L. Keener, Christian Lebœuf, Hanfried Lenz, Nicolas Puech, John Scholes, Bernulf Weißbach, and *many* others. Thanks again for help and support to Ruth Allewelt and Karl-Friedrich Koch at Springer Heidelberg, to Christoph Eyrich and Torsten Heldmann in Berlin, and to Karl H. Hofmann for some superb new drawings.

Berlin, September 2000 *Martin Aigner · Günter M. Ziegler*

Preface to the Third Edition

We would never have dreamt, when preparing the first edition of this book in 1998, of the great success this project would have, with translations into many languages, enthusiastic responses from so many readers, and so many wonderful suggestions for improvements, additions, and new topics — that could keep us busy for years.

So, this third edition offers two new chapters (on Euler's partition identities, and on card shuffling), three proofs of Euler's series appear in a separate chapter, and there is a number of other improvements, such as the Calkin-Wilf-Newman treatment of "enumerating the rationals." That's it, for now!

We thank everyone who has supported this project during the last five years, and whose input has made a difference for this new edition. This includes David Bevan, Anders Björner, Dietrich Braess, John Cosgrave, Hubert Kalf, Günter Pickert, Alistair Sinclair, and Herb Wilf.

Berlin, July 2003 *Martin Aigner · Günter M. Ziegler*

Table of Contents

Number Theory

"Irrationality and π"

Six proofs of the infinity of primes

Chapter 1

It is only natural that we start these notes with probably the oldest Book Proof, usually attributed to Euclid (*Elements* IX, 20). It shows that the sequence of primes does not end.

■ **Euclid's Proof.** For any finite set $\{p_1, \ldots, p_r\}$ of primes, consider the number $n = p_1 p_2 \cdots p_r + 1$. This n has a prime divisor p. But p is not one of the p_i: otherwise p would be a divisor of n and of the product $p_1 p_2 \cdots p_r$, and thus also of the difference $n - p_1 p_2 \ldots p_r = 1$, which is impossible. So a finite set $\{p_1, \ldots, p_r\}$ cannot be the collection of *all* prime numbers. □

Before we continue let us fix some notation. $\mathbb{N} = \{1, 2, 3, \ldots\}$ is the set of natural numbers, $\mathbb{Z} = \{\ldots, -2, -1, 0, 1, 2, \ldots\}$ the set of integers, and $\mathbb{P} = \{2, 3, 5, 7, \ldots\}$ the set of primes.

In the following, we will exhibit various other proofs (out of a much longer list) which we hope the reader will like as much as we do. Although they use different view-points, the following basic idea is common to all of them: The natural numbers grow beyond all bounds, and every natural number $n \geq 2$ has a prime divisor. These two facts taken together force $\mathbb{P}$ to be infinite. The next proof is due to Christian Goldbach (from a letter to Leonhard Euler 1730), the third proof is apparently folklore, the fourth one is by Euler himself, the fifth proof was proposed by Harry Fürstenberg, while the last proof is due to Paul Erdős.

■ **Second Proof.** Let us first look at the *Fermat numbers* $F_n = 2^{2^n} + 1$ for $n = 0, 1, 2, \ldots$. We will show that any two Fermat numbers are relatively prime; hence there must be infinitely many primes. To this end, we verify the recursion

$$\prod_{k=0}^{n-1} F_k = F_n - 2 \qquad (n \geq 1),$$

from which our assertion follows immediately. Indeed, if m is a divisor of, say, F_k and F_n $(k < n)$, then m divides 2, and hence $m = 1$ or 2. But $m = 2$ is impossible since all Fermat numbers are odd.

To prove the recursion we use induction on n. For $n = 1$ we have $F_0 = 3$ and $F_1 - 2 = 3$. With induction we now conclude

$$\prod_{k=0}^{n} F_k = \Big(\prod_{k=0}^{n-1} F_k\Big) F_n = (F_n - 2) F_n =$$
$$= (2^{2^n} - 1)(2^{2^n} + 1) = 2^{2^{n+1}} - 1 = F_{n+1} - 2. \qquad \square$$

$$\begin{aligned} F_0 &= 3 \\ F_1 &= 5 \\ F_2 &= 17 \\ F_3 &= 257 \\ F_4 &= 65537 \\ F_5 &= 641 \cdot 6700417 \end{aligned}$$

The first few Fermat numbers

Lagrange's Theorem

If G is a finite (multiplicative) group and U is a subgroup, then $|U|$ divides $|G|$.

■ **Proof.** Consider the binary relation

$$a \sim b :\Longleftrightarrow ba^{-1} \in U.$$

It follows from the group axioms that $\sim$ is an equivalence relation. The equivalence class containing an element a is precisely the coset

$$Ua = \{xa : x \in U\}.$$

Since clearly $|Ua| = |U|$, we find that G decomposes into equivalence classes, all of size $|U|$, and hence that $|U|$ divides $|G|$. □

In the special case when U is a cyclic subgroup $\{a, a^2, \ldots, a^m\}$ we find that m (the smallest positive integer such that $a^m = 1$, called the *order* of a) divides the size $|G|$ of the group.

■ **Third Proof.** Suppose $\mathbb{P}$ is finite and p is the largest prime. We consider the so-called *Mersenne number* $2^p - 1$ and show that any prime factor q of $2^p - 1$ is bigger than p, which will yield the desired conclusion. Let q be a prime dividing

$2^p - 1$, so we have $2^p \equiv 1 \pmod{q}$. Since p is prime, this means that the element 2 has order p in the multiplicative group $\mathbb{Z}_q \backslash \{0\}$ of the field $\mathbb{Z}_q$. This group has $q - 1$ elements. By Lagrange's theorem (see the box) we know that the order of every element divides the size of the group, that is, we have $p \mid q - 1$, and hence $p < q$. □

Now let us look at a proof that uses elementary calculus.

■ **Fourth Proof.** Let $\pi(x) := \#\{p \leq x : p \in \mathbb{P}\}$ be the number of primes that are less than or equal to the real number x. We number the primes $\mathbb{P} = \{p_1, p_2, p_3, \ldots\}$ in increasing order. Consider the natural logarithm $\log x$, defined as $\log x = \int_1^x \frac{1}{t} dt$.

Now we compare the area below the graph of $f(t) = \frac{1}{t}$ with an upper step function. (See also the appendix on page 10 for this method.) Thus for $n \leq x < n + 1$ we have

$$\begin{aligned} \log x \quad &\leq \quad 1 + \frac{1}{2} + \frac{1}{3} + \ldots + \frac{1}{n-1} + \frac{1}{n} \\ &\leq \quad \sum \frac{1}{m}, \text{ where the sum extends over all } m \in \mathbb{N} \text{ which have only prime divisors } p \leq x. \end{aligned}$$

Since every such m can be written in a *unique* way as a product of the form $\prod\limits_{p \leq x} p^{k_p}$, we see that the last sum is equal to

$$\prod_{\substack{p \in \mathbb{P} \\ p \leq x}} \Big(\sum_{k \geq 0} \frac{1}{p^k} \Big).$$

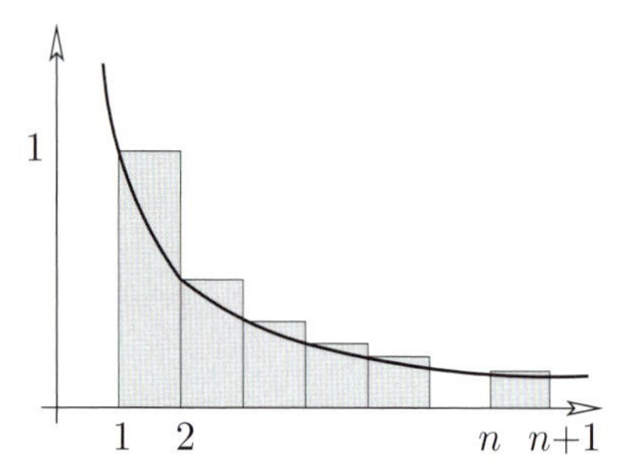

Steps above the function $f(t) = \frac{1}{t}$

The inner sum is a geometric series with ratio $\frac{1}{p}$, hence

$$\log x \;\leq\; \prod_{\substack{p \in \mathbb{P} \\ p \leq x}} \frac{1}{1 - \frac{1}{p}} \;=\; \prod_{\substack{p \in \mathbb{P} \\ p \leq x}} \frac{p}{p-1} \;=\; \prod_{k=1}^{\pi(x)} \frac{p_k}{p_k - 1}.$$

Now clearly $p_k \geq k + 1$, and thus

$$\frac{p_k}{p_k - 1} \;=\; 1 + \frac{1}{p_k - 1} \;\leq\; 1 + \frac{1}{k} \;=\; \frac{k+1}{k},$$

and therefore

$$\log x \;\leq\; \prod_{k=1}^{\pi(x)} \frac{k+1}{k} \;=\; \pi(x) + 1.$$

Everybody knows that $\log x$ is not bounded, so we conclude that $\pi(x)$ is unbounded as well, and so there are infinitely many primes. □

■ **Fifth Proof.** After analysis it's topology now! Consider the following curious topology on the set $\mathbb{Z}$ of integers. For $a, b \in \mathbb{Z}$, $b > 0$, we set

$$N_{a,b} = \{a + nb : n \in \mathbb{Z}\}.$$

Each set $N_{a,b}$ is a two-way infinite arithmetic progression. Now call a set $O \subseteq \mathbb{Z}$ *open* if either O is empty, or if to every $a \in O$ there exists some $b > 0$ with $N_{a,b} \subseteq O$. Clearly, the union of open sets is open again. If O_1, O_2 are open, and $a \in O_1 \cap O_2$ with $N_{a,b_1} \subseteq O_1$ and $N_{a,b_2} \subseteq O_2$, then $a \in N_{a,b_1 b_2} \subseteq O_1 \cap O_2$. So we conclude that any finite intersection of open sets is again open. So, this family of open sets induces a bona fide topology on $\mathbb{Z}$.

Let us note two facts:

(A) Any non-empty open set is infinite.

(B) Any set $N_{a,b}$ is closed as well.

Indeed, the first fact follows from the definition. For the second we observe

$$N_{a,b} = \mathbb{Z} \setminus \bigcup_{i=1}^{b-1} N_{a+i,b},$$

which proves that $N_{a,b}$ is the complement of an open set and hence closed.

"Pitching flat rocks, infinitely"

So far the primes have not yet entered the picture — but here they come. Since any number $n \neq 1, -1$ has a prime divisor p, and hence is contained in $N_{0,p}$, we conclude

$$\mathbb{Z} \setminus \{1, -1\} = \bigcup_{p \in \mathbb{P}} N_{0,p}.$$

Now if $\mathbb{P}$ were finite, then $\bigcup_{p \in \mathbb{P}} N_{0,p}$ would be a finite union of closed sets (by (B)), and hence closed. Consequently, $\{1, -1\}$ would be an open set, in violation of (A). □

■ **Sixth Proof.** Our final proof goes a considerable step further and demonstrates not only that there are infinitely many primes, but also that the series $\sum_{p \in \mathbb{P}} \frac{1}{p}$ diverges. The first proof of this important result was given by Euler (and is interesting in its own right), but our proof, devised by Erdős, is of compelling beauty.

Let $p_1, p_2, p_3, \ldots$ be the sequence of primes in increasing order, and assume that $\sum_{p \in \mathbb{P}} \frac{1}{p}$ converges. Then there must be a natural number k such that $\sum_{i \geq k+1} \frac{1}{p_i} < \frac{1}{2}$. Let us call $p_1, \ldots, p_k$ the *small* primes, and $p_{k+1}, p_{k+2}, \ldots$ the *big* primes. For an arbitrary natural number N we therefore find

$$\sum_{i \geq k+1} \frac{N}{p_i} < \frac{N}{2}. \qquad (1)$$

Let N_b be the number of positive integers $n \le N$ which are divisible by at least one big prime, and N_s the number of positive integers $n \le N$ which have only small prime divisors. We are going to show that for a suitable N

$$N_b + N_s \;<\; N,$$

which will be our desired contradiction, since by definition $N_b + N_s$ would have to be equal to N.

To estimate N_b note that $\lfloor \frac{N}{p_i} \rfloor$ counts the positive integers $n \le N$ which are multiples of p_i. Hence by (1) we obtain

$$N_b \;\le\; \sum_{i \ge k+1} \left\lfloor \frac{N}{p_i} \right\rfloor \;<\; \frac{N}{2}. \tag{2}$$

Let us now look at N_s. We write every $n \le N$ which has only small prime divisors in the form $n = a_n b_n^2$, where a_n is the square-free part. Every a_n is thus a product of *different* small primes, and we conclude that there are precisely 2^k different square-free parts. Furthermore, as $b_n \le \sqrt{n} \le \sqrt{N}$, we find that there are at most $\sqrt{N}$ different square parts, and so

$$N_s \;\le\; 2^k \sqrt{N}.$$

Since (2) holds for *any* N, it remains to find a number N with $2^k \sqrt{N} \le \frac{N}{2}$ or $2^{k+1} \le \sqrt{N}$, and for this $N = 2^{2k+2}$ will do. $\square$

References

[1] B. ARTMANN: *Euclid — The Creation of Mathematics,* Springer-Verlag, New York 1999.

[2] P. ERDŐS: *Über die Reihe* $\sum \frac{1}{p}$, Mathematica, Zutphen B **7** (1938), 1-2.

[3] L. EULER: *Introductio in Analysin Infinitorum,* Tomus Primus, Lausanne 1748; Opera Omnia, Ser. 1, Vol. 8.

[4] H. FÜRSTENBERG: *On the infinitude of primes,* Amer. Math. Monthly **62** (1955), 353.

Bertrand's postulate

Chapter 2

We have seen that the sequence of prime numbers $2, 3, 5, 7, \ldots$ is infinite. To see that the size of its gaps is not bounded, let $N := 2 \cdot 3 \cdot 5 \cdot \ldots \cdot p$ denote the product of all prime numbers that are smaller than $k + 2$, and note that none of the k numbers

$$N + 2, N + 3, N + 4, \ldots, N + k, N + (k + 1)$$

is prime, since for $2 \leq i \leq k + 1$ we know that i has a prime factor that is smaller than $k + 2$, and this factor also divides N, and hence also $N + i$. With this recipe, we find, for example, for $k = 10$ that none of the ten numbers

$$2312, 2313, 2314, \ldots, 2321$$

is prime.

Joseph Bertrand

But there are also upper bounds for the gaps in the sequence of prime numbers. A famous bound states that "the gap to the next prime cannot be larger than the number we start our search at." This is known as Bertrand's postulate, since it was conjectured and verified empirically for $n < 3\,000\,000$ by Joseph Bertrand. It was first proved for all n by Pafnuty Chebyshev in 1850. A much simpler proof was given by the Indian genius Ramanujan. Our Book Proof is by Paul Erdős: it is taken from Erdős' first published paper, which appeared in 1932, when Erdős was 19.

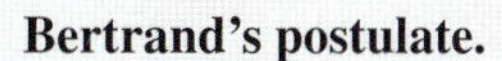

Bertrand's postulate.
For every $n \geq 1$, there is some prime number p with $n < p \leq 2n$.

Beweis eines Satzes von Tschebyschef.

Von P. Erdős in Budapest.

Für den zuerst von Tschebyschef bewiesenen Satz, laut dessen es zwischen einer natürlichen Zahl und ihrer zweifachen stets wenigstens eine Primzahl gibt, liegen in der Literatur mehrere Beweise vor. Als einfachsten kann man ohne Zweifel den Beweis von Ramanujan[1]) bezeichnen. In seinem Werk *Vorlesungen über Zahlentheorie* (Leipzig, 1927), Band I, S. 66—68 gibt Herr Landau einen besonders einfachen Beweis für einen Satz über die Anzahl der Primzahlen unter einer gegebenen Grenze, aus welchem unmittelbar folgt, daß für ein geeignetes q zwischen einer natürlichen Zahl und ihrer q-fachen stets eine Primzahl liegt. Für die augenblicklichen Zwecken des Herrn Landau kommt es nicht auf die numerische Bestimmung der im Beweis auftretenden Konstanten an; man überzeugt sich aber durch eine numerische Verfolgung des Beweises leicht, daß q jedenfalls größer als 2 ausfällt.

In den folgenden Zeilen werde ich zeigen, daß man durch eine Verschärfung der dem Landauschen Beweis zugrunde liegenden Ideen zu einem Beweis des oben erwähnten Tschebyschefschen Satzes gelangen kann, der — wie mir scheint — an Einfachkeit nicht hinter dem Ramanujanschen Beweis steht. Griechische Buchstaben sollen im Folgenden durchwegs positive, lateinische Buchstaben natürliche Zahlen bezeichnen; die Bezeichnung p ist für Primzahlen vorbehalten.

1. Der Binomialkoeffizient

$$\binom{2a}{a} = \frac{(2a)!}{(a!)^2}$$

[1]) Sr. Ramanujan, A Proof of Bertrand's Postulate, *Journal of the Indian Mathematical Society*, 11 (1919), S. 181—182 — *Collected Papers of* Srinivasa Ramanujan (Cambridge, 1927), S. 208—209.

■ **Proof.** We will estimate the size of the binomial coefficient $\binom{2n}{n}$ carefully enough to see that if it didn't have any prime factors in the range $n < p \leq 2n$, then it would be "too small." Our argument is in five steps.

(1) We first prove Bertrand's postulate for $n < 4000$. For this one does not need to check 4000 cases: it suffices (this is "Landau's trick") to check that

$$2, 3, 5, 7, 13, 23, 43, 83, 163, 317, 631, 1259, 2503, 4001$$

is a sequence of prime numbers, where each is smaller than twice the previous one. Hence every interval $\{y : n < y \leq 2n\}$, with $n \leq 4000$, contains one of these 14 primes.

(2) Next we prove that

$$\prod_{p \le x} p \;\le\; 4^{x-1} \qquad \text{for all real } x \ge 2, \tag{1}$$

where our notation — here and in the following — is meant to imply that the product is taken over all *prime* numbers $p \le x$. The proof that we present for this fact uses induction on the number of these primes. It is not from Erdős' original paper, but it is also due to Erdős (see the margin), and it is a true Book Proof. First we note that if q is the largest prime with $q \le x$, then

$$\prod_{p \le x} p = \prod_{p \le q} p \qquad \text{and} \qquad 4^{q-1} \le 4^{x-1}.$$

Thus it suffices to check (1) for the case where $x = q$ is a prime number. For $q = 2$ we get "$2 \le 4$," so we proceed to consider odd primes $q = 2m+1$. (Here we may assume, by induction, that (1) is valid for all integers x in the set $\{2, 3, \ldots, 2m\}$.) For $q = 2m+1$ we split the product and compute

$$\prod_{p \le 2m+1} p = \prod_{p \le m+1} p \cdot \prod_{m+1 < p \le 2m+1} p \;\le\; 4^m \binom{2m+1}{m} \;\le\; 4^m 2^{2m} \;=\; 4^{2m}.$$

All the pieces of this "one-line computation" are easy to see. In fact,

$$\prod_{p \le m+1} p \;\le\; 4^m$$

holds by induction. The inequality

$$\prod_{m+1 < p \le 2m+1} p \;\le\; \binom{2m+1}{m}$$

follows from the observation that $\binom{2m+1}{m} = \frac{(2m+1)!}{m!(m+1)!}$ is an integer, where the primes that we consider all are factors of the numerator $(2m+1)!$, but not of the denominator $m!(m+1)!$. Finally

$$\binom{2m+1}{m} \;\le\; 2^{2m}$$

holds since

$$\binom{2m+1}{m} \quad \text{and} \quad \binom{2m+1}{m+1}$$

are two (equal!) summands that appear in

$$\sum_{k=0}^{2m+1} \binom{2m+1}{k} = 2^{2m+1}.$$

(3) From Legendre's theorem (see the box) we get that $\binom{2n}{n} = \frac{(2n)!}{n!n!}$ contains the prime factor p exactly

$$\sum_{k \ge 1} \left(\left\lfloor \frac{2n}{p^k} \right\rfloor - 2 \left\lfloor \frac{n}{p^k} \right\rfloor \right)$$

Legendre's theorem

The number $n!$ contains the prime factor p exactly

$$\sum_{k \ge 1} \left\lfloor \frac{n}{p^k} \right\rfloor$$

times.

■ **Proof.** Exactly $\lfloor \frac{n}{p} \rfloor$ of the factors of $n! = 1 \cdot 2 \cdot 3 \cdot \ldots \cdot n$ are divisible by p, which accounts for $\lfloor \frac{n}{p} \rfloor$ p-factors. Next, $\lfloor \frac{n}{p^2} \rfloor$ of the factors of $n!$ are even divisible by p^2, which accounts for the next $\lfloor \frac{n}{p^2} \rfloor$ prime factors p of $n!$, etc. □

times. Here each summand is at most 1, since it satisfies

$$\left\lfloor \frac{2n}{p^k} \right\rfloor - 2\left\lfloor \frac{n}{p^k} \right\rfloor \;<\; \frac{2n}{p^k} - 2\left(\frac{n}{p^k} - 1\right) \;=\; 2,$$

and it is an integer. Furthermore the summands vanish whenever $p^k > 2n$. Thus $\binom{2n}{n}$ contains p exactly

$$\sum_{k\geq 1}\left(\left\lfloor \frac{2n}{p^k} \right\rfloor - 2\left\lfloor \frac{n}{p^k} \right\rfloor\right) \;\leq\; \max\{r : p^r \leq 2n\}$$

times. Hence the largest power of p that divides $\binom{2n}{n}$ is not larger than $2n$. In particular, primes $p > \sqrt{2n}$ appear at most once in $\binom{2n}{n}$.

Furthermore — and this, according to Erdős, is the key fact for his proof — primes p that satisfy $\frac{2}{3}n < p \leq n$ do not divide $\binom{2n}{n}$ at all! Indeed, $3p > 2n$ implies (for $n \geq 3$, and hence $p \geq 3$) that p and $2p$ are the only multiples of p that appear as factors in the numerator of $\frac{(2n)!}{n!n!}$, while we get two p-factors in the denominator.

Examples such as
$\binom{26}{13} = 2^3 \cdot 5^2 \cdot 7 \cdot 17 \cdot 19 \cdot 23$
$\binom{28}{14} = 2^3 \cdot 3^3 \cdot 5^2 \cdot 17 \cdot 19 \cdot 23$
$\binom{30}{15} = 2^4 \cdot 3^2 \cdot 5 \cdot 17 \cdot 19 \cdot 23 \cdot 29$
illustrate that "very small" prime factors $p < \sqrt{2n}$ can appear as higher powers in $\binom{2n}{n}$, "small" primes with $\sqrt{2n} < p \leq \frac{2}{3}n$ appear at most once, while factors in the gap with $\frac{2}{3}n < p \leq n$ don't appear at all.

(4) Now we are ready to estimate $\binom{2n}{n}$. For $n \geq 3$, using an estimate from page 12 for the lower bound, we get

$$\frac{4^n}{2n} \;\leq\; \binom{2n}{n} \;\leq\; \prod_{p\leq\sqrt{2n}} 2n \;\cdot \prod_{\sqrt{2n}<p\leq\frac{2}{3}n} p \;\cdot \prod_{n<p\leq 2n} p$$

and thus, since there are not more than $\sqrt{2n}$ primes $p \leq \sqrt{2n}$,

$$4^n \;\leq\; (2n)^{1+\sqrt{2n}} \cdot \prod_{\sqrt{2n}<p\leq\frac{2}{3}n} p \;\cdot \prod_{n<p\leq 2n} p \qquad \text{for} \quad n \geq 3. \tag{2}$$

(5) Assume now that there is no prime p with $n < p \leq 2n$, so the second product in (2) is 1. Substituting (1) into (2) we get

$$4^n \;\leq\; (2n)^{1+\sqrt{2n}}\, 4^{\frac{2}{3}n}$$

or

$$4^{\frac{1}{3}n} \;\leq\; (2n)^{1+\sqrt{2n}}, \tag{3}$$

which is false for n large enough! In fact, using $a + 1 < 2^a$ (which holds for all $a \geq 2$, by induction) we get

$$2n = \left(\sqrt[6]{2n}\right)^6 < \left(\left\lfloor \sqrt[6]{2n} \right\rfloor + 1\right)^6 < 2^{6\left\lfloor \sqrt[6]{2n} \right\rfloor} \leq 2^{6\sqrt[6]{2n}}, \tag{4}$$

and thus for $n \geq 50$ (and hence $18 < 2\sqrt{2n}$) we obtain from (3) and (4)

$$2^{2n} \leq (2n)^{3\left(1+\sqrt{2n}\right)} < 2^{\sqrt[6]{2n}\left(18+18\sqrt{2n}\right)} < 2^{20\sqrt[6]{2n}\sqrt{2n}} = 2^{20(2n)^{2/3}}.$$

This implies $(2n)^{1/3} < 20$, and thus $n < 4000$. □

One can extract even more from this type of estimates: From (2) one can derive with the same methods that

$$\prod_{n<p\leq 2n} p \geq 2^{\frac{1}{30}n} \quad \text{for} \quad n \geq 4000,$$

and thus that there are at least

$$\log_{2n}\left(2^{\frac{1}{30}n}\right) = \frac{1}{30}\frac{n}{\log_2 n + 1}$$

primes in the range between n and $2n$.

This is not that bad an estimate: the "true" number of primes in this range is roughly $n/\log n$. This follows from the "prime number theorem," which says that the limit

$$\lim_{n\to\infty} \frac{\#\{p \leq n : p \text{ is prime}\}}{n/\log n}$$

exists, and equals 1. This famous result was first proved by Hadamard and de la Vallée-Poussin in 1896; Selberg and Erdős found an elementary proof (without complex analysis tools, but still long and involved) in 1948.

On the prime number theorem itself the final word, it seems, is still not in: for example a proof of the Riemann hypothesis (see page 41), one of the major unsolved open problems in mathematics, would also give a substantial improvement for the estimates of the prime number theorem. But also for Bertrand's postulate, one could expect dramatic improvements. In fact, the following is a famous unsolved problem:

Is there always a prime between n^2 and $(n+1)^2$?

For additional information see [3, p. 19] and [4, pp. 248, 257].

Appendix: Some estimates

Estimating via integrals

There is a very simple-but-effective method of estimating sums by integrals (as already encountered on page 4). For estimating the *harmonic numbers*

$$H_n = \sum_{k=1}^{n} \frac{1}{k}$$

we draw the figure in the margin and derive from it

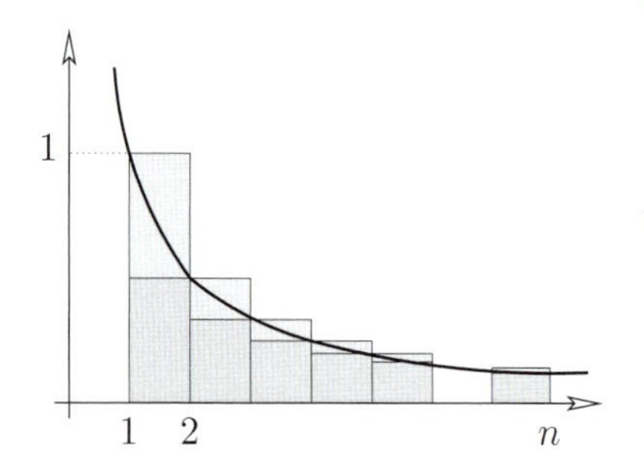

$$H_n - 1 = \sum_{k=2}^{n} \frac{1}{k} < \int_1^n \frac{1}{t}\,dt = \log n$$

by comparing the area below the graph of $f(t) = \frac{1}{t}$ $(1 \leq t \leq n)$ with the area of the dark shaded rectangles, and

$$H_n - \frac{1}{n} = \sum_{k=1}^{n-1} \frac{1}{k} > \int_1^n \frac{1}{t}\,dt = \log n$$

by comparing with the area of the large rectangles (including the lightly shaded parts). Taken together, this yields

$$\log n + \frac{1}{n} \; < \; H_n \; < \; \log n \; + \; 1.$$

In particular, $\lim\limits_{n\to\infty} H_n \to \infty$, and the order of growth of H_n is given by $\lim\limits_{n\to\infty} \frac{H_n}{\log n} = 1$. But much better estimates are known (see [2]), such as

$$H_n \;=\; \log n + \gamma + \frac{1}{2n} - \frac{1}{12n^2} + \frac{1}{120n^4} + O\left(\frac{1}{n^6}\right),$$

Here $O\left(\frac{1}{n^6}\right)$ denotes a function $f(n)$ such that $f(n) \leq c\frac{1}{n^6}$ holds for some constant c.

where $\gamma \approx 0.5772$ is "Euler's constant."

Estimating factorials — Stirling's formula

The same method applied to

$$\log(n!) \;=\; \log 2 + \log 3 + \ldots + \log n \;=\; \sum_{k=2}^{n} \log k$$

yields

$$\log((n-1)!) \; < \; \int_1^n \log t \, dt \; < \; \log(n!),$$

where the integral is easily computed:

$$\int_1^n \log t \, dt \;=\; \Big[t \log t - t\Big]_1^n \;=\; n \log n - n + 1.$$

Thus we get a lower estimate on $n!$

$$n! \; > \; e^{n \log n - n + 1} \;=\; e\left(\frac{n}{e}\right)^n$$

and at the same time an upper estimate

$$n! \;=\; n\,(n-1)! \; < \; n e^{n \log n - n + 1} \;=\; en\left(\frac{n}{e}\right)^n.$$

Here a more careful analysis is needed to get the asymptotics of $n!$, as given by *Stirling's formula*

$$n! \;\sim\; \sqrt{2\pi n}\left(\frac{n}{e}\right)^n.$$

Here $f(n) \sim g(n)$ means that $\lim\limits_{n\to\infty} \frac{f(n)}{g(n)} = 1$.

And again there are more precise versions available, such as

$$n! \;=\; \sqrt{2\pi n}\left(\frac{n}{e}\right)^n \left(1 + \frac{1}{12n} + \frac{1}{288n^2} - \frac{139}{5140n^3} + O\left(\frac{1}{n^4}\right)\right).$$

Estimating binomial coefficients

Just from the definition of the binomial coefficients $\binom{n}{k}$ as the number of k-subsets of an n-set, we know that the sequence $\binom{n}{0}, \binom{n}{1}, \ldots, \binom{n}{n}$ of binomial coefficients

- sums to $\sum_{k=0}^{n} \binom{n}{k} = 2^n$
- is symmetric: $\binom{n}{k} = \binom{n}{n-k}$.

							1							
						1		1						
					1		2		1					
				1		3		3		1				
			1		4		6		4		1			
		1		5		10		10		5		1		
	1		6		15		20		15		6		1	
1		7		21		35		35		21		7		1

Pascal's triangle

From the functional equation $\binom{n}{k} = \frac{n-k+1}{k}\binom{n}{k-1}$ one easily finds that for every n the binomial coefficients $\binom{n}{k}$ form a sequence that is symmetric and *unimodal*: it increases towards the middle, so that the middle binomial coefficients are the largest ones in the sequence:

$$1 = \binom{n}{0} < \binom{n}{1} < \ldots < \binom{n}{\lfloor n/2 \rfloor} = \binom{n}{\lceil n/2 \rceil} > \ldots > \binom{n}{n-1} > \binom{n}{n} = 1.$$

Here $\lfloor x \rfloor$ resp. $\lceil x \rceil$ denotes the number x rounded down resp. rounded up to the nearest integer.

From the asymptotic formulas for the factorials mentioned above one can obtain very precise estimates for the sizes of binomial coefficients. However, we will only need very weak and simple estimates in this book, such as the following: $\binom{n}{k} \leq 2^n$ for all k, while for $n \geq 2$ we have

$$\binom{n}{\lfloor n/2 \rfloor} \geq \frac{2^n}{n},$$

with equality only for $n = 2$. In particular, for $n \geq 1$,

$$\binom{2n}{n} \geq \frac{4^n}{2n}.$$

This holds since $\binom{n}{\lfloor n/2 \rfloor}$, a middle binomial coefficient, is the largest entry in the sequence $\binom{n}{0}+\binom{n}{n}, \binom{n}{1}, \binom{n}{2}, \ldots, \binom{n}{n-1}$, whose sum is 2^n, and whose average is thus $\frac{2^n}{n}$.

On the other hand, we note the upper bound for binomial coefficients

$$\binom{n}{k} = \frac{n(n-1)\cdots(n-k+1)}{k!} \leq \frac{n^k}{k!} \leq \frac{n^k}{2^{k-1}},$$

which is a reasonably good estimate for the "small" binomial coefficients at the tails of the sequence, when n is large (compared to k).

References

[1] P. ERDŐS: *Beweis eines Satzes von Tschebyschef,* Acta Sci. Math. (Szeged) **5** (1930-32), 194-198.

[2] R. L. GRAHAM, D. E. KNUTH & O. PATASHNIK: *Concrete Mathematics. A Foundation for Computer Science,* Addison-Wesley, Reading MA 1989.

[3] G. H. HARDY & E. M. WRIGHT: *An Introduction to the Theory of Numbers,* fifth edition, Oxford University Press 1979.

[4] P. RIBENBOIM: *The New Book of Prime Number Records,* Springer-Verlag, New York 1989.

Binomial coefficients are (almost) never powers

Chapter 3

There is an epilogue to Bertrand's postulate which leads to a beautiful result on binomial coefficients. In 1892 Sylvester strengthened Bertrand's postulate in the following way:

If $n \geq 2k$, then at least one of the numbers $n, n-1, \ldots, n-k+1$ has a prime divisor p greater than k.

Note that for $n = 2k$ we obtain precisely Bertrand's postulate. In 1934, Erdős gave a short and elementary Book Proof of Sylvester's result, running along the lines of his proof of Bertrand's postulate. There is an equivalent way of stating Sylvester's theorem:

The binomial coefficient

$$\binom{n}{k} = \frac{n(n-1)\cdots(n-k+1)}{k!} \qquad (n \geq 2k)$$

always has a prime factor $p > k$.

With this observation in mind, we turn to another one of Erdős' jewels. When is $\binom{n}{k}$ equal to a power m^ℓ? It is easy to see that there are infinitely many solutions for $k = \ell = 2$, that is, of the equation $\binom{n}{2} = m^2$. Indeed, if $\binom{n}{2}$ is a square, then so is $\binom{(2n-1)^2}{2}$. To see this, set $n(n-1) = 2m^2$. It follows that

$$(2n-1)^2((2n-1)^2-1) = (2n-1)^2 4n(n-1) = 2(2m(2n-1))^2,$$

and hence

$$\binom{(2n-1)^2}{2} = (2m(2n-1))^2.$$

Beginning with $\binom{9}{2} = 6^2$ we thus obtain infinitely many solutions — the next one is $\binom{289}{2} = 204^2$. However, this does not yield all solutions. For example, $\binom{50}{2} = 35^2$ starts another series, as does $\binom{1682}{2} = 1189^2$. For $k = 3$ it is known that $\binom{n}{3} = m^2$ has the unique solution $n = 50$, $m = 140$. But now we are at the end of the line. For $k \geq 4$ and any $\ell \geq 2$ no solutions exist, and this is what Erdős proved by an ingenious argument.

$\binom{50}{3} = 140^2$
is the only solution for $k = 3$, $\ell = 2$

Theorem. *The equation $\binom{n}{k} = m^\ell$ has no integer solutions with $\ell \geq 2$ and $4 \leq k \leq n-4$.*

■ **Proof.** Note first that we may assume $n \geq 2k$ because of $\binom{n}{k} = \binom{n}{n-k}$. Suppose the theorem is false, and that $\binom{n}{k} = m^\ell$. The proof, by contradiction, proceeds in the following four steps.

(1) By Sylvester's theorem, there is a prime factor p of $\binom{n}{k}$ greater than k, hence p^ℓ divides $n(n-1)\cdots(n-k+1)$. Clearly, only one of the factors $n-i$ can be a multiple of p (because of $p > k$), and we conclude $p^\ell \mid n-i$, and therefore

$$n \;\geq\; p^\ell \;>\; k^\ell \;\geq\; k^2.$$

(2) Consider any factor $n-j$ of the numerator and write it in the form $n-j = a_j m_j^\ell$, where a_j is not divisible by any nontrivial ℓ-th power. We note by **(1)** that a_j has only prime divisors less than or equal to k. We want to show next that $a_i \neq a_j$ for $i \neq j$. Assume to the contrary that $a_i = a_j$ for some $i < j$. Then $m_i \geq m_j + 1$ and

$$\begin{aligned} k \;&>\; (n-i)-(n-j) \;=\; a_j(m_i^\ell - m_j^\ell) \;\geq\; a_j((m_j+1)^\ell - m_j^\ell) \\ &>\; a_j \ell m_j^{\ell-1} \;\geq\; \ell(a_j m_j^\ell)^{1/2} \;\geq\; \ell(n-k+1)^{1/2} \\ &\geq\; \ell(\tfrac{n}{2}+1)^{1/2} \;>\; n^{1/2}, \end{aligned}$$

which contradicts $n > k^2$ from above.

(3) Next we prove that the a_i's are the integers $1, 2, \ldots, k$ in some order. (According to Erdős, this is the crux of the proof.) Since we already know that they are all distinct, it suffices to prove that

$$a_0 a_1 \cdots a_{k-1} \quad \text{divides} \quad k!.$$

Substituting $n-j = a_j m_j^\ell$ into the equation $\binom{n}{k} = m^\ell$, we obtain

$$a_0 a_1 \cdots a_{k-1}(m_0 m_1 \cdots m_{k-1})^\ell \;=\; k! m^\ell.$$

Cancelling the common factors of $m_0 \cdots m_{k-1}$ and m yields

$$a_0 a_1 \cdots a_{k-1} u^\ell \;=\; k! v^\ell$$

with $\gcd(u,v) = 1$. It remains to show that $v = 1$. If not, then v contains a prime divisor p. Since $\gcd(u,v) = 1$, p must be a prime divisor of $a_0 a_1 \cdots a_{k-1}$ and hence is less than or equal to k. By the theorem of Legendre (see page 8) we know that $k!$ contains p to the power $\sum_{i\geq 1} \lfloor \frac{k}{p^i} \rfloor$. We now estimate the exponent of p in $n(n-1)\cdots(n-k+1)$. Let i be a positive integer, and let $b_1 < b_2 < \ldots < b_s$ be the multiples of p^i among $n, n-1, \ldots, n-k+1$. Then $b_s = b_1 + (s-1)p^i$ and hence

$$(s-1)p^i \;=\; b_s - b_1 \;\leq\; n-(n-k+1) \;=\; k-1,$$

which implies

$$s \;\leq\; \left\lfloor \frac{k-1}{p^i} \right\rfloor + 1 \;\leq\; \left\lfloor \frac{k}{p^i} \right\rfloor + 1.$$

So for each i the number of multiples of p^i among $n, \dots, n-k+1$, and hence among the a_j's, is bounded by $\lfloor \frac{k}{p^i} \rfloor + 1$. This implies that the exponent of p in $a_0 a_1 \cdots a_{k-1}$ is at most

$$\sum_{i=1}^{\ell-1} \left(\left\lfloor \frac{k}{p^i} \right\rfloor + 1 \right)$$

with the reasoning that we used for Legendre's theorem in Chapter 2. The only difference is that this time the sum stops at $i = \ell - 1$, since the a_j's contain no ℓ-th powers.

Taking both counts together, we find that the exponent of p in v^ℓ is at most

$$\sum_{i=1}^{\ell-1} \left(\left\lfloor \frac{k}{p^i} \right\rfloor + 1 \right) - \sum_{i \geq 1} \left\lfloor \frac{k}{p^i} \right\rfloor \;\leq\; \ell - 1,$$

and we have our desired contradiction, since v^ℓ is an ℓ-th power.

This suffices already to settle the case $\ell = 2$. Indeed, since $k \geq 4$ one of the a_i's must be equal to 4, but the a_i's contain no squares. So let us now assume that $\ell \geq 3$.

We see that our analysis so far agrees with $\binom{50}{3} = 140^2$, as

$$\begin{aligned} 50 &= 2 \cdot 5^2 \\ 49 &= 1 \cdot 7^2 \\ 48 &= 3 \cdot 4^2 \end{aligned}$$

and $5 \cdot 7 \cdot 4 = 140$.

(4) Since $k \geq 4$, we must have $a_{i_1} = 1, a_{i_2} = 2, a_{i_3} = 4$ for some i_1, i_2, i_3, that is,

$$n - i_1 = m_1^\ell, \;\; n - i_2 = 2m_2^\ell, \;\; n - i_3 = 4m_3^\ell.$$

We claim that $(n - i_2)^2 \neq (n - i_1)(n - i_3)$. If not, put $b = n - i_2$ and $n - i_1 = b - x$, $n - i_3 = b + y$, where $0 < |x|, |y| < k$. Hence

$$b^2 = (b - x)(b + y) \quad \text{or} \quad (y - x)b = xy,$$

where $x = y$ is plainly impossible. Now we have by part **(1)**

$$|xy| \;=\; b|y - x| \;\geq\; b \;>\; n - k \;>\; (k-1)^2 \;\geq\; |xy|,$$

which is absurd.

So we have $m_2^2 \neq m_1 m_3$, where we assume $m_2^2 > m_1 m_3$ (the other case being analogous), and proceed to our last chains of inequalities. We obtain

$$\begin{aligned} 2(k-1)n &> n^2 - (n-k+1)^2 \;>\; (n - i_2)^2 - (n - i_1)(n - i_3) \\ &= 4[m_2^{2\ell} - (m_1 m_3)^\ell] \;\geq\; 4[(m_1 m_3 + 1)^\ell - (m_1 m_3)^\ell] \\ &\geq 4\ell m_1^{\ell-1} m_3^{\ell-1}. \end{aligned}$$

Since $\ell \geq 3$ and $n > k^\ell \geq k^3 > 6k$, this yields

$$\begin{aligned} 2(k-1)n m_1 m_3 &> 4\ell m_1^\ell m_3^\ell \;=\; \ell(n - i_1)(n - i_3) \\ &> \ell(n-k+1)^2 \;>\; 3(n - \frac{n}{6})^2 \;>\; 2n^2. \end{aligned}$$

Now since $m_i \leq n^{1/\ell} \leq n^{1/3}$ we finally obtain

$$kn^{2/3} \geq km_1m_3 > (k-1)m_1m_3 > n,$$

or $k^3 > n$. With this contradiction, the proof is complete. □

References

[1] P. ERDŐS: *A theorem of Sylvester and Schur,* J. London Math. Soc. **9** (1934), 282-288.

[2] P. ERDŐS: *On a diophantine equation,* J. London Math. Soc. **26** (1951), 176-178.

[3] J. J. SYLVESTER: *On arithmetical series,* Messenger of Math. **21** (1892), 1-19, 87-120; Collected Mathematical Papers Vol. 4, 1912, 687-731.

Representing numbers as sums of two squares

Chapter 4

Which numbers can be written as sums of two squares?

Pierre de Fermat

This question is as old as number theory, and its solution is a classic in the field. The "hard" part of the solution is to see that every prime number of the form $4m + 1$ is a sum of two squares. G. H. Hardy writes that this *two square theorem* of Fermat "is ranked, very justly, as one of the finest in arithmetic." Nevertheless, one of our Book Proofs below is quite recent.

Let's start with some "warm-ups." First, we need to distinguish between the prime $p = 2$, the primes of the form $p = 4m + 1$, and the primes of the form $p = 4m + 3$. Every prime number belongs to exactly one of these three classes. At this point we may note (using a method "à la Euclid") that there are infinitely many primes of the form $4m + 3$. In fact, if there were only finitely many, then we could take p_k to be the largest prime of this form. Setting

$$N_k \ := \ 2^2 \cdot 3 \cdot 5 \cdots p_k - 1$$

(where $p_1 = 2$, $p_2 = 3$, $p_3 = 5$, ... denotes the sequence of all primes), we find that N_k is congruent to $3 \pmod 4$, so it must have a prime factor of the form $4m + 3$, and this prime factor is larger than p_k — contradiction. At the end of this chapter we will also derive that there are infinitely many primes of the other kind, $p = 4m + 1$.

Our first lemma is a special case of the famous "law of reciprocity": It characterizes the primes for which -1 is a square in the field $\mathbb{Z}_p$ (which is reviewed in the box on the next page).

Lemma 1. *For primes $p = 4m + 1$ the equation $s^2 \equiv -1 \pmod p$ has two solutions $s \in \{1, 2, \ldots, p-1\}$, for $p = 2$ there is one such solution, while for primes of the form $p = 4m + 3$ there is no solution.*

■ **Proof.** For $p = 2$ take $s = 1$. For odd p, we construct the equivalence relation on $\{1, 2, \ldots, p - 1\}$ that is generated by identifying every element with its additive inverse and with its multiplicative inverse in $\mathbb{Z}_p$. Thus the "general" equivalence classes will contain four elements

$$\{x, -x, \overline{x}, -\overline{x}\}$$

since such a 4-element set contains both inverses for all its elements. However, there are smaller equivalence classes if some of the four numbers are not distinct:

- $x \equiv -x$ is impossible for odd p.
- $x \equiv \overline{x}$ is equivalent to $x^2 \equiv 1$. This has two solutions, namely $x = 1$ and $x = p-1$, leading to the equivalence class $\{1, p-1\}$ of size 2.
- $x \equiv -\overline{x}$ is equivalent to $x^2 \equiv -1$. This equation may have no solution or two distinct solutions $x_0, p-x_0$: in this case the equivalence class is $\{x_0, p-x_0\}$.

For $p = 11$ the partition is $\{1,10\}, \{2,9,6,5\}, \{3,8,4,7\}$; for $p = 13$ it is $\{1,12\}, \{2,11,7,6\}, \{3,10,9,4\}, \{5,8\}$: the pair $\{5,8\}$ yields the two solutions of $s^2 \equiv -1 \bmod 13$.

The set $\{1, 2, \ldots, p-1\}$ has $p-1$ elements, and we have partitioned it into quadruples (equivalence classes of size 4), plus one or two pairs (equivalence classes of size 2). For $p-1 = 4m+2$ we find that there is only the one pair $\{1, p-1\}$, the rest is quadruples, and thus $s^2 \equiv -1 \pmod p$ has no solution. For $p-1 = 4m$ there has to be the second pair, and this contains the two solutions of $s^2 \equiv -1$ that we were looking for. □

Prime fields

If p is a prime, then the set $\mathbb{Z}_p = \{0, 1, \ldots, p-1\}$ with addition and multiplication defined "modulo p" forms a finite field. We will need the following simple properties:

- For $x \in \mathbb{Z}_p$, $x \neq 0$, the additive inverse (for which we usually write $-x$) is given by $p - x \in \{1, 2, \ldots, p-1\}$. If $p > 2$, then x and $-x$ are different elements of $\mathbb{Z}_p$.
- Each $x \in \mathbb{Z}_p \backslash \{0\}$ has a unique multiplicative inverse $\overline{x} \in \mathbb{Z}_p \backslash \{0\}$, with $x\overline{x} \equiv 1 \pmod p$.
  The definition of primes implies that the map $\mathbb{Z}_p \to \mathbb{Z}_p$, $z \mapsto xz$ is injective for $x \neq 0$. Thus on the finite set $\mathbb{Z}_p \backslash \{0\}$ it must be surjective as well, and hence for each x there is a unique $\overline{x} \neq 0$ with $x\overline{x} \equiv 1 \pmod p$.
- The squares $0^2, 1^2, 2^2, \ldots, h^2$ define different elements of $\mathbb{Z}_p$, for $h = \lfloor \frac{p}{2} \rfloor$.
  This is since $x^2 \equiv y^2$, or $(x+y)(x-y) \equiv 0$, implies that $x \equiv y$ or that $x \equiv -y$. The $1 + \lfloor \frac{p}{2} \rfloor$ elements $0^2, 1^2, \ldots, h^2$ are called the *squares* in $\mathbb{Z}_p$.

+	0	1	2	3	4
0	0	1	2	3	4
1	1	2	3	4	0
2	2	3	4	0	1
3	3	4	0	1	2
4	4	0	1	2	3

·	0	1	2	3	4
0	0	0	0	0	0
1	0	1	2	3	4
2	0	2	4	1	3
3	0	3	1	4	2
4	0	4	3	2	1

Addition and multiplication in $\mathbb{Z}_5$

At this point, let us note "on the fly" that for *all* primes there are solutions for $x^2 + y^2 \equiv -1 \pmod p$. In fact, there are $\lfloor \frac{p}{2} \rfloor + 1$ distinct squares x^2 in $\mathbb{Z}_p$, and there are $\lfloor \frac{p}{2} \rfloor + 1$ distinct numbers of the form $-(1 + y^2)$. These two sets of numbers are too large to be disjoint, since $\mathbb{Z}_p$ has only p elements, and thus there must exist x and y with $x^2 \equiv -(1+y^2) \pmod p$.

Lemma 2. *No number $n = 4m+3$ is a sum of two squares.*

■ **Proof.** The square of any even number is $(2k)^2 = 4k^2 \equiv 0 \pmod 4$, while squares of odd numbers yield $(2k+1)^2 = 4(k^2+k)+1 \equiv 1 \pmod 4$. Thus any sum of two squares is congruent to $0, 1$ or $2 \pmod 4$. □

This is enough evidence for us that the primes $p = 4m+3$ are "bad." Thus, we proceed with "good" properties for primes of the form $p = 4m+1$. On the way to the main theorem, the following is the key step.

Proposition. *Every prime of the form* $p = 4m+1$ *is a sum of two squares, that is, it can be written as* $p = x^2+y^2$ *for some natural numbers* $x, y \in \mathbb{N}$.

We shall present here two proofs of this result — both of them elegant and surprising. The first proof features a striking application of the "pigeon-hole principle" (which we have already used "on the fly" before Lemma 2; see Chapter 22 for more), as well as a clever move to arguments "modulo p" and back. The idea is due to the Norwegian number theorist Axel Thue.

■ **Proof.** Consider the pairs (x', y') of integers with $0 \le x', y' \le \sqrt{p}$, that is, $x', y' \in \{0, 1, \dots, \lfloor\sqrt{p}\rfloor\}$. There are $(\lfloor\sqrt{p}\rfloor + 1)^2$ such pairs. Using the estimate $\lfloor x\rfloor + 1 > x$ for $x = \sqrt{p}$, we see that we have more than p such pairs of integers. Thus for any $s \in \mathbb{Z}$, it is impossible that all the values $x' - sy'$ produced by the pairs (x', y') are distinct modulo p. That is, for every s there are two distinct pairs

$$(x', y'),\ (x'', y'') \ \in\ \{0, 1, \dots, \lfloor\sqrt{p}\rfloor\}^2$$

with

$$x' - sy' \ \equiv\ x'' - sy'' \pmod{p}.$$

Now we take differences: We have $x' - x'' \equiv s(y' - y'') \pmod{p}$. Thus if we define

$$x := |x' - x''|, \quad y := |y' - y''|,$$

then we get

$$(x, y) \in \{0, 1, \dots, \lfloor\sqrt{p}\rfloor\}^2 \quad \text{with} \quad x \equiv \pm sy \pmod{p}.$$

Also we know that not both x and y can be zero, because the pairs (x', y') and (x'', y'') are distinct.

Now let s be a solution of $s^2 \equiv -1 \pmod{p}$, which exists by Lemma 1. Then $x^2 \equiv s^2y^2 \equiv -y^2 \pmod{p}$, and so we have produced

$$(x, y) \in \mathbb{Z}^2 \quad \text{with} \quad 0 < x^2 + y^2 < 2p \quad \text{and} \quad x^2 + y^2 \equiv 0 \pmod{p}.$$

But p is the only number between 0 and $2p$ that is divisible by p. Thus $x^2 + y^2 = p$: done! □

For $p = 13$, $\lfloor\sqrt{p}\rfloor = 3$ we consider $x', y' \in \{0, 1, 2, 3\}$. For $s = 5$, the sum $x' - sy'$ (mod 13) assumes the following values:

x' \ y'	0	1	2	3
0	0	8	3	11
1	1	9	4	12
2	2	10	5	0
3	3	11	6	1

Our second proof for the proposition — also clearly a Book Proof — was discovered by Roger Heath-Brown in 1971 and appeared in 1984. (A condensed "one-sentence version" was given by Don Zagier.) It is so elementary that we don't even need to use Lemma 1.

Heath-Brown's argument features three linear involutions: a quite obvious one, a hidden one, and a trivial one that gives "the final blow." The second, unexpected, involution corresponds to some hidden structure on the set of integral solutions of the equation $4xy + z^2 = p$.

■ **Proof.** We study the set

$$S \ := \ \{(x,y,z) \in \mathbb{Z}^3 : 4xy + z^2 = p, \quad x > 0, \quad y > 0\}.$$

This set is finite. Indeed, $x \geq 1$ and $y \geq 1$ implies $y \leq \frac{p}{4}$ and $x \leq \frac{p}{4}$. So there are only finitely many possible values for x and y, and given x and y, there are at most two values for z.

1. The first linear involution is given by

$$f : S \longrightarrow S, \quad (x,y,z) \longmapsto (y,x,-z),$$

that is, "interchange x and y, and negate z." This clearly maps S to itself, and it is an *involution*: Applied twice, it yields the identity. Also, f has no fixed points, since $z = 0$ would imply $p = 4xy$, which is impossible. Furthermore, f maps the solutions in

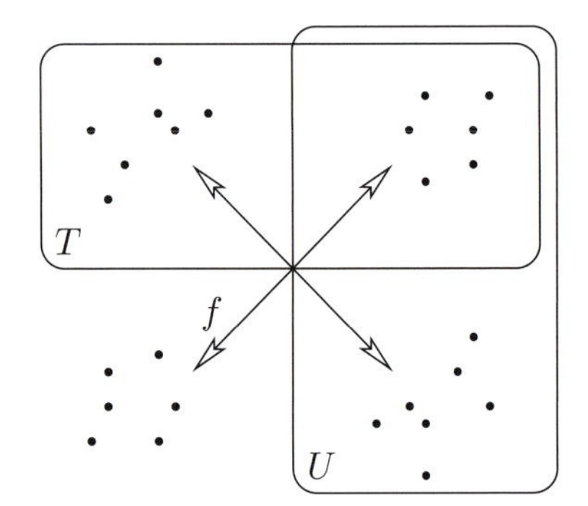

$$T \ := \ \{(x,y,z) \in S : z > 0\}$$

to the solutions in $S \backslash T$, which satisfy $z < 0$. Also, f reverses the signs of $x - y$ and of z, so it maps the solutions in

$$U \ := \ \{(x,y,z) \in S : (x-y) + z > 0\}$$

to the solutions in $S \backslash U$. For this we have to see that there is no solution with $(x-y)+z = 0$, but there is none since this would give $p = 4xy + z^2 = 4xy + (x-y)^2 = (x+y)^2$.

What do we get from the study of f? The main observation is that since f maps the sets T and U to their complements, it also interchanges the elements in $T \backslash U$ with these in $U \backslash T$. That is, there is the same number of solutions in U that are not in T as there are solutions in T that are not in U — *so T and U have the same cardinality.*

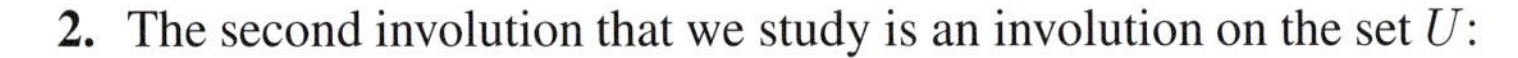

2. The second involution that we study is an involution on the set U:

$$g : U \longrightarrow U, \quad (x,y,z) \longmapsto (x - y + z, y, 2y - z).$$

First we check that indeed this is a well-defined map: If $(x,y,z) \in U$, then $x - y + z > 0, y > 0$ and $4(x - y + z)y + (2y - z)^2 = 4xy + z^2$, so $g(x,y,z) \in S$. By $(x - y + z) - y + (2y - z) = x > 0$ we find that indeed $g(x,y,z) \in U$.

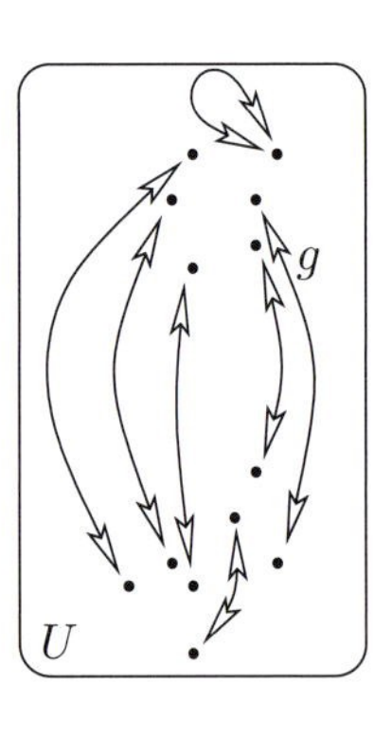

Also g is an involution: $g(x,y,z) = (x - y + z, y, 2y - z)$ is mapped by g to $((x - y + z) - y + (2y - z), y, 2y - (2y - z)) = (x,y,z)$.

And finally: g has exactly one fixed point:

$$(x,y,z) \ = \ g(x,y,z) \ = \ (x - y + z, y, 2y - z)$$

holds exactly if $y = z$: But then $p = 4xy + y^2 = (4x + y)y$, which holds only for $y = 1 = z$, and $x = \frac{p-1}{4}$.

But if g is an involution on U that has exactly one fixed point, then *the cardinality of U is odd.*

3. The third, trivial, involution that we study is the involution on T that interchanges x and y:

$$h : T \longrightarrow T, \quad (x, y, z) \longmapsto (y, x, z).$$

This map is clearly well-defined, and an involution. We combine now our knowledge derived from the other two involutions: The cardinality of T is equal to the cardinality of U, which is odd. But if h is an involution on a finite set of odd cardinality, then it *has a fixed point*: There is a point $(x, y, z) \in T$ with $x = y$, that is, a solution of

$$p = 4x^2 + z^2 = (2x)^2 + z^2. \qquad \square$$

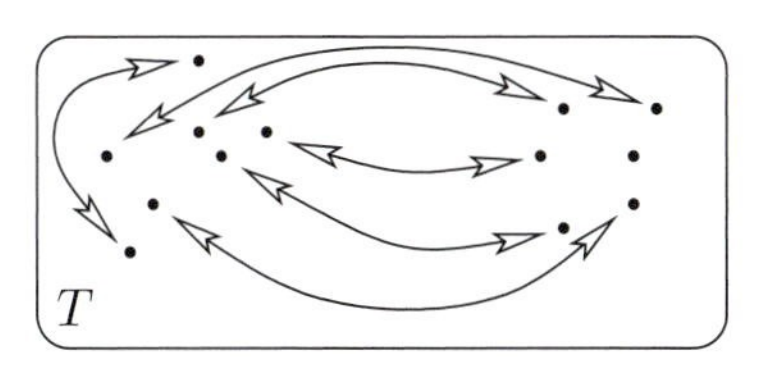

On a finite set of odd cardinality, every involution has at least one fixed point.

Note that this proof yields more — the number of representations of p in the form $p = x^2 + (2y)^2$ is *odd* for all primes of the form $p = 4m+1$. (The representation is actually unique, see [3].) Also note that both proofs are not effective: Try to find x and y for a ten digit prime! Efficient ways to find such representations as sums of two squares are discussed in [1] and [7].

The following theorem completely answers the question which started this chapter.

Theorem. *A natural number n can be represented as a sum of two squares if and only if every prime factor of the form $p = 4m + 3$ appears with an even exponent in the prime decomposition of n.*

■ **Proof.** Call a number n *representable* if it is a sum of two squares, that is, if $n = x^2 + y^2$ for some $x, y \in \mathbb{N}_0$. The theorem is a consequence of the following five facts.

(1) $1 = 1^2 + 0^2$ and $2 = 1^2 + 1^2$ are representable. Every prime of the form $p = 4m + 1$ is representable.

(2) The product of any two representable numbers $n_1 = x_1^2 + y_1^2$ and $n_2 = x_2^2 + y_2^2$ is representable: $n_1 n_2 = (x_1 x_2 + y_1 y_2)^2 + (x_1 y_2 - x_2 y_1)^2$.

(3) If n is representable, $n = x^2 + y^2$, then also nz^2 is representable, by $nz^2 = (xz)^2 + (yz)^2$.

Facts (1), (2) and (3) together yield the "if" part of the theorem.

(4) If $p = 4m + 3$ is a prime that divides a representable number $n = x^2 + y^2$, then p divides both x and y, and thus p^2 divides n. In fact, if we had $x \not\equiv 0 \pmod{p}$, then we could find $\overline{x}$ such that $x\overline{x} \equiv 1 \pmod{p}$, multiply the equation $x^2 + y^2 \equiv 0$ by $\overline{x}^2$, and thus obtain $1 + y^2\overline{x}^2 = 1 + (\overline{x}y)^2 \equiv 0 \pmod{p}$, which is impossible for $p = 4m + 3$ by Lemma 1.

(5) If n is representable, and $p = 4m + 3$ divides n, then p^2 divides n, and n/p^2 is representable. This follows from (4), and completes the proof. $\square$

As a corollary, we obtain that there are infinitely many primes of the form $p = 4m + 1$. For this, we consider

$$M_k = (3 \cdot 5 \cdot 7 \cdots p_k)^2 + 2^2,$$

a number that is congruent to $1 \pmod 4$. All its prime factors are larger than p_k, and by fact (4) of the previous proof, it has no prime factors of the form $4m + 3$. Thus M_k has a prime factor of the form $4m + 1$ that is larger than p_k.

Two remarks close our discussion:

- If a and b are two natural numbers that are relatively prime, then there are infinitely many primes of the form $am + b$ $(m \in \mathbb{N})$ — this is a famous (and difficult) theorem of Dirichlet. More precisely, one can show that the number of primes $p \leq x$ of the form $p = am + b$ is described very accurately for large x by the function $\frac{1}{\varphi(a)} \frac{x}{\log x}$, where $\varphi(x)$ denotes the number of b with $1 \leq b < a$ that are relatively prime to a. (This is a substantial refinement of the prime number theorem, which we had discussed on page 10.)
- This means that the primes for fixed a and varying b appear essentially at the same rate. Nevertheless, for example for $a = 4$ one can observe a rather subtle, but nevertheless noticable and persistent tendency towards "more" primes of the form $4m+3$: If you look for a large random x, then chances are that there are more primes $p \leq x$ of the form $p = 4m + 3$ than of the form $p = 4m + 1$. This effect is known as "Chebyshev's bias"; see Riesel [4] and Rubinstein and Sarnak [5].

References

[1] F. W. Clarke, W. N. Everitt, L. L. Littlejohn & S. J. R. Vorster: *H. J. S. Smith and the Fermat Two Squares Theorem,* Amer. Math. Monthly **106** (1999), 652-665.

[2] D. R. Heath-Brown: *Fermat's two squares theorem,* Invariant (1984), 2-5.

[3] I. Niven & H. S. Zuckerman: *An Introduction to the Theory of Numbers,* Fifth edition, Wiley, New York 1972.

[4] H. Riesel: *Prime Numbers and Computer Methods for Factorization,* Second edition, Progress in Mathematics **126**, Birkhäuser, Boston MA 1994.

[5] M. Rubinstein & P. Sarnak: *Chebyshev's bias,* Experimental Mathematics **3** (1994), 173-197.

[6] A. Thue: *Et par antydninger til en taltheoretisk metode,* Kra. Vidensk. Selsk. Forh. **7** (1902), 57-75.

[7] S. Wagon: *Editor's corner: The Euclidean algorithm strikes again,* Amer. Math. Monthly **97** (1990), 125-129.

[8] D. Zagier: *A one-sentence proof that every prime* $p \equiv 1 \pmod 4$ *is a sum of two squares,* Amer. Math. Monthly **97** (1990), 144.

Every finite division ring is a field

Chapter 5

Rings are important structures in modern algebra. If a ring R has a multiplicative unit element 1 and every nonzero element has a multiplicative inverse, then R is called a *division ring*. So, all that is missing in R from being a field is the commutativity of multiplication. The best-known example of a non-commutative division ring is the ring of quaternions discovered by Hamilton. But, as the chapter title says, every such division ring must of necessity be infinite. If R is finite, then the axioms force the multiplication to be commutative.

This result which is now a classic has caught the imagination of many mathematicians, because, as Herstein writes: "It is so unexpectedly interrelating two seemingly unrelated things, the number of elements in a certain algebraic system and the multiplication of that system."

Ernst Witt

Theorem. *Every finite division ring R is commutative.*

This beautiful theorem which is usually attributed to MacLagan Wedderburn has been proved by many people using a variety of different ideas. Wedderburn himself gave three proofs in 1905, and another proof was given by Leonard E. Dickson in the same year. More proofs were later given by Emil Artin, Hans Zassenhaus, Nicolas Bourbaki, and many others. One proof stands out for its simplicity and elegance. It was found by Ernst Witt in 1931 and combines two elementary ideas towards a glorious finish.

■ **Proof.** Our first ingredient comes from a blend of linear algebra and basic group theory. For an arbitrary element $s \in R$, let C_s be the set $\{x \in R : xs = sx\}$ of elements which commute with s; C_s is called the *centralizer* of s. Clearly, C_s contains 0 and 1 and is a sub-division ring of R. The *center* Z is the set of elements which commute with all elements of R, thus $Z = \bigcap_{s \in R} C_s$. In particular, all elements of Z commute, 0 and 1 are in Z, and so Z is a *finite field*. Let us set $|Z| = q$.

We can regard R and C_s as vector spaces over the field Z and deduce that $|R| = q^n$, where n is the dimension of the vector space R over Z, and similarly $|C_s| = q^{n_s}$ for suitable integers $n_s \geq 1$.

Now let us assume that R is not a field. This means that for *some* $s \in R$ the centralizer C_s is not all of R, or, what is the same, $n_s < n$.

On the set $R^* := R \backslash \{0\}$ we consider the relation

$$r' \sim r \quad :\Longleftrightarrow \quad r' = x^{-1}rx \quad \text{for some } x \in R^*.$$

It is easy to check that $\sim$ is an equivalence relation. Let

$$A_s := \{x^{-1}sx : x \in R^*\}$$

be the equivalence class containing s. We note that $|A_s| = 1$ precisely when s is in the center Z. So by our assumption, there are classes A_s with $|A_s| \geq 2$. Consider now for $s \in R^*$ the map $f_s : x \longmapsto x^{-1}sx$ from R^* onto A_s. For $x, y \in R^*$ we find

$$\begin{aligned} x^{-1}sx = y^{-1}sy \quad &\iff \quad (yx^{-1})s = s(yx^{-1}) \\ &\iff \quad yx^{-1} \in C_s^* \quad \iff \quad y \in C_s^* x, \end{aligned}$$

for $C_s^* := C_s \backslash \{0\}$, where $C_s^* x = \{zx : z \in C_s^*\}$ has size $|C_s^*|$. Hence any element $x^{-1}sx$ is the image of precisely $|C_s^*| = q^{n_s} - 1$ elements in R^* under the map f_s, and we deduce $|R^*| = |A_s|\,|C_s^*|$. In particular, we note that

$$\frac{|R^*|}{|C_s^*|} = \frac{q^n - 1}{q^{n_s} - 1} = |A_s| \quad \text{is an } \textit{integer} \text{ for all } s.$$

We know that the equivalence classes partition R^*. We now group the central elements Z^* together and denote by $A_1, \dots, A_t$ the equivalence classes containing more than one element. By our assumption we know $t \geq 1$. Since $|R^*| = |Z^*| + \sum_{k=1}^{t} |A_k|$, we have proved the so-called *class formula*

$$q^n - 1 \;=\; q - 1 + \sum_{k=1}^{t} \frac{q^n - 1}{q^{n_k} - 1}, \tag{1}$$

where we have $1 < \frac{q^n - 1}{q^{n_k} - 1} \in \mathbb{N}$ for all k.

With (1) we have left abstract algebra and are back to the natural numbers. Next we claim that $q^{n_k} - 1 \mid q^n - 1$ implies $n_k \mid n$. Indeed, write $n = an_k + r$ with $0 \leq r < n_k$, then $q^{n_k} - 1 \mid q^{an_k + r} - 1$ implies

$$q^{n_k} - 1 \mid (q^{an_k + r} - 1) - (q^{n_k} - 1) = q^{n_k}(q^{(a-1)n_k + r} - 1),$$

and thus $q^{n_k} - 1 \mid q^{(a-1)n_k + r} - 1$, since q^{n_k} and $q^{n_k} - 1$ are relatively prime. Continuing in this way we find $q^{n_k} - 1 \mid q^r - 1$ with $0 \leq r < n_k$, which is only possible for $r = 0$, that is, $n_k \mid n$. In summary, we note

$$n_k \mid n \quad \text{for all } k. \tag{2}$$

Now comes the second ingredient: the complex numbers $\mathbb{C}$. Consider the polynomial $x^n - 1$. Its roots in $\mathbb{C}$ are called the *n-th roots of unity*. Since $\lambda^n = 1$, all these roots λ have $|\lambda| = 1$ and lie therefore on the unit circle of the complex plane. In fact, they are precisely the numbers $\lambda_k = e^{\frac{2k\pi i}{n}} = \cos(2k\pi/n) + i\sin(2k\pi/n)$, $0 \leq k \leq n-1$ (see the box on the next page). Some of the roots λ satisfy $\lambda^d = 1$ for $d < n$; for example, the root $\lambda = -1$ satisfies $\lambda^2 = 1$. For a root λ, let d be the smallest positive exponent with $\lambda^d = 1$, that is, d is the order of λ in the group of the roots of unity. Then $d \mid n$, by Lagrange's theorem ("the order of every element of a group divides the order of the group" — see the box in Chapter 1). Note that there are roots of order n, such as $\lambda_1 = e^{\frac{2\pi i}{n}}$.

Roots of unity

Any complex number $z = x + iy$ may be written in the "polar" form

$$z = re^{i\varphi} = r(\cos\varphi + i\sin\varphi),$$

where $r = |z| = \sqrt{x^2 + y^2}$ is the distance of z to the origin, and φ is the angle measured from the positive x-axis. The n-th roots of unity are therefore of the form

$$\lambda_k = e^{\frac{2k\pi i}{n}} = \cos(2k\pi/n) + i\sin(2k\pi/n), \qquad 0 \le k \le n-1,$$

since for all k

$$\lambda_k^n = e^{2k\pi i} = \cos(2k\pi) + i\sin(2k\pi) = 1.$$

We obtain these roots geometrically by inscribing a regular n-gon into the unit circle. Note that $\lambda_k = \zeta^k$ for all k, where $\zeta = e^{\frac{2\pi i}{n}}$. Thus the n-th roots of unity form a cyclic group $\{\zeta, \zeta^2, \ldots, \zeta^{n-1}, \zeta^n = 1\}$ of order n.

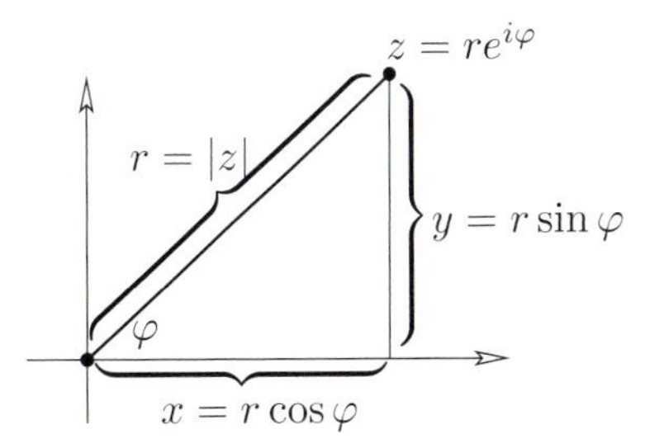

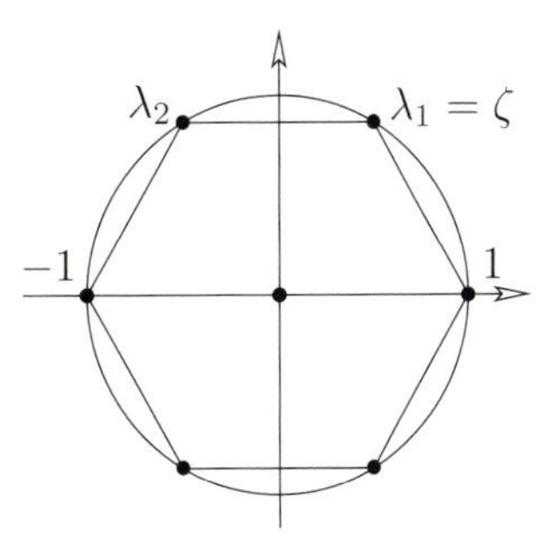

The roots of unity for $n = 6$

Now we group all roots of order d together and set

$$\phi_d(x) := \prod_{\lambda \text{ of order } d} (x - \lambda).$$

Note that the definition of $\phi_d(x)$ is independent of n. Since every root has some order d, we conclude that

$$x^n - 1 = \prod_{d\,|\,n} \phi_d(x). \tag{3}$$

Here is the crucial observation: The *coefficients* of the polynomials $\phi_n(x)$ are *integers* (that is, $\phi_n(x) \in \mathbb{Z}[x]$ for all n), where in addition the constant coefficient is either 1 or -1.

Let us carefully verify this claim. For $n = 1$ we have 1 as the only root, and so $\phi_1(x) = x - 1$. Now we proceed by induction, where we assume $\phi_d(x) \in \mathbb{Z}[x]$ for all $d < n$, and that the constant coefficient of $\phi_d(x)$ is 1 or -1. By (3),

$$x^n - 1 = p(x)\,\phi_n(x) \tag{4}$$

where $p(x) = \sum_{j=0}^{\ell} p_j x^j$, $\phi_n(x) = \sum_{k=0}^{n-\ell} a_k x^k$, with $p_0 = 1$ or $p_0 = -1$.

Since $-1 = p_0 a_0$, we see $a_0 \in \{1, -1\}$. Suppose we already know that $a_0, a_1, \ldots, a_{k-1} \in \mathbb{Z}$. Computing the coefficient of x^k on both sides of (4) we find

$$\sum_{j=0}^{k} p_j a_{k-j} \;=\; \sum_{j=1}^{k} p_j a_{k-j} + p_0 a_k \;\in\; \mathbb{Z}.$$

By assumption, all $a_0, \ldots, a_{k-1}$ (and all p_j) are in $\mathbb{Z}$. Thus $p_0 a_k$ and hence a_k must also be integers, since p_0 is 1 or -1.

We are ready for the *coup de grâce*. Let $n_k \mid n$ be one of the numbers appearing in (1). Then

$$x^n - 1 \;=\; \prod_{d \mid n} \phi_d(x) \;=\; (x^{n_k} - 1)\phi_n(x) \prod_{d \mid n,\, d \nmid n_k,\, d \neq n} \phi_d(x).$$

We conclude that in $\mathbb{Z}$ we have the divisibility relations

$$\phi_n(q) \mid q^n - 1 \qquad \text{and} \qquad \phi_n(q) \mid \frac{q^n - 1}{q^{n_k} - 1}. \tag{5}$$

Since (5) holds for all k, we deduce from the class formula (1)

$$\phi_n(q) \mid q - 1,$$

but this cannot be. Why? We know $\phi_n(x) = \prod(x - \lambda)$ where λ runs through all roots of $x^n - 1$ of order n. Let $\widetilde{\lambda} = a + ib$ be one of those roots. By $n > 1$ (because of $R \neq Z$) we have $\widetilde{\lambda} \neq 1$, which implies that the real part a is smaller than 1. Now $|\widetilde{\lambda}|^2 = a^2 + b^2 = 1$, and hence

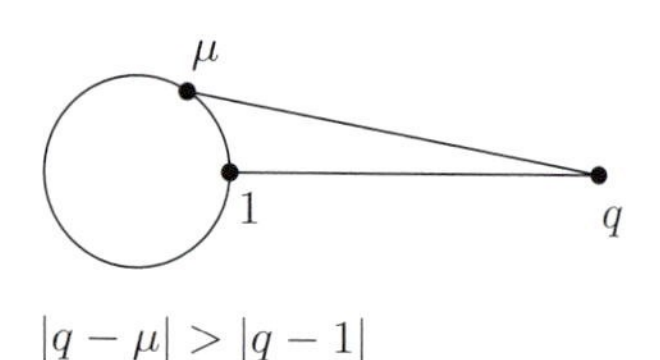

$|q - \mu| > |q - 1|$

$$\begin{aligned}
|q - \widetilde{\lambda}|^2 &= |q - a - ib|^2 \;=\; (q - a)^2 + b^2 \\
&= q^2 - 2aq + a^2 + b^2 \;=\; q^2 - 2aq + 1 \\
&> q^2 - 2q + 1 \qquad \text{(because of } a < 1\text{)} \\
&= (q - 1)^2,
\end{aligned}$$

and so $|q - \widetilde{\lambda}| > q - 1$ holds for *all* roots of order n. This implies

$$|\phi_n(q)| \;=\; \prod_{\lambda} |q - \lambda| > q - 1,$$

which means that $\phi_n(q)$ cannot be a divisor of $q - 1$, contradiction and end of proof. □

References

[1] L. E. DICKSON: *On finite algebras,* Nachrichten der Akad. Wissenschaften Göttingen Math.-Phys. Klasse (1905), 1-36; Collected Mathematical Papers Vol. III, Chelsea Publ. Comp, The Bronx, NY 1975, 539-574.

[2] J. H. M. WEDDERBURN: *A theorem on finite algebras,* Trans. Amer. Math. Soc. **6** (1905), 349-352.

[3] E. WITT: *Über die Kommutativität endlicher Schiefkörper,* Abh. Math. Sem. Univ. Hamburg **8** (1931), 413.

Some irrational numbers

Chapter 6

> *"π is irrational"*

This was already conjectured by Aristotle, when he claimed that diameter and circumference of a circle are not commensurable. The first proof of this fundamental fact was given by Johann Heinrich Lambert in 1766. Our Book Proof is due to Ivan Niven, 1947: an extremely elegant one-page proof that needs only elementary calculus. Its idea is powerful, and quite a bit more can be derived from it, as was shown by Iwamoto and Koksma, respectively:

- π^2 is irrational and
- e^r is irrational for rational $r \neq 0$.

Charles Hermite

Niven's method does, however, have its roots and predecessors: It can be traced back to the classical paper by Charles Hermite from 1873 which first established that e is transcendental, that is, that e is not a zero of a polynomial with rational coefficients.

$$e := 1 + \tfrac{1}{1} + \tfrac{1}{2} + \tfrac{1}{6} + \tfrac{1}{24} + \dots = 2.718281828\dots$$

Before we treat π we will look at e and its powers, and see that these are irrational. This is much easier, and we thus also follow the historical order in the development of the results.

To start with, it is rather easy to see (as did Fourier in 1815) that $e = \sum_{k\geq 0} \frac{1}{k!}$ is irrational. Indeed, if we had $e = \frac{a}{b}$ for integers a and $b > 0$, then we would get

$$n!be = n!a$$

for *every* $n \geq 0$. But this cannot be true, because on the right-hand side we have an integer, while the left-hand side with

$$e = \Big(1+\frac{1}{1!}+\frac{1}{2!}+\dots+\frac{1}{n!}\Big) + \Big(\frac{1}{(n+1)!}+\frac{1}{(n+2)!}+\frac{1}{(n+3)!}+\dots\Big)$$

decomposes into an integral part

$$bn!\Big(1 + \frac{1}{1!} + \frac{1}{2!} + \dots + \frac{1}{n!}\Big)$$

and a second part

$$b\Big(\frac{1}{n+1} + \frac{1}{(n+1)(n+2)} + \frac{1}{(n+1)(n+2)(n+3)} + \dots\Big)$$

Geometric series

For the infinite geometric series
$$Q = \tfrac{1}{q} + \tfrac{1}{q^2} + \tfrac{1}{q^3} + \ldots$$
with $q > 1$ we clearly have
$$qQ = 1 + \tfrac{1}{q} + \tfrac{1}{q^2} + \ldots = 1 + Q$$
and thus
$$Q = \frac{1}{q-1}.$$

which is *approximately* $\frac{b}{n}$, so that for large n it certainly cannot be integral: It is larger than $\frac{b}{n+1}$ and smaller than $\frac{b}{n}$, as one can see from a comparison with a geometric series:

$$\begin{aligned}\frac{1}{n+1} \;<\; & \frac{1}{n+1} + \frac{1}{(n+1)(n+2)} + \frac{1}{(n+1)(n+2)(n+3)} + \ldots \\ <\; & \frac{1}{n+1} + \frac{1}{(n+1)^2} + \frac{1}{(n+1)^3} + \ldots \;=\; \frac{1}{n}.\end{aligned}$$

Now one might be led to think that this simple multiply–by–$n!$ trick is not even sufficient to show that e^2 is irrational. This is a stronger statement: $\sqrt{2}$ is an example of a number which is irrational, but whose square is not.

From John Cosgrave we have learned that with two nice ideas/observations (let's call them "tricks") one can get two steps further nevertheless: Each of the tricks is sufficient to show that e^2 is irrational, the combination of both of them even yields the same for e^4. The first trick may be found in a one page paper by J. Liouville from 1840 — and the second one in a two page "addendum" which Liouville published on the next two journal pages.

Why is e^2 irrational? What can we derive from $e^2 = \frac{a}{b}$? According to Liouville we should write this as

$$be = ae^{-1},$$

substitute the series

$$e = 1 + \frac{1}{1} + \frac{1}{2} + \frac{1}{6} + \frac{1}{24} + \frac{1}{120} + \ldots$$

and

$$e^{-1} = 1 - \frac{1}{1} + \frac{1}{2} - \frac{1}{6} + \frac{1}{24} - \frac{1}{120} \pm \ldots,$$

and then multiply by $n!$, for a sufficiently large even n. Then we see that $n!be$ is nearly integral:

$$n!b\Big(1 + \frac{1}{1} + \frac{1}{2} + \frac{1}{6} + \ldots + \frac{1}{n!}\Big)$$

is an integer, and the rest

$$n!b\Big(\frac{1}{(n+1)!} + \frac{1}{(n+2)!} + \ldots\Big)$$

is approximately $\frac{b}{n}$: It is larger than $\frac{b}{n+1}$ but smaller than $\frac{b}{n}$, as we have seen above.

At the same time $n!ae^{-1}$ is nearly integral as well: Again we get a large integral part, and then a rest

$$(-1)^{n+1} n!a\Big(\frac{1}{(n+1)!} - \frac{1}{(n+2)!} + \frac{1}{(n+3)!} \mp \ldots\Big),$$

192 JOURNAL DE MATHÉMATIQUES

SUR L'IRRATIONNALITÉ DU NOMBRE

$e = 2{,}718\ldots;$

PAR J. LIOUVILLE.

On prouve dans les éléments que le nombre e, base des logarithmes népériens, n'a pas une valeur rationnelle. On devrait, ce me semble, ajouter que la même méthode prouve aussi que e ne peut pas être racine d'une équation du second degré à coefficients rationnels, en sorte que l'on ne peut pas avoir $ae + \frac{b}{e} = c$, a étant un entier positif et b, c, des entiers positifs ou négatifs. En effet, si l'on remplace dans cette équation e et $\frac{1}{e}$ ou e^{-1} par leurs développements déduits de celui de e^x, puis qu'on multiplie les deux membres par $1.2.3\ldots n$, on trouvera aisément

$$\frac{a}{n+1}\Big(1 + \frac{1}{n+2} + \ldots\Big) \pm \frac{b}{n+1}\Big(1 - \frac{1}{n+2} + \ldots\Big) = \mu,$$

μ étant un entier. On peut toujours faire en sorte que le facteur

$$\pm\frac{b}{n+1}$$

soit positif; il suffira de supposer n pair si b est <0 et n impair si b est >0; en prenant de plus n très grand, l'équation que nous venons d'écrire conduira dès lors à une absurdité; car son premier membre étant essentiellement positif et très petit, sera compris entre 0 et 1, et ne pourra pas être égal à un entier μ. Donc, etc.

Liouville's paper

and this is approximately $(-1)^{n+1}\frac{a}{n}$. More precisely: for even n the rest is larger than $-\frac{a}{n}$, but smaller than

$$-a\Big(\frac{1}{n+1}-\frac{1}{(n+1)^2}-\frac{1}{(n+1)^3}-\dots\Big) = -\frac{a}{n+1}\Big(1-\frac{1}{n}\Big) < 0.$$

But this cannot be true, since for large even n it would imply that $n!ae^{-1}$ is just a bit smaller than an integer, while $n!be$ is a bit larger than an integer, so $n!ae^{-1} = n!be$ cannot hold. □

In order to show that e^4 is irrational, we now courageously assume that $e^4 = \frac{a}{b}$ were rational, and write this as

$$be^2 = ae^{-2}.$$

We could now try to multiply this by $n!$ for some large n, and collect the non-integral summands, but this leads to nothing useful: The sum of the remaining terms on the left-hand side will be approximately $b\frac{2^{n+1}}{n}$, on the right side $(-1)^{n+1}a\frac{2^{n+1}}{n}$, and both will be very large if n gets large.
So one has to examine the situation a bit more carefully, and make two little adjustments to the strategy: First we will not take an *arbitrary* large n, but a large power of two, $n = 2^m$; and secondly we will not multiply by $n!$, but by $\frac{n!}{2^{n-1}}$. Then we need a little lemma, a special case of Legendre's theorem (see page 8): For any $n \geq 1$ the integer $n!$ contains the prime factor 2 at most $n-1$ times — with equality if (and only if) n is a power of two, $n = 2^m$.
This lemma is not hard to show: $\lfloor\frac{n}{2}\rfloor$ of the factors of $n!$ are even, $\lfloor\frac{n}{4}\rfloor$ of them are divisible by 4, and so on. So if 2^k is the largest power of two which satisfies $2^k \leq n$, then $n!$ contains the prime factor 2 exactly

$$\Big\lfloor\frac{n}{2}\Big\rfloor+\Big\lfloor\frac{n}{4}\Big\rfloor+\dots+\Big\lfloor\frac{n}{2^k}\Big\rfloor \leq \frac{n}{2}+\frac{n}{4}+\dots+\frac{n}{2^k} = n\Big(1-\frac{1}{2^k}\Big) \leq n-1$$

times, with equality in both inequalities exactly if $n = 2^k$.
Let's get back to $be^2 = ae^{-2}$. We are looking at

$$b\frac{n!}{2^{n-1}}e^2 = a\frac{n!}{2^{n-1}}e^{-2} \tag{1}$$

and substitute the series

$$e^2 = 1+\frac{2}{1}+\frac{4}{2}+\frac{8}{6}+\dots+\frac{2^r}{r!}+\dots$$

and

$$e^{-2} = 1-\frac{2}{1}+\frac{4}{2}-\frac{8}{6}\pm\dots+(-1)^r\frac{2^r}{r!}+\dots$$

For $r \leq n$ we get integral summands on both sides, namely

$$b\frac{n!}{2^{n-1}}\frac{2^r}{r!} \quad \text{resp.} \quad (-1)^r a\frac{n!}{2^{n-1}}\frac{2^r}{r!},$$

where for $r > 0$ the denominator $r!$ contains the prime factor 2 *at most* $r-1$ times, while $n!$ contains it *exactly* $n-1$ times. (So for $r > 0$ the summands are even.)

And since n is even (we assume that $n = 2^m$), the series that we get for $r \geq n+1$ are

$$2b\Big(\frac{2}{n+1} + \frac{4}{(n+1)(n+2)} + \frac{8}{(n+1)(n+2)(n+3)} + \dots\Big)$$

resp.

$$2a\Big(-\frac{2}{n+1} + \frac{4}{(n+1)(n+2)} - \frac{8}{(n+1)(n+2)(n+3)} \pm \dots\Big).$$

These series will for large n be roughly $\frac{4b}{n}$ resp. $-\frac{4a}{n}$, as one sees again by comparison with geometric series. For large $n = 2^m$ this means that the left-hand side of (1) is *a bit* larger than an integer, while the right-hand side is *a bit* smaller — contradiction! □

So we know that e^4 is irrational; to show that e^3, e^5 etc. are irrational as well, we need heavier machinery (that is, a bit of calculus), and a new idea — which essentially goes back to Charles Hermite, and for which the key is hidden in the following simple lemma.

Lemma. *For some fixed $n \geq 1$, let*

$$f(x) \quad = \quad \frac{x^n(1-x)^n}{n!}.$$

(i) *The function $f(x)$ is a polynomial of the form $f(x) = \frac{1}{n!}\sum_{i=n}^{2n} c_i x^i$, where the coefficients c_i are integers.*

(ii) *For $0 < x < 1$ we have $0 < f(x) < \frac{1}{n!}$.*

(iii) *The derivatives $f^{(k)}(0)$ and $f^{(k)}(1)$ are integers for all $k \geq 0$.*

■ **Proof.** Parts (i) and (ii) are clear.
For (iii) note that by (i) the k-th derivative $f^{(k)}$ vanishes at $x = 0$ unless $n \leq k \leq 2n$, and in this range $f^{(k)}(0) = \frac{k!}{n!}c_k$ is an integer. From $f(x) = f(1-x)$ we get $f^{(k)}(x) = (-1)^k f^{(k)}(1-x)$ for all x, and hence $f^{(k)}(1) = (-1)^k f^{(k)}(0)$, which is an integer. □

Theorem 1. *e^r is irrational for every $r \in \mathbb{Q}\backslash\{0\}$.*

■ **Proof.** It suffices to show that e^s cannot be rational for a positive integer s (if $e^{\frac{s}{t}}$ were rational, then $\big(e^{\frac{s}{t}}\big)^t = e^s$ would be rational, too). Assume that $e^s = \frac{a}{b}$ for integers $a, b > 0$, and let n be so large that $n! > as^{2n+1}$. Put

The estimate $n! > e(\frac{n}{e})^n$ yields an explicit n that is "large enough."

$$F(x) \quad := \quad s^{2n}f(x) - s^{2n-1}f'(x) + s^{2n-2}f''(x) \mp \dots + f^{(2n)}(x),$$

where $f(x)$ is the function of the lemma.

$F(x)$ may also be written as an infinite sum

$$F(x) \;=\; s^{2n}f(x) - s^{2n-1}f'(x) + s^{2n-2}f''(x) \mp \ldots,$$

since the higher derivatives $f^{(k)}(x)$, for $k > 2n$, vanish. From this we see that the polynomial $F(x)$ satisfies the identity

$$F'(x) \;=\; -s\,F(x) + s^{2n+1}f(x).$$

Thus differentiation yields

$$\frac{d}{dx}\left[e^{sx}F(x)\right] \;=\; se^{sx}F(x) + e^{sx}F'(x) \;=\; s^{2n+1}e^{sx}f(x)$$

and hence

$$N := b\int_0^1 s^{2n+1}e^{sx}f(x)dx \;=\; b\left[e^{sx}F(x)\right]_0^1 \;=\; aF(1) - bF(0).$$

This is an integer, since part (iii) of the lemma implies that $F(0)$ and $F(1)$ are integers. However, part (ii) of the lemma yields estimates for the size of N from below and from above,

$$0 < N = b\int_0^1 s^{2n+1}e^{sx}f(x)dx < bs^{2n+1}e^s\frac{1}{n!} \;=\; \frac{as^{2n+1}}{n!} < 1,$$

which shows that N cannot be an integer: contradiction. □

Now that this trick was so successful, we use it once more.

Theorem 2. *π^2 is irrational.*

■ **Proof.** Assume that $\pi^2 = \frac{a}{b}$ for integers $a, b > 0$. We now use the polynomial

$$F(x) \;:=\; b^n\Big(\pi^{2n}f(x) - \pi^{2n-2}f^{(2)}(x) + \pi^{2n-4}f^{(4)}(x) \mp \ldots\Big),$$

which satisfies $F''(x) = -\pi^2F(x) + b^n\pi^{2n+2}f(x)$.

From part (iii) of the lemma we get that $F(0)$ and $F(1)$ are integers. Elementary differentiation rules yield

$$\begin{aligned}\frac{d}{dx}\big[F'(x)\sin\pi x - \pi F(x)\cos\pi x\big] \;&=\; \big(F''(x) + \pi^2F(x)\big)\sin\pi x\\ &=\; b^n\pi^{2n+2}f(x)\sin\pi x\\ &=\; \pi^2a^nf(x)\sin\pi x,\end{aligned}$$

and thus we obtain

$$\begin{aligned}N := \pi\int_0^1 a^nf(x)\sin\pi x\,dx \;&=\; \left[\frac{1}{\pi}F'(x)\sin\pi x - F(x)\cos\pi x\right]_0^1\\ &=\; F(0) + F(1),\end{aligned}$$

which is an integer. Furthermore N is positive since it is defined as the

π is not rational, but it does have "good approximations" by rationals — some of these were known since antiquity:

$$\begin{aligned}\tfrac{22}{7} &= 3.142857142857...\\ \tfrac{355}{113} &= 3.141592920353...\\ \tfrac{104348}{33215} &= 3.141592653921...\\ \pi &= 3.141592653589...\end{aligned}$$

integral of a function that is positive (except on the boundary). However, if we choose n so large that $\frac{\pi a^n}{n!} < 1$, then from part (ii) of the lemma we obtain

$$0 < N = \pi \int_0^1 a^n f(x) \sin \pi x \, dx < \frac{\pi a^n}{n!} < 1,$$

a contradiction. □

Here comes our final irrationality result.

Theorem 3. *For every odd integer $n \geq 3$, the number*

$$A(n) := \frac{1}{\pi} \arccos\left(\frac{1}{\sqrt{n}}\right)$$

is irrational.

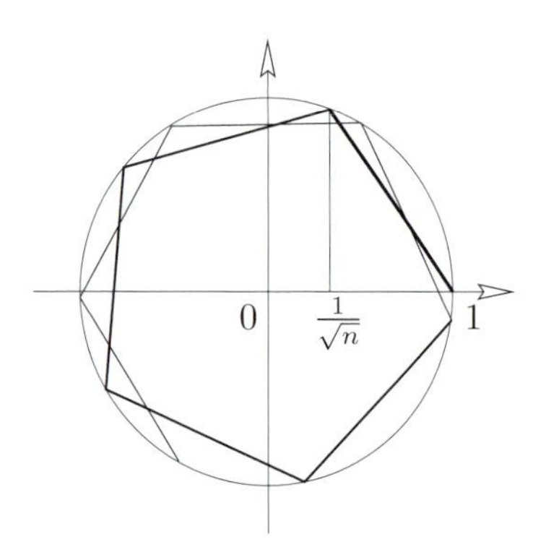

We will need this result for Hilbert's third problem (see Chapter 8) in the cases $n = 3$ and $n = 9$. For $n = 2$ and $n = 4$ we have $A(2) = \frac{1}{4}$ and $A(4) = \frac{1}{3}$, so the restriction to odd integers is essential. These values are easily derived by appealing to the diagram in the margin, in which the statement "$\frac{1}{\pi} \arccos\left(\frac{1}{\sqrt{n}}\right)$ is irrational" is equivalent to saying that the polygonal arc constructed from $\frac{1}{\sqrt{n}}$, all of whose chords have the same length, never closes into itself.

We leave it as an exercise for the reader to show that $A(n)$ is rational *only* for $n \in \{1, 2, 4\}$. For that, distinguish the cases when $n = 2^r$, and when n is not a power of 2.

■ **Proof.** We use the addition theorem

$$\cos\alpha + \cos\beta = 2\cos\tfrac{\alpha+\beta}{2}\cos\tfrac{\alpha-\beta}{2}$$

from elementary trigonometry, which for $\alpha = (k+1)\varphi$ and $\beta = (k-1)\varphi$ yields

$$\cos(k+1)\varphi = 2\cos\varphi\,\cos k\varphi - \cos(k-1)\varphi. \tag{2}$$

For the angle $\varphi_n = \arccos\left(\frac{1}{\sqrt{n}}\right)$, which is defined by $\cos\varphi_n = \frac{1}{\sqrt{n}}$ and $0 \leq \varphi_n \leq \pi$, this yields representations of the form

$$\cos k\varphi_n = \frac{A_k}{\sqrt{n}^k},$$

where A_k is an integer that is not divisible by n, for all $k \geq 0$. In fact, we have such a representation for $k = 0, 1$ with $A_0 = A_1 = 1$, and by induction on k using (2) we get for $k \geq 1$

$$\cos(k+1)\varphi_n = 2\frac{1}{\sqrt{n}}\frac{A_k}{\sqrt{n}^k} - \frac{A_{k-1}}{\sqrt{n}^{k-1}} = \frac{2A_k - nA_{k-1}}{\sqrt{n}^{k+1}}.$$

Thus we obtain $A_{k+1} = 2A_k - nA_{k-1}$. If $n \geq 3$ is odd, and A_k is not divisible by n, then we find that A_{k+1} cannot be divisible by n, either.

Now assume that

$$A(n) \;=\; \frac{1}{\pi}\varphi_n \;=\; \frac{k}{\ell}$$

is rational (with integers $k, \ell > 0$). Then $\ell\varphi_n = k\pi$ yields

$$\pm 1 \;=\; \cos k\pi \;=\; \frac{A_\ell}{\sqrt{n}^{\,\ell}}.$$

Thus $\sqrt{n}^{\,\ell} = \pm A_\ell$ is an integer, with $\ell \geq 2$, and hence $n \mid \sqrt{n}^{\,\ell}$. With $\sqrt{n}^{\,\ell} \mid A_\ell$ we find that n divides A_ℓ, a contradiction. □

References

[1] C. HERMITE: *Sur la fonction exponentielle,* Comptes rendus de l'Académie des Sciences (Paris) **77** (1873), 18-24; Œuvres de Charles Hermite, Vol. III, Gauthier-Villars, Paris 1912, pp. 150-181.

[2] Y. IWAMOTO: *A proof that π^2 is irrational,* J. Osaka Institute of Science and Technology **1** (1949), 147-148.

[3] J. F. KOKSMA: *On Niven's proof that π is irrational,* Nieuw Archief voor Wiskunde (2) **23** (1949), 39.

[4] J. LIOUVILLE: *Sur l'irrationalité du nombre $e = 2{,}718\ldots$,* Journal de Mathématiques Pures et Appl. (1) **5** (1840), 192; *Addition*, 193-194.

[5] I. NIVEN: *A simple proof that π is irrational,* Bulletin Amer. Math. Soc. **53** (1947), 509.

Three times $\pi^2/6$

Chapter 7

We know that the infinite series $\sum_{n\geq 1} \frac{1}{n}$ does not converge. Indeed, in Chapter 1 we have seen that even the series $\sum_{p\in\mathbb{P}} \frac{1}{p}$ diverges.

However, the sum of the reciprocals of the squares converges (although very slowly, as we will also see), and it produces an interesting value.

Euler's series.

$$\sum_{n\geq 1} \frac{1}{n^2} = \frac{\pi^2}{6}.$$

This is a classical, famous and important result by Leonhard Euler from 1734. One of its key interpretations is that it yields the first non-trivial value $\zeta(2)$ of Riemann's zeta function (see the appendix on page 41). This value is irrational, as we have seen in Chapter 6.

But not only the result has a prominent place in mathematics history, there are also a number of extremely elegant and clever proofs that have their history: For some of these the joy of discovery and rediscovery has been shared by many. In this chapter, we present three such proofs.

1	$=$	1.000000
$1+\frac{1}{4}$	$=$	1.250000
$1+\frac{1}{4}+\frac{1}{9}$	$=$	1.361111
$1+\frac{1}{4}+\frac{1}{9}+\frac{1}{6}$	$=$	1.423611
$1+\frac{1}{4}+\frac{1}{9}+\frac{1}{6}+\frac{1}{25}$	$=$	1.463611
$1+\frac{1}{4}+\frac{1}{9}+\frac{1}{6}+\frac{1}{25}+\frac{1}{36}$	$=$	1.491388
$\pi^2/6$	$=$	$1.644934.$

■ **Proof.** The first proof appears as an exercise in William J. LeVeque's number theory textbook from 1956. But he says: "I haven't the slightest idea where that problem came from, but I'm pretty certain that it wasn't original with me."

The proof consists in two different evaluations of the double integral

$$I := \int_0^1\int_0^1 \frac{1}{1-xy}\,dx\,dy.$$

For the first one, we expand $\frac{1}{1-xy}$ as a geometric series, decompose the summands as products, and integrate effortlessly:

$$\begin{aligned} I &= \int_0^1\int_0^1 \sum_{n\geq 0}(xy)^n\,dx\,dy = \sum_{n\geq 0}\int_0^1\int_0^1 x^n y^n\,dx\,dy \\ &= \sum_{n\geq 0}\Big(\int_0^1 x^n dx\Big)\Big(\int_0^1 y^n dy\Big) = \sum_{n\geq 0}\frac{1}{n+1}\,\frac{1}{n+1} \\ &= \sum_{n\geq 0}\frac{1}{(n+1)^2} = \sum_{n\geq 1}\frac{1}{n^2} = \zeta(2). \end{aligned}$$

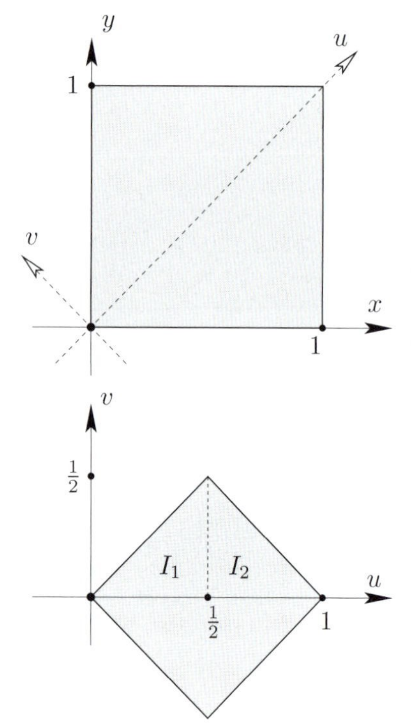

This evaluation also shows that the double integral (over a positive function with a pole at $x = y = 1$) is finite. Note that the computation is also easy and straightforward if we read it backwards — thus the evaluation of $\zeta(2)$ leads one to the double integral I.

The second way to evaluate I comes from a change of coordinates: in the new coordinates given by $u := \frac{y+x}{2}$ and $v := \frac{y-x}{2}$ the domain of integration is a square of side length $\frac{1}{2}\sqrt{2}$, which we get from the old domain by first rotating it by $45°$ and then shrinking it by a factor of $\sqrt{2}$. Substitution of $x = u - v$ and $y = u + v$ yields

$$\frac{1}{1-xy} \quad = \quad \frac{1}{1-u^2+v^2}.$$

To transform the integral, we have to replace $dx\,dy$ by $2\,du\,dv$, to compensate for the fact that our coordinate transformation reduces areas by a constant factor of 2 (which is the Jacobi determinant of the transformation; see the box on the next page). The new domain of integration, and the function to be integrated, are symmetric with respect to the u-axis, so we just need to compute two times (another factor of 2 arises here!) the integral over the upper half domain, which we split into two parts in the most natural way:

$$I \;=\; 4\int_0^{1/2}\Big(\int_0^{u}\frac{dv}{1-u^2+v^2}\Big)du \;+\; 4\int_{1/2}^{1}\Big(\int_0^{1-u}\frac{dv}{1-u^2+v^2}\Big)du.$$

Using $\displaystyle\int\frac{dx}{a^2+x^2} = \frac{1}{a}\arctan\frac{x}{a} + C$, this becomes

$$\begin{aligned} I \;=\; & \;4\int_0^{1/2}\frac{1}{\sqrt{1-u^2}}\arctan\Big(\frac{u}{\sqrt{1-u^2}}\Big)du \\ & + \;4\int_{1/2}^{1}\frac{1}{\sqrt{1-u^2}}\arctan\Big(\frac{1-u}{\sqrt{1-u^2}}\Big)du. \end{aligned}$$

These integrals can be simplified and finally evaluated by substituting $u = \sin\theta$ resp. $u = \cos\theta$. But we proceed more directly, by computing that the derivative of $g(u) := \arctan\big(\frac{u}{\sqrt{1-u^2}}\big)$ is $g'(u) = \frac{1}{\sqrt{1-u^2}}$, while the derivative of $h(u) := \arctan\big(\frac{1-u}{\sqrt{1-u^2}}\big) = \arctan\big(\sqrt{\frac{1-u}{1+u}}\big)$ is $h'(u) = -\frac{1}{2}\frac{1}{\sqrt{1-u^2}}$. So we may use $\int_a^b f'(x)f(x)dx = \big[\frac{1}{2}f(x)^2\big]_a^b = \frac{1}{2}f(b)^2 - \frac{1}{2}f(a)^2$ and get

$$\begin{aligned} I \;&=\; 4\int_0^{1/2} g'(u)g(u)\,du \;+\; 4\int_{1/2}^{1} -2h'(u)h(u)\,du \\ &=\; 2\Big[g(u)^2\Big]_0^{1/2} \;-\; 4\Big[h(u)^2\Big]_{1/2}^{1} \\ &=\; 2g(\tfrac{1}{2})^2 \;-\; 2g(0)^2 \;-\; 4h(1)^2 \;+\; 4h(\tfrac{1}{2})^2 \\ &=\; 2\big(\tfrac{\pi}{6}\big)^2 - 0 - 0 + 4\big(\tfrac{\pi}{6}\big)^2 \;=\; \tfrac{\pi^2}{6}. \end{aligned}$$

□

This proof extracted the value of Euler's series from an integral via a rather simple coordinate transformation. An ingenious proof of this type — with an entirely non-trivial coordinate transformation — was later discovered by Beukers, Calabi and Kolk. The point of departure for that proof is to split the sum $\sum_{n\geq 1}\frac{1}{n^2}$ into the even terms and the odd terms. Clearly the even terms $\frac{1}{2^2}+\frac{1}{4^2}+\frac{1}{6^2}+\ldots=\sum_{k\geq 1}\frac{1}{(2k)^2}$ sum to $\frac{1}{4}\zeta(2)$, so the odd terms $\frac{1}{1^2}+\frac{1}{3^2}+\frac{1}{5^2}+\ldots=\sum_{k\geq 0}\frac{1}{(2k+1)^2}$ make up three quarters of the total sum $\zeta(2)$. Thus Euler's series is equivalent to

$$\sum_{k\geq 0}\frac{1}{(2k+1)^2} \;=\; \frac{\pi^2}{8}.$$

The Substitution Formula

To compute a double integral

$$I \;=\; \int_S f(x,y)\,dx\,dy.$$

we may perform a substitution of variables

$$x = x(u,v) \quad y = y(u,v),$$

if the correspondence of $(u,v)\in T$ to $(x,y)\in S$ is bijective and continuously differentiable. Then I equals

$$\int_T f(x(u,v),y(u,v))\left|\frac{d(x,y)}{d(u,v)}\right| du\,dv,$$

where $\frac{d(x,y)}{d(u,v)}$ is the Jacobi determinant:

$$\frac{d(x,y)}{d(u,v)} \;=\; \det\begin{pmatrix}\frac{dx}{du} & \frac{dx}{dv}\\ \frac{dy}{du} & \frac{dy}{dv}\end{pmatrix}.$$

■ **Proof.** As above, we may express this as a double integral, namely

$$J \;=\; \int_0^1\!\!\int_0^1 \frac{1}{1-x^2y^2}\,dx\,dy \;=\; \sum_{k\geq 0}\frac{1}{(2k+1)^2}.$$

So we have to compute this integral J. And for this Beukers, Calabi and Kolk proposed the new coordinates

$$u \;:=\; \arccos\sqrt{\frac{1-x^2}{1-x^2y^2}} \qquad v \;:=\; \arccos\sqrt{\frac{1-y^2}{1-x^2y^2}}.$$

To compute the double integral, we may ignore the boundary of the domain, and consider x, y in the range $0<x<1$ and $0<y<1$. Then u, v will lie in the triangle $u>0$, $v>0$, $u+v<\pi/2$. The coordinate transformation can be inverted explicitly, which leads one to the substitution

$$x \;=\; \frac{\sin u}{\cos v} \qquad \text{and} \qquad y \;=\; \frac{\sin v}{\cos u}.$$

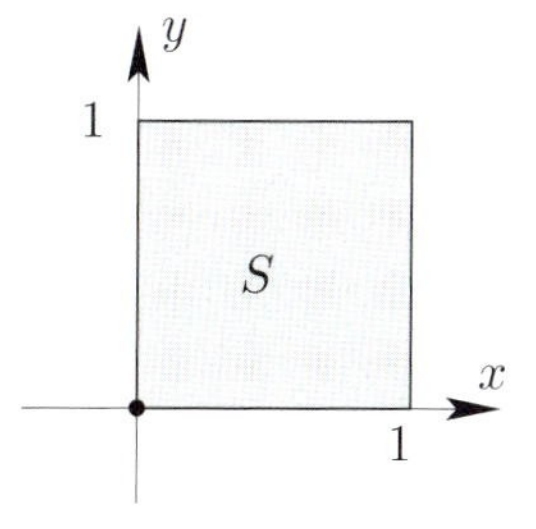

It is easy to check that these formulas define a bijective coordinate transformation between the interior of the unit square $S=\{(x,y): 0\leq x,y\leq 1\}$ and the interior of the triangle $T=\{(u,v): u,v\geq 0,\; u+v\leq \pi/2\}$.

Now we have to compute the Jacobi determinant of the coordinate transformation, and magically it turns out to be

$$\det\begin{pmatrix}\frac{\cos u}{\cos v} & \frac{\sin u\sin v}{\cos^2 v}\\ \frac{\sin u\sin v}{\cos^2 u} & \frac{\cos v}{\cos u}\end{pmatrix} \;=\; 1-\frac{\sin^2 u\sin^2 v}{\cos^2 u\cos^2 v} \;=\; 1-x^2y^2.$$

But this means that the integral that we want to compute is transformed into

$$J \;=\; \int_0^{\pi/2}\!\!\int_0^{\pi/2-u} 1\,du\,dv,$$

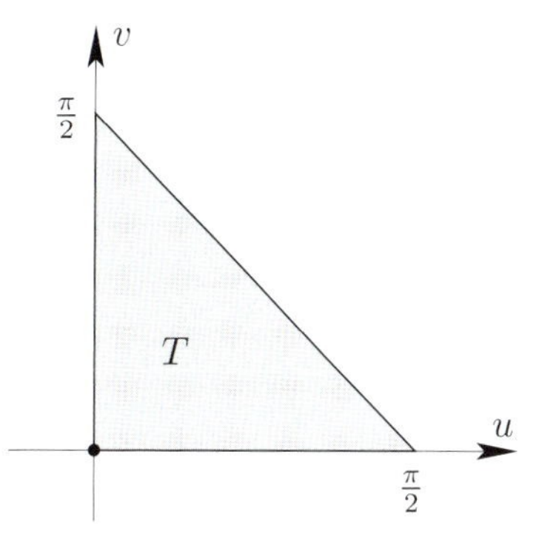

which is just the area $\frac{1}{2}(\frac{\pi}{2})^2=\frac{\pi^2}{8}$ of the triangle T. □

Beautiful — even more so, as the same method of proof extends to the computation of $\zeta(2k)$ in terms of a $2k$-dimensional integral, for all $k \geq 1$. We refer to the original paper of Beuker, Calabi and Kolk [2], and to Chapter 20, where we'll achieve this on a different path, using the Herglotz trick and Euler's original approach.

After these two proofs via coordinate transformation we can't resist the temptation to present another, entirely different and completely elementary proof for $\sum_{n\geq 1} \frac{1}{n^2} = \frac{\pi^2}{6}$. It appears in a sequence of exercises in the problem book by the twin brothers Akiva and Isaak Yaglom, whose Russian original edition appeared in 1954. Versions of this beautiful proof were rediscovered and presented by F. Holme (1970), I. Papadimitriou (1973), and by Ransford (1982) who attributed it to John Scholes.

■ **Proof.** The first step is to establish a remarkable relation between values of the (squared) cotangent function. Namely, for all $m \geq 1$ one has

$$\cot^2\left(\tfrac{\pi}{2m+1}\right) + \cot^2\left(\tfrac{2\pi}{2m+1}\right) + \ldots + \cot^2\left(\tfrac{m\pi}{2m+1}\right) \;=\; \tfrac{2m(2m-1)}{6}. \quad (1)$$

For $m = 1, 2, 3$ this yields
$\cot^2 \frac{\pi}{3} = \frac{1}{3}$
$\cot^2 \frac{\pi}{5} + \cot^2 \frac{2\pi}{5} = 2$
$\cot^2 \frac{\pi}{7} + \cot^2 \frac{2\pi}{7} + \cot^2 \frac{3\pi}{7} = 5$

To establish this, we start with the relation

$$\cos nx + i \sin nx \;=\; (\cos x + i \sin x)^n$$

and take its imaginary part, which is

$$\sin nx \;=\; \binom{n}{1} \sin x \cos^{n-1} x - \binom{n}{3} \sin^3 x \cos^{n-3} x \pm \ldots \quad (2)$$

Now we let $n = 2m+1$, while for x we will consider the m different values $x = \frac{r\pi}{2m+1}$, for $r = 1, 2, \ldots, m$. For each of these values we have $nx = r\pi$, and thus $\sin nx = 0$, while $0 < x < \frac{\pi}{2}$ implies that for $\sin x$ we get m distinct positive values.

In particular, we can divide (2) by $\sin^n x$, which yields

$$0 \;=\; \binom{n}{1} \cot^{n-1} x - \binom{n}{3} \cot^{n-3} x \pm \ldots,$$

that is,

$$0 \;=\; \binom{2m+1}{1} \cot^{2m} x - \binom{2m+1}{3} \cot^{2m-2} x \pm \ldots$$

for each of the m distinct values of x. Thus for the polynomial of degree m

$$p(t) \;:=\; \binom{2m+1}{1} t^m - \binom{2m+1}{3} t^{m-1} \pm \ldots + (-1)^m \binom{2m+1}{2m+1}$$

we know m distinct roots

$$a_r \;=\; \cot^2\left(\tfrac{r\pi}{2m+1}\right) \quad \text{for} \quad r = 1, 2, \ldots, m.$$

Hence the polynomial coincides with

$$p(t) \;=\; \binom{2m+1}{1} \left(t - \cot^2\left(\tfrac{\pi}{2m+1}\right)\right) \cdot \ldots \cdot \left(t - \cot^2\left(\tfrac{m\pi}{2m+1}\right)\right).$$

Comparison of the coefficients of t^{m-1} in $p(t)$ now yields that the sum of the roots is

$$a_1 + \ldots + a_r \;=\; \frac{\binom{2m+1}{3}}{\binom{2m+1}{1}} \;=\; \frac{2m(2m-1)}{6},$$

which proves (1).

Comparison of coefficients: If $p(t) = c(t-a_1)\cdots(t-a_m)$, then the coefficient of t^{m-1} is $-c(a_1 + \ldots + a_m)$.

We also need a second identity, of the same type,

$$\csc^2\left(\tfrac{\pi}{2m+1}\right) + \csc^2\left(\tfrac{2\pi}{2m+1}\right) + \ldots + \csc^2\left(\tfrac{m\pi}{2m+1}\right) = \frac{2m(2m+2)}{6}, \qquad (3)$$

for the cosecant function $\csc x = \frac{1}{\sin x}$. But

$$\csc^2 x = \frac{1}{\sin^2 x} = \frac{\cos^2 x + \sin^2 x}{\sin^2 x} = \cot^2 x + 1,$$

so we can derive (3) from (1) by adding m to both sides of the equation.

Now the stage is set, and everything falls into place. We use that in the range $0 < y < \frac{\pi}{2}$ we have

$$0 \;<\; \sin y \;<\; y \;<\; \tan y,$$

and thus

$$0 \;<\; \cot y \;<\; \tfrac{1}{y} \;<\; \csc y,$$

which implies

$$\cot^2 y \;<\; \tfrac{1}{y^2} \;<\; \csc^2 y.$$

$0 < a < b < c$ implies $0 < \frac{1}{c} < \frac{1}{b} < \frac{1}{a}$

Now we take this double inequality, apply it to each of the m distinct values of x, and add the results. Using (1) for the left-hand side, and (3) for the right-hand side, we obtain

$$\frac{2m(2m-1)}{6} \;<\; \left(\tfrac{2m+1}{\pi}\right)^2 + \left(\tfrac{2m+1}{2\pi}\right)^2 + \ldots + \left(\tfrac{2m+1}{m\pi}\right)^2 \;<\; \frac{2m(2m+2)}{6},$$

that is,

$$\frac{\pi^2}{6}\frac{2m}{2m+1}\frac{2m-1}{2m+1} \;<\; \frac{1}{1^2} + \frac{1}{2^2} + \ldots + \frac{1}{m^2} \;<\; \frac{\pi^2}{6}\frac{2m}{2m+1}\frac{2m+2}{2m+1}.$$

Both the left-hand and the right-hand side converge to $\frac{\pi^2}{6}$ for $m \longrightarrow \infty$: end of proof. □

So how fast does $\sum \frac{1}{n^2}$ converge to $\pi^2/6$? For this we have to estimate the difference

$$\frac{\pi^2}{6} - \sum_{n=1}^{\infty} \frac{1}{n^2} \;=\; \sum_{n=m+1}^{\infty} \frac{1}{n^2}.$$

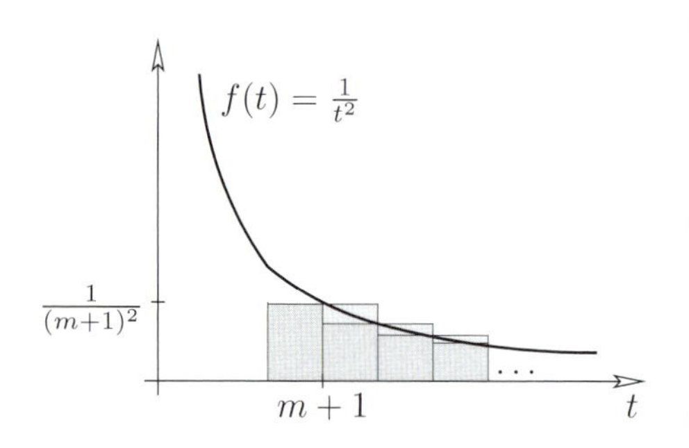

This is very easy with the technique of "comparing with an integral" that we have reviewed already in the appendix to Chapter 2 (page 10). It yields

$$\sum_{n=m+1}^{\infty} \frac{1}{n^2} \;<\; \int_m^{\infty} \frac{1}{t^2} dt \;=\; \frac{1}{m}$$

for an upper bound and

$$\sum_{n=m+1}^{\infty} \frac{1}{n^2} \;>\; \int_{m+1}^{\infty} \frac{1}{t^2} dt \;=\; \frac{1}{m+1}$$

for a lower bound on the "remaining summands" — or even

$$\sum_{n=m+1}^{\infty} \frac{1}{n^2} \;>\; \int_{m+\frac{1}{2}}^{\infty} \frac{1}{t^2} dt \;=\; \frac{1}{m+\frac{1}{2}}$$

if you are willing to do a slightly more careful estimate, using that the function $f(t) = \frac{1}{t^2}$ is convex.

This means that our series does not converge too well; if we sum the first one thousand summands, then we expect an error in the third digit after the decimal point, while for the sum of the first one million summands, $m = 1000000$, we expect to get an error in the sixth decimal digit, and we do. However, then comes a big surprise: to an accuracy of 45 digits,

$$\pi^2/6 \;=\; 1.644934066848226436472415166646025189218949901,$$

$$\sum_{n=1}^{10^6} \frac{1}{n^2} \;=\; 1.644933066848726436305748499979391855885616544.$$

So the sixth digit after the comma is wrong (too small by 1), but *the next six digits are right*! And then one digit is wrong (too large by 5), then again five are correct. This surprising discovery is quite recent, due to Roy D. North from Colorado Springs, 1988. (In 1982, Martin R. Powell, a school teacher from Amersham, Bucks, England, failed to notice the full effect due to the insufficient computing power available at the time.) It is too strange to be purely coincidental … A look at the error term, which again to 45 digits reads

$$\sum_{n=10^6+1}^{\infty} \frac{1}{n^2} = 0.000000999999500000166666666666633333333333357,$$

reveals that clearly there is a pattern. You might try to rewrite this last number as

$$+\,10^{-6} - \tfrac{1}{2}10^{-12} + \tfrac{1}{6}10^{-18} - \tfrac{1}{30}10^{-30} + \tfrac{1}{42}10^{-42} + \ldots$$

where the coefficients $(1, -\frac{1}{2}, \frac{1}{6}, 0, -\frac{1}{30}, 0, \frac{1}{42})$ of 10^{-6i} form the beginning of the sequence of *Bernoulli numbers* that we'll meet again in Chapter 20. We refer our readers to the article by Borwein, Borwein & Dilcher [3] for more such surprising "coincidences" — and for proofs.

Appendix: The Riemann zeta function

The *Riemann zeta function* $\zeta(s)$ is defined for real $s > 1$ by

$$\zeta(s) \ := \ \sum_{n\geq 1} \frac{1}{n^s}.$$

Our estimates for H_n (see page 10) imply that the series for $\zeta(1)$ diverges, but for any real $s > 1$ it does converge. The zeta function has a canonical continuation to the entire complex plane (with one simple pole at $s = 1$), which can be constructed using power series expansions. The resulting complex function is of utmost importance for the theory of prime numbers. Let us mention three diverse connections:

(1) The remarkable identity

$$\zeta(s) \ = \ \prod_p \frac{1}{1 - p^{-s}}$$

is due to Euler. It encodes the basic fact that every natural number has a unique (!) decomposition into prime factors; using this, Euler's identity is a simple consequence of the geometric series expansion

$$\frac{1}{1 - p^{-s}} \ = \ 1 + \frac{1}{p^s} + \frac{1}{p^{2s}} + \frac{1}{p^{3s}} + \dots$$

(2) The location of the complex zeros of the zeta function is the subject of the "Riemann hypothesis": one of the most famous and important unresolved conjectures in all of mathematics. It claims that all the non-trivial zeros $s \in \mathbb{C}$ of the zeta function satisfy $\text{Re}(s) = \frac{1}{2}$. (The zeta function vanishes at all the negative even integers, which are referred to as the "trivial zeros.")

Very recently, Jeff Lagarias showed that, surprisingly, the Riemann hypothesis is equivalent to the following elementary statement: For all $n \geq 1$,

$$\sum_{d\,|\,n} d \ \leq \ H_n \ + \ \exp(H_n)\log(H_n),$$

with equality only for $n = 1$, where H_n is again the n-th harmonic number.

(3) It has been known for a long time that $\zeta(s)$ is a rational multiple of π^s, and hence irrational, if s is an *even* integer $s \geq 2$; see Chapter 20. In contrast, the irrationality of $\zeta(3)$ was proved by Roger Apéry only in 1979. Despite considerable effort the picture is rather incomplete about $\zeta(s)$ for the other odd integers, $s = 2t+1 \geq 5$. Very recently, Keith Ball and Tanguy Rivoal proved that infinitely many of the values $\zeta(2t+1)$ are irrational. And indeed, although it is not known for any single odd value $s \geq 5$ that $\zeta(s)$ is irrational, Wadim Zudilin has proved that at least one of the four values $\zeta(5)$, $\zeta(7)$, $\zeta(9)$, and $\zeta(11)$ is irrational. We refer to the beautiful survey by Fischler [4].

References

[1] K. BALL & T. RIVOAL: *Irrationalité d'une infinité de valeurs de la fonction zêta aux entiers impairs,* Inventiones math. **146** (2001), 193-207.

[2] F. BEUKERS, J. A. C. KOLK & E. CALABI: *Sums of generalized harmonic series and volumes,* Nieuw Archief voor Wiskunde (4) **11** (1993), 217-224.

[3] J. M. BORWEIN, P. B. BORWEIN & K. DILCHER: *Pi, Euler numbers, and asymptotic expansions,* Amer. Math. Monthly **96** (1989), 681-687.

[4] S. FISCHLER: *Irrationalité de valeurs de zêta (d'après Apéry, Rivoal, ...),* Bourbaki Seminar, No. 910, November 2002; to appear in Astérisque; Preprint `arXiv:math.NT/0303066`, March 2003, 45 pages.

[5] J. C. LAGARIAS: *An elementary problem equivalent to the Riemann hypothesis,* Amer. Math. Monthly **109** (2002), 534-543.

[6] W. J. LEVEQUE: *Topics in Number Theory,* Vol. I, Addison-Wesley, Reading MA 1956.

[7] A. M. YAGLOM & I. M. YAGLOM: *Challenging mathematical problems with elementary solutions,* Vol. II, Holden-Day, Inc., San Francisco, CA 1967.

[8] W. ZUDILIN: *Arithmetic of linear forms involving odd zeta values,* Preprint, August 2001, 42 pages; `arXiv:math.NT/0206176`.

Geometry

"Platonic solids — child's play!"

Hilbert's third problem: decomposing polyhedra

Chapter 8

In his legendary address to the International Congress of Mathematicians at Paris in 1900 David Hilbert asked — as the third of his twenty-three problems — to specify

> *"two tetrahedra of equal bases and equal altitudes which can in no way be split into congruent tetrahedra, and which cannot be combined with congruent tetrahedra to form two polyhedra which themselves could be split up into congruent tetrahedra."*

David Hilbert

This problem can be traced back to two letters of Carl Friedrich Gauss from 1844 (published in Gauss' collected works in 1900). If tetrahedra of equal volume could be split into congruent pieces, then this would give one an "elementary" proof of Euclid's theorem XII.5 that pyramids with the same base and height have the same volume. It would thus provide an elementary definition of the volume for polyhedra (that would not depend on analysis, and hence on continuity arguments). A similar statement is true in plane geometry: the Bolyai-Gerwien Theorem [1, Sect. 2.7] states that planar polygons are both *equidecomposable* (can be dissected into congruent triangles) and *equicomplementable* (can be made congruent by adding congruent triangles) if and only if they have the same area.

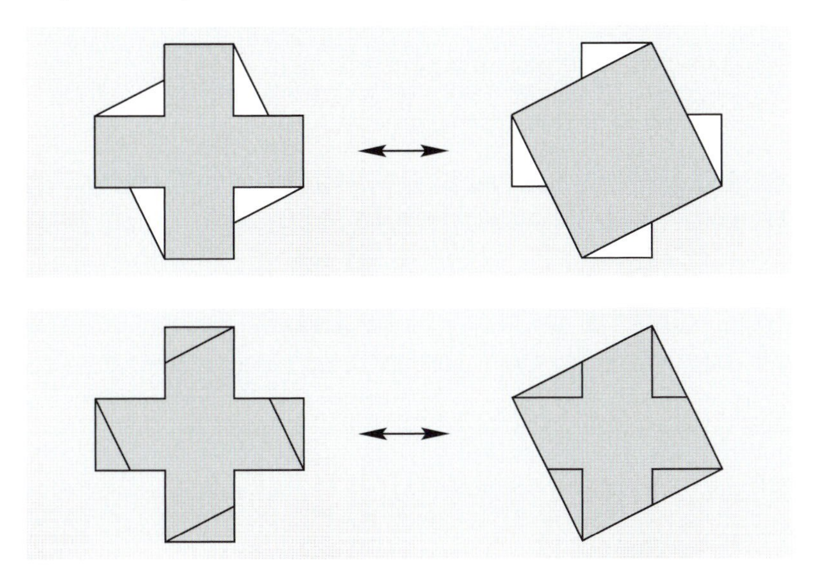

The cross is equicomplementable with a square of the same area.

In fact, they are even equidecomposable.

Hilbert — as we can see from his wording of the problem — did expect that there is no analogous theorem in dimension 3, and he was right. In fact, the problem was completely solved by Hilbert's student Max Dehn in two papers: the first one, exhibiting non-equidecomposable tetrahedra of equal base and height, appeared already in 1900, the second one, also covering equicomplementability, appeared in 1902. However, Dehn's papers are not easy to understand, and it takes effort to see whether Dehn did not fall into a subtle trap which ensnared others: a very-elegant-but-unfortunately-wrong proof was found by Bricard (in 1896!), by Meschkowski (1960), and probably by others. Luckily, Dehn's proof was reworked and redone, and after combined efforts of V. F. Kagan (1903/1930), Hugo Hadwiger (1949/54) and Vladimir G. Boltianskii, we now have a Book Proof — as follows. (The appendix to this chapter provides some basics about polyhedra.)

(1) A little linear algebra

For every finite set of real numbers $M = \{m_1, \ldots, m_k\} \subseteq \mathbb{R}$, we define $V(M)$ as the set of all linear combinations of numbers in M with rational coefficients, that is,

$$V(M) := \Big\{ \sum_{i=1}^{k} q_i m_i : q_i \in \mathbb{Q} \Big\} \subseteq \mathbb{R}.$$

The first (trivial, but important) observation is that $V(M)$ is a finite dimensional vector space over the field $\mathbb{Q}$ of rational numbers. In fact, $V(M)$ is clearly closed under taking sums and under multiplication with rationals, and the field axioms for $\mathbb{R}$ make $V(M)$ into a vector space. The dimension of $V(M)$ is the size of any minimal generating set. Since M generates $V(M)$ by definition, we see that it contains a minimal generating set, and hence

$$\dim_{\mathbb{Q}} V(M) \leq k = |M|.$$

In the following, we shall need and use $\mathbb{Q}$-*linear functions*

$$f : V(M) \to \mathbb{Q}$$

which we interpret as linear maps of $\mathbb{Q}$-vector spaces. The key property is that for every rational linear dependence $\sum_{i=1}^{k} q_i m_i = 0$ with $q_i \in \mathbb{Q}$, we must have $\sum_{i=1}^{k} q_i f(m_i) = f(0) = 0$. Here is the simple lemma that gets things going.

Lemma. *For any finite subsets $M \subseteq M'$ of $\mathbb{R}$, the $\mathbb{Q}$-vector space $V(M)$ is a subspace of the $\mathbb{Q}$-vector space $V(M')$. Thus if $f : V(M) \to \mathbb{Q}$ is a $\mathbb{Q}$-linear function, then f can be extended to a $\mathbb{Q}$-linear function $f' : V(M') \to \mathbb{Q}$ so that $f'(m) = f(m)$ for all $m \in M$.*

■ **Proof.** Any $\mathbb{Q}$-linear function $V(M) \to \mathbb{Q}$ is determined as soon as its values on a $\mathbb{Q}$-basis of $V(M)$ are fixed. Since every basis of $V(M)$ can be extended to a basis of $V(M')$, the rest follows. □

(2) Dehn invariants

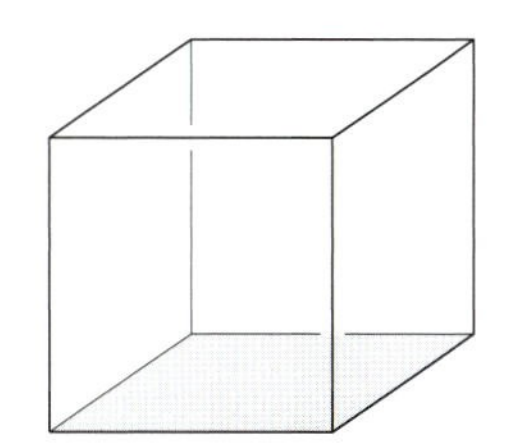

$M_C = \{\frac{\pi}{2}, \pi\}$

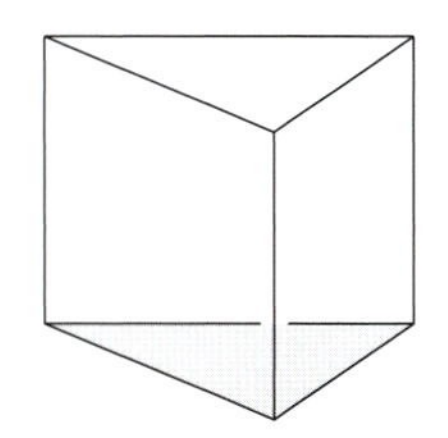

$M_Q = \{\frac{\pi}{3}, \frac{\pi}{2}, \pi\}$

For a 3-dimensional polyhedron P, let M_P denote the set of all angles between adjacent facets (*dihedral angles*), together with the number π. Thus for a cube C we get $M_C = \{\frac{\pi}{2}, \pi\}$, while for an orthogonal prism Q over an equilateral triangle we get $M_Q = \{\frac{\pi}{3}, \frac{\pi}{2}, \pi\}$.

Given any finite set $M \subseteq \mathbb{R}$ that contains M_P, and any $\mathbb{Q}$-linear function

$$f : V(M) \rightarrow \mathbb{Q}$$

that satisfies $f(\pi) = 0$, we define the *Dehn invariant* of P (with respect to f) to be the real number

$$D_f(P) := \sum_{e \in P} \ell(e) \cdot f(\alpha(e)),$$

where the sum extends over all edges e of the polyhedron, $\ell(e)$ denotes the length of e, and $\alpha(e)$ is the angle between the two facets that meet in e.

We will calculate various Dehn invariants later. For now just note that $f\left(\frac{\pi}{2}\right) = \frac{1}{2}f(\pi) = 0$ must hold for *any* such $\mathbb{Q}$-linear function f, and thus

$$D_f(C) = 0,$$

that is, the Dehn invariant of a cube is zero with respect to *any* f.

(3) The Dehn-Hadwiger theorem

As above we call two polyhedra P, Q *equidecomposable* if they can be decomposed into finite sets of polyhedra $P_1, \ldots, P_n$ and $Q_1, \ldots, Q_n$ such that P_i and Q_i are congruent for all i ($1 \leq i \leq n$). Two polyhedra are *equicomplementable* if there are polyhedra $P_1, \ldots, P_m$ and $Q_1, \ldots, Q_m$ so that the interiors of the P_i are disjoint from each other and from P, and similarly for the Q_i and Q, such that P_i is congruent to Q_i for all i, and such that $\widetilde{P} := P \cup P_1 \cup P_2 \cup \ldots \cup P_m$ and $\widetilde{Q} := Q \cup Q_1 \cup Q_2 \cup \ldots \cup Q_m$ are equidecomposable. A theorem of Gerling from 1844 implies that it does not matter whether we admit reflections when considering congruences, or not.

Clearly equidecomposable polyhedra are equicomplementable, but the converse is far from clear. The following theorem of Hadwiger (in the version of Boltianskii) provides our tool to find — as Hilbert proposed — tetrahedra of equal volume that are not equicomplementable, and thus not equidecomposable.

Theorem. *Let P and Q be polyhedra with dihedral angles $\alpha_1, \ldots, \alpha_p$ resp. $\beta_1, \ldots, \beta_q$ at their edges, and let M be a finite set of real numbers with*

$$\{\alpha_1, \ldots, \alpha_p, \beta_1, \ldots, \beta_q, \pi\} \subseteq M.$$

If $f : V(M) \rightarrow \mathbb{Q}$ is any $\mathbb{Q}$-linear function with $f(\pi) = 0$ such that

$$D_f(P) \neq D_f(Q),$$

then P and Q are not equicomplementable.

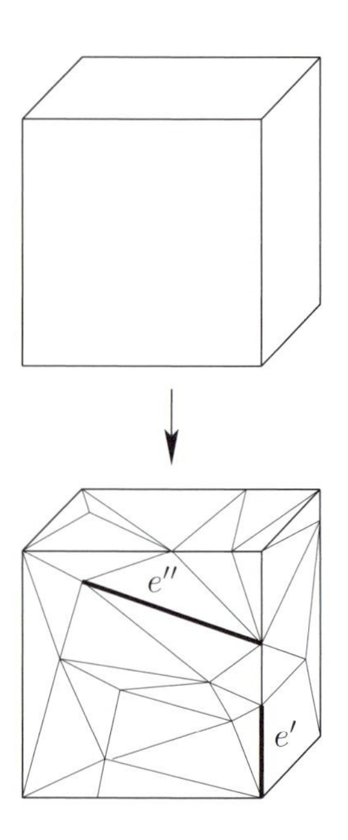

■ **Proof.** The argument comes in two parts.

(1) If a polyhedron P has a decomposition into finitely many polyhedral pieces $P_1, \ldots, P_n$, and if all the dihedral angles of the pieces $P_1, \ldots, P_n$ are contained in the set M, then for every $\mathbb{Q}$-linear $f : V(M) \to \mathbb{Q}$, the Dehn invariants add up:

$$D_f(P) \;=\; D_f(P_1) + \ldots + D_f(P_n).$$

For this, we associate a *mass* to any part of an edge of a polyhedron: if $e' \subseteq e$ is a part of an edge e of P, then its mass will be

$$m_f(e') \;:=\; \ell(e')\, f(\alpha(e')),$$

its length times the f-value of its dihedral angle.

Now if P is decomposed into $P_1, \ldots, P_n$, consider the union of all the edges of the pieces P_i. Along the edges e' that are contained in edges of P, we see that the dihedral angles of the pieces add up to the dihedral angle of P at e', and hence the masses add up.

At any other edge e'' of one of the P_i's which is contained in the interior of a face of P or in the interior of P, the angles add up to π or to 2π, so the f-values of the angles in the pieces add up to $f(\pi) = 0$ resp. to $f(2\pi) = 0$. Thus for the sum of the masses we get the same value that we had attached to these edges for P in the first place, namely 0.

(2) Assuming that P and Q are equicomplementable, we can enlarge M to a superset M' that also includes all the dihedral angles appearing in any of the pieces involved. M' is finite, since we only consider finite decompositions. Then our lemma above allows us to extend f to $f' : V(M') \to \mathbb{Q}$, and hence part **(1)** yields an equation of the type

$$D_{f'}(P)+D_{f'}(P_1)+\ldots+D_{f'}(P_m) = D_{f'}(Q)+D_{f'}(Q_1)+\ldots+D_{f'}(Q_m)$$

where $D_{f'}(P_i) \;=\; D_{f'}(Q_i)$ since P_i and Q_i are congruent. Hence we conclude $D_f(P) = D_f(Q)$, a contradiction. □

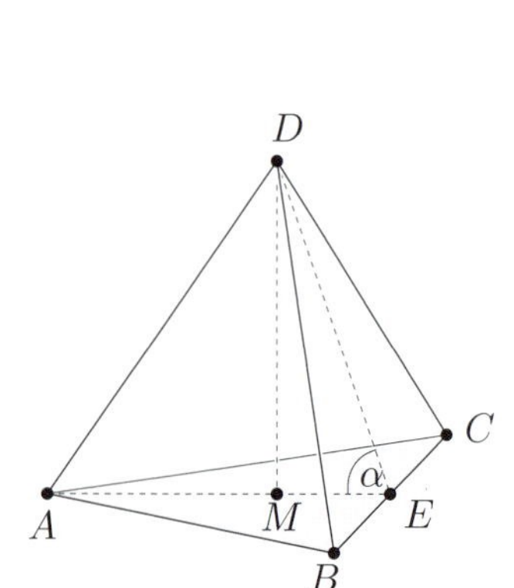

Example 1. For a regular tetrahedron T_0 with edge lengths ℓ, we calculate the dihedral angle from the sketch. The midpoint M of the base triangle divides the height AE of the base triangle by 1:2, and since $|AE| = |DE|$, we find $\cos\alpha = \frac{1}{3}$, and thus

$$\alpha \;=\; \arccos\tfrac{1}{3}.$$

Setting $M := \{\alpha, \pi\}$ we note that the ratio

$$\frac{\alpha}{\pi} \;=\; \frac{1}{\pi}\arccos\tfrac{1}{3}$$

is irrational, according to Theorem 3 of Chapter 6 (taking $n = 9$). Thus the $\mathbb{Q}$-vector space $V(M)$ is 2-dimensional with basis M, and there is a $\mathbb{Q}$-linear function $f : V(M) \to \mathbb{Q}$ with

$$f(\alpha) := 1,\; f(\pi) := 0.$$

For this f we have

$$D_f(T_0) \;=\; 6\ell f(\alpha) \;=\; 6\ell \;\neq\; 0,$$

and thus a *regular tetrahedron cannot be equidecomposable or equicomplementable with a cube*, since the Dehn invariant of a cube vanishes for any f.

Example 2. Let T_1 be a tetrahedron spanned by three orthogonal edges AB, AC, AD of length u. This tetrahedron has three dihedral angles that are right angles, and three more dihedral angles of equal size φ, which we calculate from the sketch as

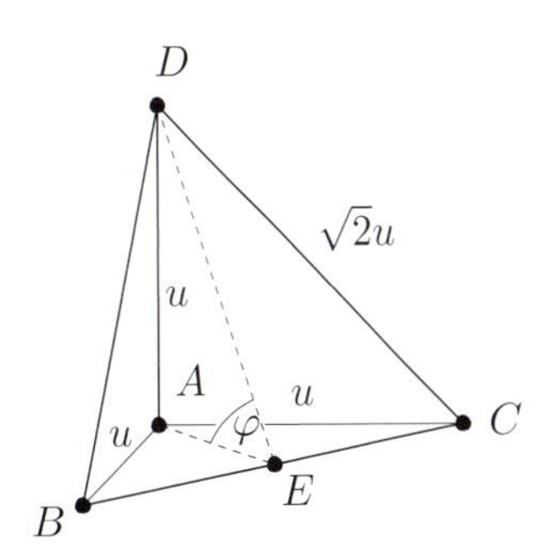

$$\cos\varphi \;=\; \frac{|AE|}{|DE|} \;=\; \frac{\frac{1}{2}\sqrt{2}u}{\frac{1}{2}\sqrt{3}\,\sqrt{2}u} \;=\; \frac{1}{\sqrt{3}}.$$

It follows that

$$\varphi \;=\; \arccos\frac{1}{\sqrt{3}}.$$

For $M := \{\frac{\pi}{2}, \arccos\frac{1}{\sqrt{3}}, \pi\}$, the $\mathbb{Q}$-vector space $V(M)$ has dimension 2. In fact, π and $\frac{\pi}{2}$ are linearly dependent, so $V(M) = V(\{\arccos\frac{1}{\sqrt{3}}, \pi\})$, but there is no rational relation between $\arccos\frac{1}{\sqrt{3}}$ and π — equivalently, $\frac{1}{\pi}\arccos\frac{1}{\sqrt{3}}$ is irrational, as we proved in Chapter 6 (take $n = 3$ in Thm. 3). Thus we may construct a $\mathbb{Q}$-linear map f by setting

$$f(\pi) := 0 \quad\text{and}\quad f\big(\arccos\tfrac{1}{\sqrt{3}}\big) := 1,$$

from which we obtain $f(\frac{\pi}{2}) = 0$ and hence

$$D_f(T_1) \;=\; 3uf\big(\tfrac{\pi}{2}\big) + 3(\sqrt{2}u)f\big(\arccos\tfrac{1}{\sqrt{3}}\big) \;=\; 3\sqrt{2}u \;\neq\; 0.$$

This proves that T_1 is not equidecomposable or equicomplementable with a cube C of the same volume, since $D_f(C) = 0$ holds for *any* f.

Example 3. Finally, let T_2 be a tetrahedron with three consecutive edges AB, BC and CD that are mutually orthogonal (an "orthoscheme") and of the same length u.

We will *not* calculate the angles in such a tetrahedron (they are $\frac{\pi}{2}$, $\frac{\pi}{3}$, and $\frac{\pi}{4}$), but rather argue that — using the midpoints of edges and faces, and the center — a cube of edge length u can be decomposed into 6 tetrahedra of this type (3 congruent copies, and 3 mirror images).

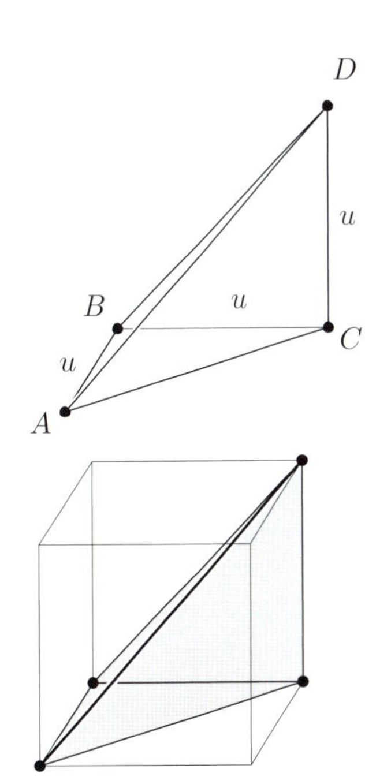

All these congruent copies and mirror images have the same Dehn invariants, and hence for every suitable functional f we will obtain

$$D_f(T_2) = \frac{1}{6}D_f(C) = 0$$

so all Dehn invariants of such a tetrahedron vanish! This solves Hilbert's third problem, since we have before constructed a different tetrahedron, T_1, with congruent bases and the same height, and with $D_f(T_1) \neq 0$. By the Dehn-Hadwiger theorem T_1 and T_2 are not equidecomposable, and not even equicomplementable.

Appendix: Polytopes and polyhedra

A *convex polytope* in $\mathbb{R}^d$ is the convex hull of a finite set $S = \{\boldsymbol{s}_1, \ldots, \boldsymbol{s}_n\}$, that is, a set of the form

$$P = \operatorname{conv}(S) := \Big\{\sum_{i=1}^{n} \lambda_i \boldsymbol{s}_i : \lambda_i \geq 0, \sum_{i=1}^{n} \lambda_i = 1\Big\}.$$

Polytopes are certainly familiar objects: prime examples are given by convex *polygons* (2-dimensional convex polytopes) and by convex *polyhedra* (3-dimensional convex polytopes).

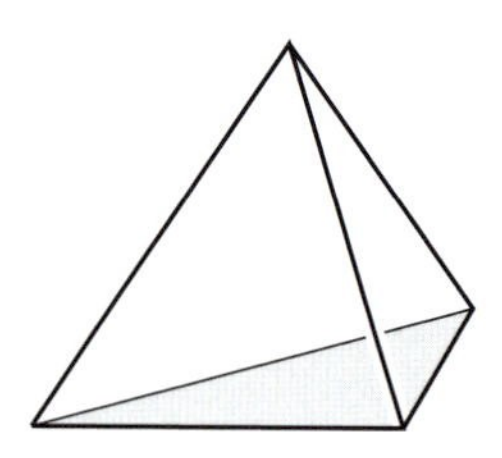

There are several types of polyhedra that generalize to higher dimensions in a natural way. For example, if the set S is affinely independent of cardinality $d+1$, then $\operatorname{conv}(S)$ is a d-dimensional *simplex* (or d-*simplex*). For $d = 2$ this yields a triangle, for $d = 3$ we obtain a tetrahedron. Similarly, squares and cubes are special cases of d-cubes, such as the *unit d-cube* given by $C_d = [0,1]^d \subseteq \mathbb{R}^d$.

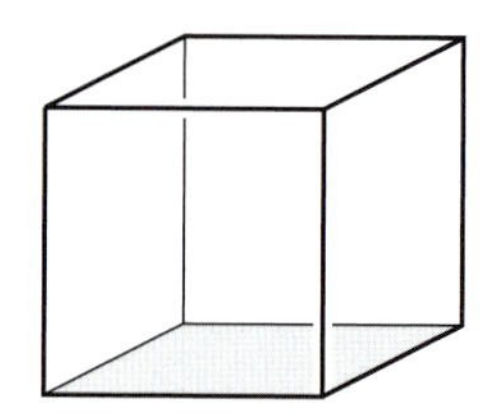

General polytopes are defined as finite unions of convex polytopes. In this book non-convex polyhedra will appear in connection with Cauchy's rigidity theorem in Chapter 12, and non-convex polygons in connection with Pick's theorem in Chapter 11, and again when we discuss the art gallery theorem in Chapter 31.

Convex polytopes can, equivalently, be defined as the bounded solution sets of finite systems of linear inequalities. Thus every convex polytope $P \subseteq \mathbb{R}^d$ has a representation of the form

$$P = \{\boldsymbol{x} \in \mathbb{R}^d : A\boldsymbol{x} \leq \boldsymbol{b}\}$$

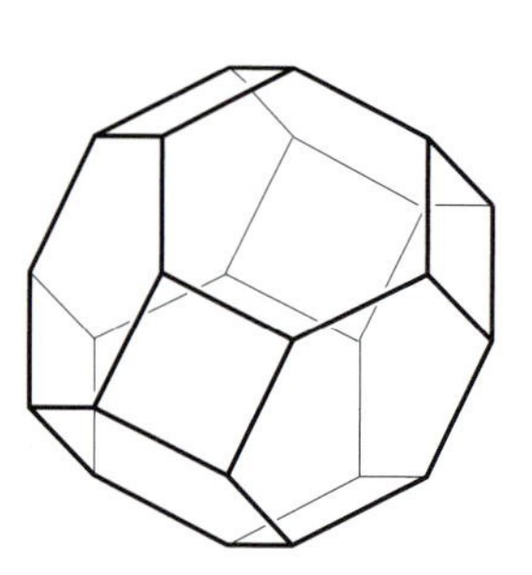

Some familiar polytopes: tetrahedron, cube and permutahedron

for some matrix $A \in \mathbb{R}^{m \times d}$ and a vector $\boldsymbol{b} \in \mathbb{R}^m$. In other words, P is the solution set of a system of m linear inequalities $\boldsymbol{a}_i^T \boldsymbol{x} \leq b_i$, where $\boldsymbol{a}_i^T$ is the i-th row of A. Conversely, every bounded such solution set is a convex polytope, and can thus be represented as the convex hull of a finite set of points.

For polygons and polyhedra, we have the familiar concepts of *vertices*, *edges*, and 2-*faces*. For higher-dimensional convex polytopes, we can define their faces as follows: a *face* of P is a subset $F \subseteq P$ of the form $P \cap \{\boldsymbol{x} \in \mathbb{R}^d : \boldsymbol{a}^T\boldsymbol{x} = b\}$, where $\boldsymbol{a}^T\boldsymbol{x} \leq b$ is a linear inequality that is valid for all points $\boldsymbol{x} \in P$.

All the faces of a polytope are themselves polytopes. The set V of vertices (0-dimensional faces) of a convex polytope is also the inclusion-minimal set such that $\operatorname{conv}(V) = P$. Assuming that $P \subseteq \mathbb{R}^d$ is a d-dimensional convex polytope, the *facets* (the $(d-1)$-dimensional faces) determine a minimal set of hyperplanes and thus of halfspaces that contain P, and whose intersection is P. In particular, this implies the following fact that we will need later: Let F be a facet of P, denote by H_F the hyperplane it determines, and by H_F^+ and H_F^- the two closed half-spaces bounded by H_F. Then one of these two halfspaces contains P (and the other one doesn't).

The *graph* $G(P)$ of the convex polytope P is given by the set V of vertices, and by the edge set E of 1-dimensional faces. If P has dimension 3, then this graph is planar, and gives rise to the famous "Euler polyhedron formula" (see Chapter 11).

Two polytopes $P, P' \subseteq \mathbb{R}^d$ are *congruent* if there is some length-preserving affine map that takes P to P'. Such a map may reverse the orientation of space, as does the reflection of P in a hyperplane, which takes P to a *mirror image* of P. They are *combinatorially equivalent* if there is a bijection from the faces of P to the faces of P' that preserves dimension and inclusions between the faces. This notion of combinatorial equivalence is much weaker than congruence: for example, our figure shows a unit cube and a "skew" cube that are combinatorially equivalent (and thus we would call any one of them "a cube"), but they are certainly not congruent.

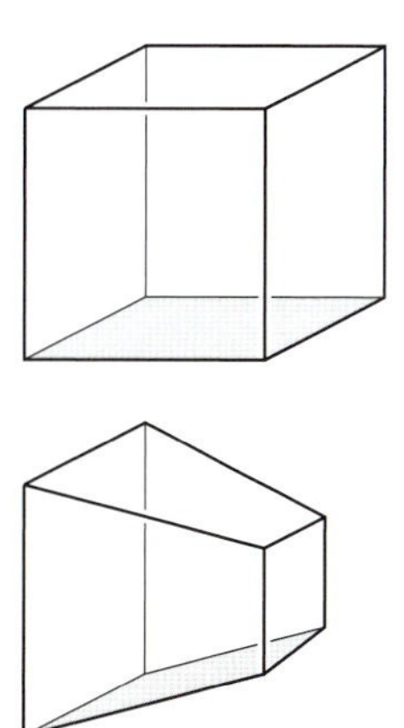
Combinatorially equivalent polytopes

A polytope (or a more general subset of $\mathbb{R}^d$) is called *centrally symmetric* if there is some point $\boldsymbol{x}_0 \in \mathbb{R}^d$ such that

$$\boldsymbol{x}_0 + \boldsymbol{x} \in P \iff \boldsymbol{x}_0 - \boldsymbol{x} \in P.$$

In this situation we call $\boldsymbol{x}_0$ the *center* of P.

References

[1] V. G. Boltianskii: *Hilbert's Third Problem,* V. H. Winston & Sons (Halsted Press, John Wiley & Sons), Washington DC 1978.

[2] M. Dehn: *Ueber raumgleiche Polyeder,* Nachrichten von der Königl. Gesellschaft der Wissenschaften, Mathematisch-physikalische Klasse (1900), 345-354.

[3] M. Dehn: *Ueber den Rauminhalt,* Mathematische Annalen **55** (1902), 465-478.

[4] C. F. Gauss: *"Congruenz und Symmetrie": Briefwechsel mit Gerling,* pp. 240-249 in: Werke, Band VIII, Königl. Gesellschaft der Wissenschaften zu Göttingen; B. G. Teubner, Leipzig 1900.

[5] D. Hilbert: *Mathematical Problems,* Lecture delivered at the International Congress of Mathematicians at Paris in 1900, Bulletin Amer. Math. Soc. **8** (1902), 437-479.

[6] G. M. Ziegler: *Lectures on Polytopes,* Graduate Texts in Mathematics **152**, Springer-Verlag, New York 1995/1998.

Lines in the plane and decompositions of graphs

Chapter 9

Perhaps the best-known problem on configurations of lines was raised by Sylvester in 1893 in a column of mathematical problems.

J. J. Sylvester

QUESTIONS FOR SOLUTION.

11851. (Professor Sylvester.)—**Prove that it is not possible to arrange any finite number of real points so that a right line through every two of them shall pass through a third, unless they all lie in the same right line.**

Whether Sylvester himself had a proof is in doubt, but a correct proof was given by Tibor Gallai [Grünwald] some 40 years later. Therefore the following theorem is commonly attributed to Sylvester and Gallai. Subsequent to Gallai's proof several others appeared, but the following argument due to L. M. Kelly may be "simply the best."

Theorem 1. *In any configuration of n points in the plane, not all on a line, there is a line which contains exactly two of the points.*

■ **Proof.** Let $\mathcal{P}$ be the given set of points and consider the set $\mathcal{L}$ of all lines which pass through at least two points of $\mathcal{P}$. Among all pairs (P, ℓ) with P not on ℓ, choose a pair (P_0, ℓ_0) such that P_0 has the smallest distance to ℓ_0, with Q being the point on ℓ_0 closest to P_0 (that is, on the line through P_0 vertical to ℓ_0).

Claim. *This line ℓ_0 does it!*

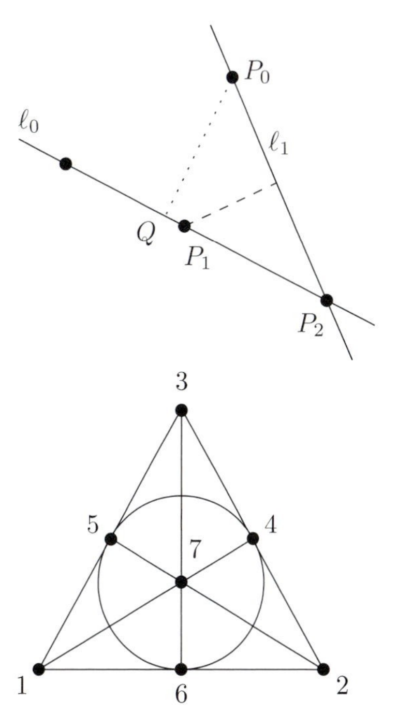

If not, then ℓ_0 contains at least three points of $\mathcal{P}$, and thus two of them, say P_1 and P_2, lie on the same side of Q. Let us assume that P_1 lies between Q and P_2, where P_1 possibly coincides with Q. The figure on the right shows the configuration. It follows that the distance of P_1 to the line ℓ_1 determined by P_0 and P_2 is smaller than the distance of P_0 to ℓ_0, and this contradicts our choice for ℓ_0 and P_0. □

In the proof we have used metric axioms (shortest distance) and order axioms (P_1 lies between Q and P_2) of the real plane. Do we really need these properties beyond the usual incidence axioms of points and lines? Well, some additional condition is required, as the famous Fano plane depicted in the margin demonstrates. Here $\mathcal{P} = \{1, 2, \ldots, 7\}$ and $\mathcal{L}$ consists of the 7 three-point lines as indicated in the figure, including the "line" $\{4, 5, 6\}$. Any two points determine a unique line, so the incidence axioms are satisfied, but there is no 2-point line. The Sylvester-Gallai theorem therefore shows that the Fano configuration cannot be embedded into the real plane such that the seven collinear triples lie on real lines: there must always be a "crooked" line.

However, it was shown by Coxeter that the order axioms will suffice for a proof of the Sylvester-Gallai theorem. Thus one can devise a proof that does not use any metric properties — see also the proof that we will give in Chapter 11, using Euler's formula.

The Sylvester-Gallai theorem directly implies another famous result on points and lines in the plane, due to Paul Erdős and Nicolaas G. de Bruijn. But in this case the result holds more generally for arbitrary point-line systems, as was observed already by Erdős and de Bruijn. We will discuss the more general result in a moment.

Theorem 2. *Let $\mathcal{P}$ be a set of $n \geq 3$ points in the plane, not all on a line. Then the set $\mathcal{L}$ of lines passing through at least two points contains at least n lines.*

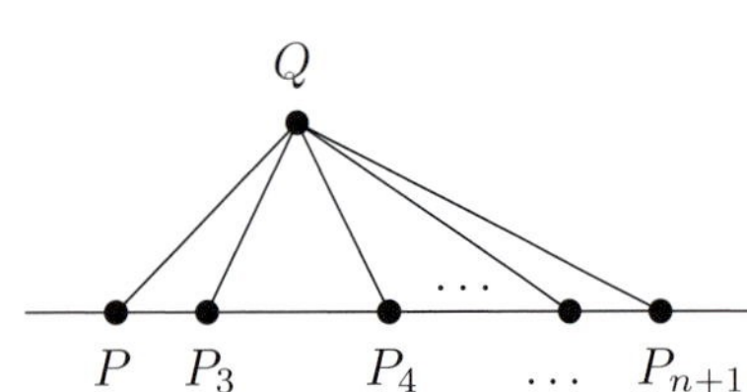

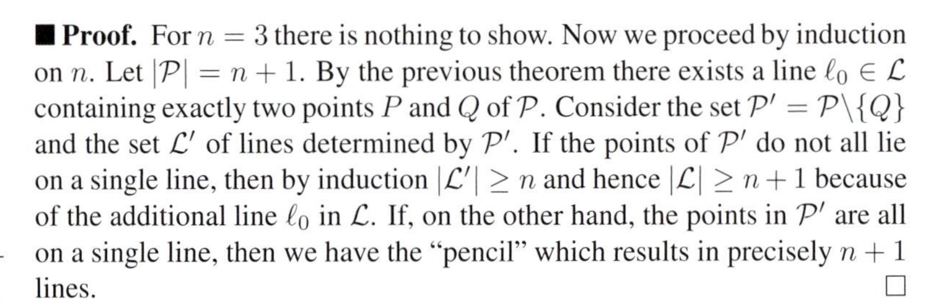

■ **Proof.** For $n = 3$ there is nothing to show. Now we proceed by induction on n. Let $|\mathcal{P}| = n + 1$. By the previous theorem there exists a line $\ell_0 \in \mathcal{L}$ containing exactly two points P and Q of $\mathcal{P}$. Consider the set $\mathcal{P}' = \mathcal{P} \backslash \{Q\}$ and the set $\mathcal{L}'$ of lines determined by $\mathcal{P}'$. If the points of $\mathcal{P}'$ do not all lie on a single line, then by induction $|\mathcal{L}'| \geq n$ and hence $|\mathcal{L}| \geq n + 1$ because of the additional line ℓ_0 in $\mathcal{L}$. If, on the other hand, the points in $\mathcal{P}'$ are all on a single line, then we have the "pencil" which results in precisely $n + 1$ lines. □

Now, as promised, here is the general result, which applies to much more general "incidence geometries."

Theorem 3. *Let X be a set of $n \geq 3$ elements, and let $A_1, \dots, A_m$ be proper subsets of X, such that every pair of elements of X is contained in precisely one set A_i. Then $m \geq n$ holds.*

■ **Proof.** The following proof, variously attributed to Motzkin or Conway, is almost one-line and truly inspired. For $x \in X$ let r_x be the number of sets A_i containing x. (Note that $2 \leq r_x < m$ by the assumptions.) Now if $x \notin A_i$, then $r_x \geq |A_i|$ because the $|A_i|$ sets containing x and an element of A_i must be distinct. Suppose $m < n$, then $m|A_i| < n\,r_x$ and thus $m(n - |A_i|) > n(m - r_x)$ for $x \notin A_i$, and we find

$$1 = \sum_{x \in X} \tfrac{1}{n} = \sum_{x \in X} \sum_{A_i : x \notin A_i} \tfrac{1}{n(m - r_x)} > \sum_{A_i} \sum_{x : x \notin A_i} \tfrac{1}{m(n - |A_i|)} = \sum_{A_i} \tfrac{1}{m} = 1,$$

which is absurd. □

There is another very short proof for this theorem that uses linear algebra. Let B be the *incidence matrix* of $(X; A_1, \dots, A_m)$, that is, the rows in B are indexed by the elements of X, the columns by $A_1, \dots, A_m$, where

$$B_{xA} := \begin{cases} 1 & \text{if } x \in A \\ 0 & \text{if } x \notin A. \end{cases}$$

Consider the product BB^T. For $x \neq x'$ we have $(BB^T)_{xx'} = 1$, since x

and x' are contained in precisely one set A_i, hence

$$BB^T = \begin{pmatrix} r_{x_1}-1 & 0 & \dots & 0 \\ 0 & r_{x_2}-1 & & \vdots \\ \vdots & & \ddots & 0 \\ 0 & \dots & 0 & r_{x_n}-1 \end{pmatrix} + \begin{pmatrix} 1 & 1 & \dots & 1 \\ 1 & 1 & & \vdots \\ \vdots & & \ddots & 1 \\ 1 & \dots & 1 & 1 \end{pmatrix}$$

where r_x is defined as above. Since the first matrix is positive definite (it has only positive eigenvalues) and the second matrix is positive semi-definite (it has the eigenvalues n and 0), we deduce that BB^T is positive definite and thus, in particular, invertible, implying $\text{rank}(BB^T) = n$. It follows that the rank of the $(n \times m)$-matrix B is at least n, and we conclude that indeed $n \leq m$, since the rank cannot exceed the number of columns.

Let us go a little beyond and turn to graph theory. (We refer to the review of basic graph concepts in the appendix to this chapter.) A moment's thought shows that the following statement is really the same as Theorem 3:

If we decompose a complete graph K_n into m cliques different from K_n, such that every edge is in a unique clique, then $m \geq n$.

Indeed, let X correspond to the vertex set of K_n and the sets A_i to the vertex sets of the cliques, then the statements are identical.

Our next task is to decompose K_n into complete bipartite graphs such that again every edge is in exactly one of these graphs. There is an easy way to do this. Number the vertices $\{1, 2, \dots, n\}$. First take the complete bipartite graph joining 1 to all other vertices. Thus we obtain the graph $K_{1,n-1}$ which is called a *star*. Next join 2 to $3, \dots, n$, resulting in a star $K_{1,n-2}$. Going on like this, we decompose K_n into stars $K_{1,n-1}, K_{1,n-2}, \dots, K_{1,1}$. This decomposition uses $n-1$ complete bipartite graphs. Can we do better, that is, use fewer graphs? No, as the following result of Ron Graham and Henry O. Pollak says:

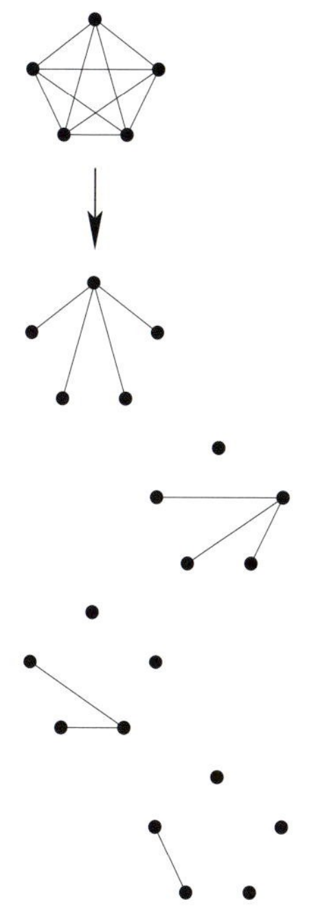

A decomposition of K_5 into 4 complete bipartite subgraphs

Theorem 4. *If K_n is decomposed into complete bipartite subgraphs $H_1, \dots, H_m$, then $m \geq n - 1$.*

The interesting thing is that, in contrast to the Erdős-de Bruijn theorem, no combinatorial proof for this result is known! All of them use linear algebra in one way or another. Of the various more or less equivalent ideas let us look at the proof due to Tverberg, which may be the most transparent.

■ **Proof.** Let the vertex set of K_n be $\{1, \dots, n\}$, and let L_j, R_j be the defining vertex sets of the complete bipartite graph H_j, $j = 1, \dots, m$. To every vertex i we associate a variable x_i. Since $H_1, \dots, H_m$ decompose K_n, we find

$$\sum_{i<j} x_i x_j \;=\; \sum_{k=1}^{m} \Big(\sum_{a \in L_k} x_a \cdot \sum_{b \in R_k} x_b \Big). \qquad (1)$$

Now suppose the theorem is false, $m < n - 1$. Then the system of linear equations

$$\begin{aligned} x_1 + \ldots + x_n &= 0, \\ \sum_{a \in L_k} x_a &= 0 \qquad (k = 1, \ldots, m) \end{aligned}$$

has fewer equations than variables, hence there exists a non-trivial solution $c_1, \ldots, c_n$. From (1) we infer

$$\sum_{i<j} c_i c_j = 0.$$

But this implies

$$0 = (c_1 + \ldots + c_n)^2 = \sum_{i=1}^{n} c_i^2 + 2\sum_{i<j} c_i c_j = \sum_{i=1}^{n} c_i^2 > 0,$$

a contradiction, and the proof is complete. □

Appendix: Basic graph concepts

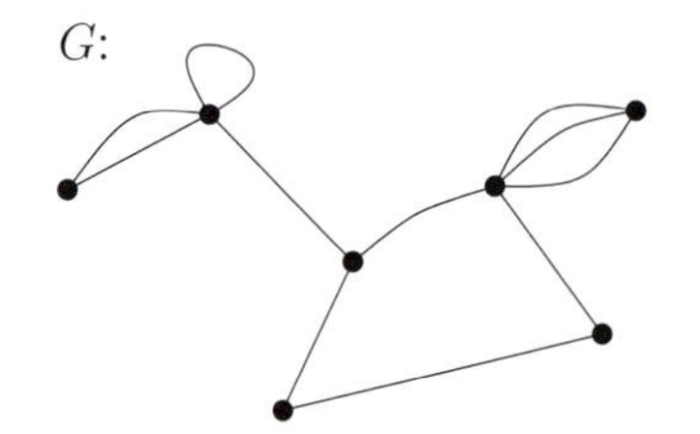

A graph G with 7 vertices and 11 edges. It has one loop, one double edge and one triple edge.

Graphs are among the most basic of all mathematical structures. Correspondingly, they have many different versions, representations, and incarnations. Abstractly, a *graph* is a pair $G = (V, E)$, where V is the set of *vertices*, E is the set of *edges*, and each edge $e \in E$ "connects" two vertices $v, w \in V$. We consider only finite graphs, where V and E are finite.

Usually, we deal with *simple graphs*: Then we do not admit *loops*, i. e., edges for which both ends coincide, and no *multiple edges* that have the same set of endvertices. Vertices of a graph are called *adjacent* or *neighbors* if they are the endvertices of an edge. A vertex and an edge are called *incident* if the edge has the vertex as an endvertex.

Here is a little picture gallery of important (simple) graphs:

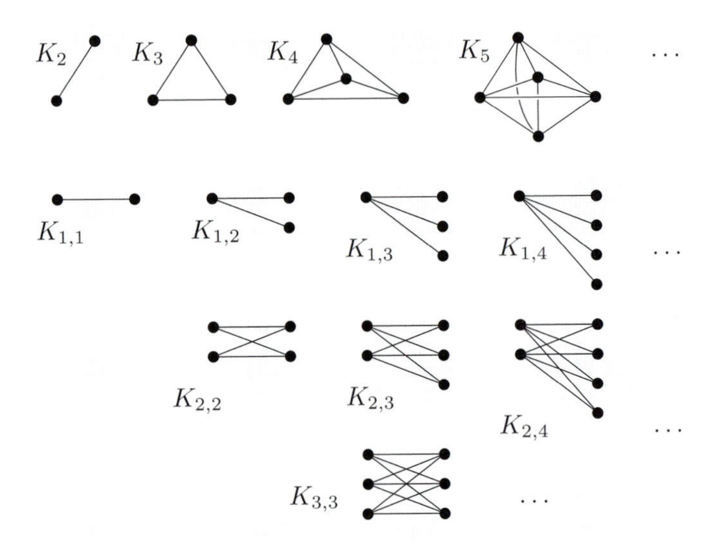

The *complete graphs* K_n on n vertices and $\binom{n}{2}$ edges

The *complete bipartite graphs* $K_{m,n}$ with $m + n$ vertices and mn edges

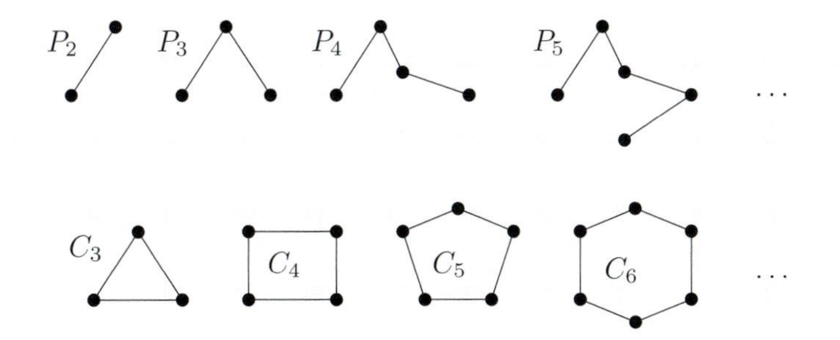

The *paths* P_n with n vertices

The *cycles* C_n with n vertices

Two graphs $G = (V, E)$ and $G' = (V', E')$ are considered *isomorphic* if there are bijections $V \to V'$ and $E \to E'$ that preserve the incidences between edges and their endvertices. (It is a major unsolved problem whether there is an efficient test to decide whether two given graphs are isomorphic.) This notion of isomorphism allows us to talk about *the* complete graph K_5 on 5 vertices, etc.

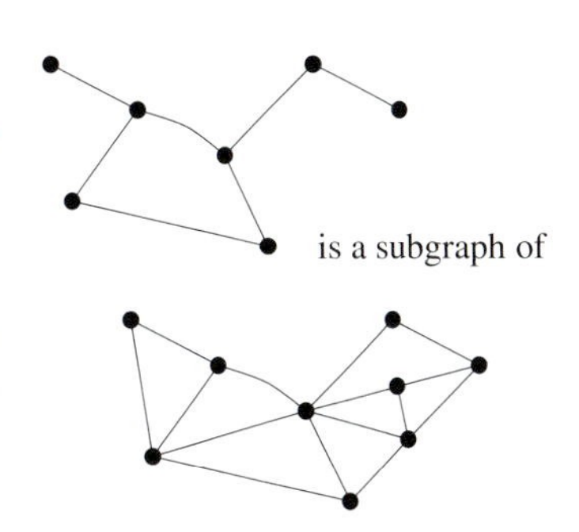

$G' = (V', E')$ is a *subgraph* of $G = (V, E)$ if $V' \subseteq V, E' \subseteq E$, and every edge $e \in E'$ has the same endvertices in G' as in G. G' is an *induced subgraph* if, additionally, *all* edges of G that connect vertices of G' are also edges of G'.

Many notions about graphs are quite intuitive: for example, a graph G is *connected* if every two distinct vertices are connected by a path in G, or equivalently, if G cannot be split into two nonempty subgraphs whose vertex sets are disjoint.

We end this survey of basic graph concepts with a few more pieces of terminology: A *clique* in G is a complete subgraph. An *independent set* in G is an induced subgraph without edges, that is, a subset of the vertex set such that no two vertices are connected by an edge of G. A graph is a *forest* if it does not contain any cycles. A *tree* is a connected forest. Finally, a graph $G = (V, E)$ is *bipartite* if it is isomorphic to a subgraph of a complete bipartite graph, that is, if its vertex set can be written as a union $V = V_1 \cup V_2$ of two independent sets.

References

[1] N. G. DE BRUIJN & P. ERDŐS: *On a combinatorial problem,* Proc. Kon. Ned. Akad. Wetensch. **51** (1948), 1277-1279.

[2] H. S. M. COXETER: *A problem of collinear points,* Amer. Math. Monthly **55** (1948), 26-28 (contains Kelly's proof).

[3] P. ERDŐS: *Problem 4065 — Three point collinearity,* Amer. Math. Monthly **51** (1944), 169-171 (contains Gallai's proof).

[4] R. L. GRAHAM & H. O. POLLAK: *On the addressing problem for loop switching,* Bell System Tech. J. **50** (1971), 2495-2519.

[5] J. J. SYLVESTER: *Mathematical Question 11851,* The Educational Times **46** (1893), 156.

[6] H. TVERBERG: *On the decomposition of K_n into complete bipartite graphs,* J. Graph Theory **6** (1982), 493-494.

The slope problem

Chapter 10

Try for yourself — before you read much further — to construct configurations of points in the plane that determine "relatively few" slopes. For this we assume, of course, that the $n \geq 3$ points do not all lie on one line. Recall from Chapter 9 on "Lines in the plane" the theorem of Erdős and de Bruijn: the n points will determine at least n different lines. But of course many of these lines may be parallel, and thus determine the same slope.

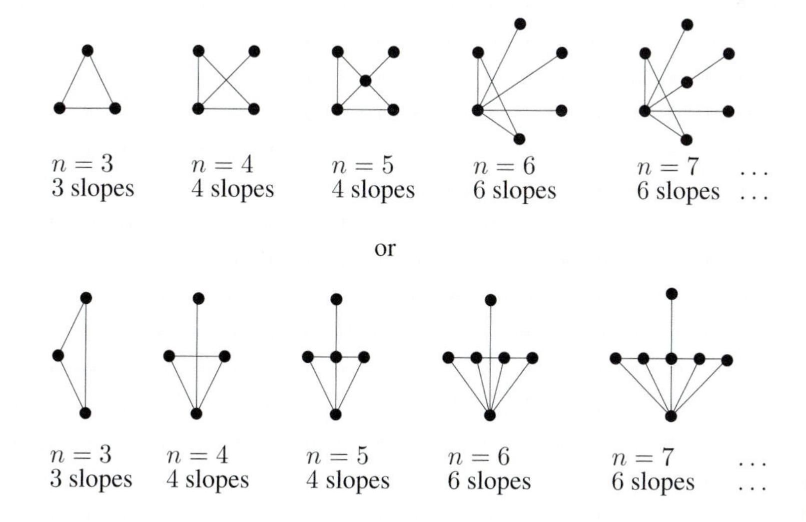

A little experimentation for small n will probably lead you to a sequence such as the two depicted here.

After some attempts at finding configurations with fewer slopes you might conjecture — as Scott did in 1970 — the following theorem.

Theorem. *If $n \geq 3$ points in the plane do not lie on one single line, then they determine at least $n-1$ different slopes, where equality is possible only if n is odd and $n \geq 5$.*

Our examples above — the drawings represent the first few configurations in two infinite sequences of examples — show that the theorem as stated is *best possible*: for any odd $n \geq 5$ there is a configuration with n points that determines exactly $n-1$ different slopes, and for any other $n \geq 3$ we have a configuration with exactly n slopes.

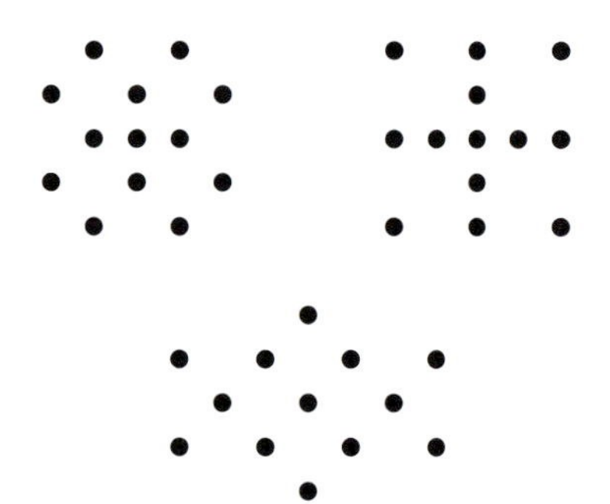

Three pretty sporadic examples from the Jamison-Hill catalogue

However, the configurations that we have drawn above are by far not the only ones. For example, Jamison and Hill described four infinite families of configurations, each of them consisting of configurations with an odd number n of points that determine only $n-1$ slopes ("slope-critical configurations"). Furthermore, they listed 102 "sporadic" examples that do not seem to fit into an infinite family, most of them found by extensive computer searches.

Conventional wisdom might say that extremal problems tend to be very difficult to solve exactly if the extreme configurations are so diverse and irregular. Indeed, there is a lot that can be said about the structure of slope-critical configurations (see [2]), but a classification seems completely out of reach. However, the theorem above has a simple proof, which has two main ingredients: a reduction to an efficient combinatorial model due to Eli Goodman and Ricky Pollack, and a beautiful argument in this model by which Peter Ungar completed the proof in 1982.

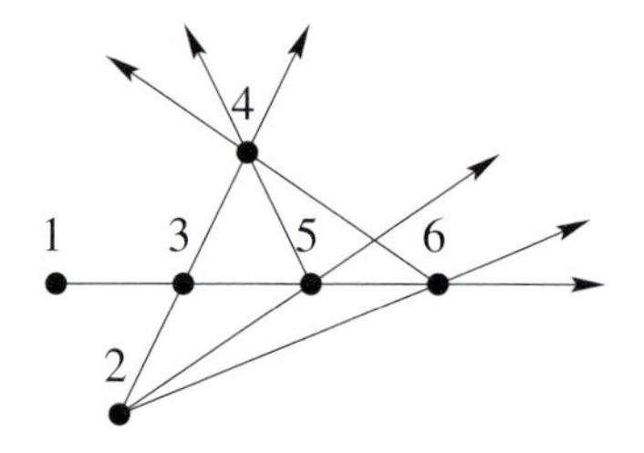

This configuration of $n = 6$ points determines $t = 6$ different slopes.

■ **Proof.** **(1)** First we notice that it suffices to show that every "even" set of $n = 2m$ points in the plane $(m \geq 2)$ determines at least n slopes. This is so since the case $n = 3$ is trivial, and for any set of $n = 2m + 1 \geq 5$ points (not all on a line) we can find a subset of $n - 1 = 2m$ points, not all on a line, which already determines $n - 1$ slopes.

Thus for the following we consider a configuration of $n = 2m$ points in the plane that determines $t \geq 2$ different slopes.

(2) The combinatorial model is obtained by constructing a periodic sequence of permutations. For this we start with some direction in the plane that is not one of the configuration's slopes, and we number the points $1, \ldots, n$ in the order in which they appear in the 1-dimensional projection in this direction. Thus the permutation $\pi_0 = 123...n$ represents the order of the points for our starting direction.

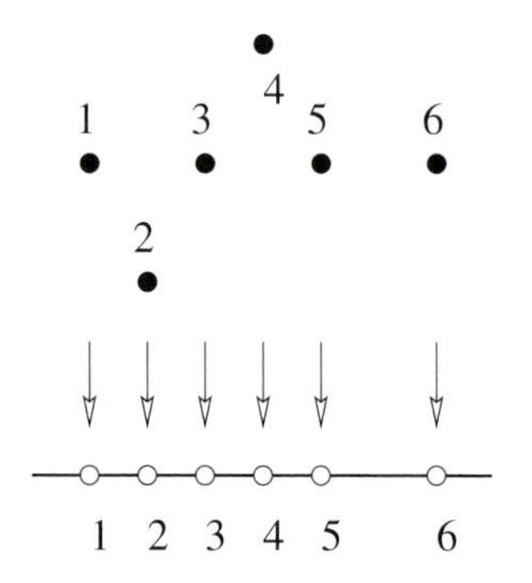

Here a vertical starting direction yields $\pi_0 = 123456$.

Next let the direction perform a counterclockwise motion, and watch how the projection and its permutation change. Changes in the order of the projected points appear exactly when the direction passes over one of the configuration's slopes.

But the changes are far from random or arbitrary: By performing a $180°$ rotation of the direction, we obtain a sequence of permutations

$$\pi_0 \to \pi_1 \to \pi_2 \to \ldots \to \pi_{t-1} \to \pi_t$$

which has the following special properties:

- The sequence starts with $\pi_0 = 123...n$ and ends with $\pi_t = n...321$.
- The length t of the sequence is the number of slopes of the point configuration.
- In the course of the sequence, every pair $i < j$ is switched exactly once. This means that on the way from $\pi_0 = 123...n$ to $\pi_t = n...321$, only *increasing* substrings are reversed.

- Every move consists in the reversal of one or more disjoint increasing substrings (corresponding to the one or more lines that have the direction which we pass at this point).

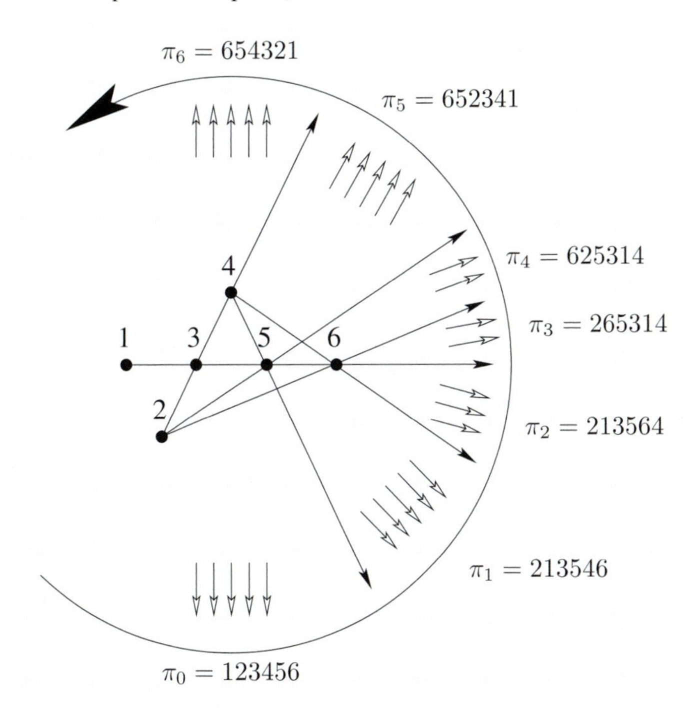

Getting the sequence of permutations for our small example

By continuing the circular motion around the configuration, one can view the sequence as a part of a two-way infinite, periodic sequence of permutations

$$\ldots \longrightarrow \pi_{-1} \longrightarrow \pi_0 \longrightarrow \ldots \longrightarrow \pi_t \longrightarrow \pi_{t+1} \longrightarrow \ldots \longrightarrow \pi_{2t} \longrightarrow \ldots$$

where π_{i+t} is the reverse of π_i for all i, and thus $\pi_{i+2t} = \pi_i$ for all $i \in \mathbb{Z}$.

We will show that *every* sequence with the above properties (and $t \geq 2$) must have length $t \geq n$.

(3) The proof's key is to divide each permutation into a "left half" and a "right half" of equal size $m = \frac{n}{2}$, and to count the letters that cross the imaginary *barrier* between the left half and the right half.

Call $\pi_i \longrightarrow \pi_{i+1}$ a *crossing move* if one of the substrings it reverses does involve letters from both sides of the barrier. The crossing move has *order* d if it moves $2d$ letters across the barrier, that is, if the crossing string has exactly d letters on one side and at least d letters on the other side. Thus in our example

$$\pi_2 = 2\underline{13{:}564} \longrightarrow \overline{265{:}314} = \pi_3$$

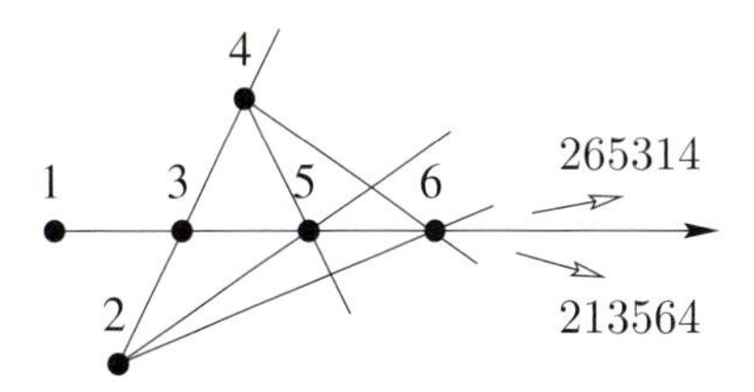

A crossing move

is a crossing move of order $d = 2$ (it moves $1, 3, 5, 6$ across the barrier, which we mark by ":"),

$$65\underline{2{:}34}1 \longrightarrow 65\overline{4{:}32}1$$

is crossing of order $d = 1$, while for example

$$6\underline{25}{:}3\underline{14} \longrightarrow 6\overline{52}{:}3\overline{41}$$

is not a crossing move.

In the course of the sequence $\pi_0 \to \pi_1 \to \ldots \to \pi_t$, each of the letters $1, 2, \ldots, n$ has to cross the barrier at least once. This implies that, if the orders of the c crossing moves are $d_1, d_2, \ldots, d_c$, then we have

$$\sum_{i=1}^{c} 2d_i \;=\; \#\{\text{letters that cross the barrier}\} \;\geq\; n.$$

This also implies that we have at least two crossing moves, since a crossing move with $2d_i = n$ occurs only if all the points are on one line, i. e. for $t = 1$. Geometrically, a crossing move corresponds to the direction of a line of the configuration that has less than m points on each side.

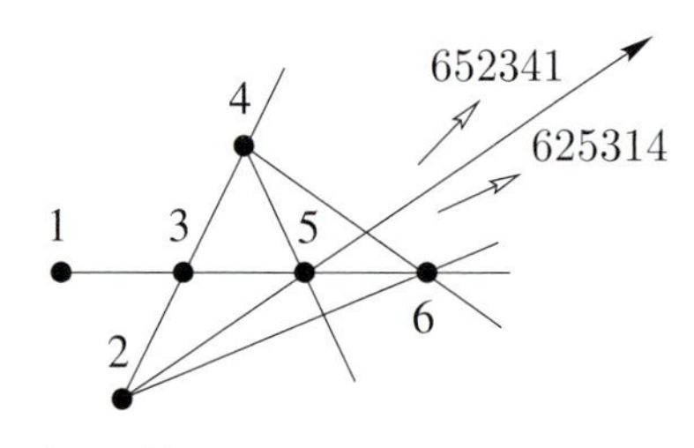

A touching move

(4) A *touching move* is a move that reverses some string that is adjacent to the central barrier, but does not cross it. For example,

$$\pi_4 \;=\; 6\underline{25}{:}3\underline{14} \longrightarrow 6\overline{52}{:}3\overline{41} \;=\; \pi_5$$

is a touching move. Geometrically, a touching move corresponds to the slope of a line of the configuration that has exactly m points on one side, and hence at most $m-2$ points on the other side.

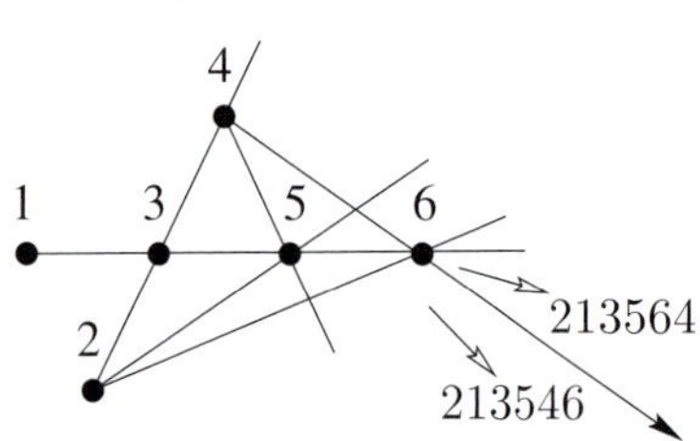

An ordinary move

Moves that are neither touching nor crossing will be called *ordinary moves*. For this

$$\pi_1 \;=\; 213{:}5\underline{46} \longrightarrow 213{:}5\overline{64} \;=\; \pi_2$$

is an example. So every move is either crossing, or touching, or ordinary, and we can use the letters T, C, O to denote the types of moves. $C(d)$ will denote a crossing move of order d. Thus for our small example we get

$$\pi_0 \xrightarrow{T} \pi_1 \xrightarrow{O} \pi_2 \xrightarrow{C(2)} \pi_3 \xrightarrow{O} \pi_4 \xrightarrow{T} \pi_5 \xrightarrow{C(1)} \pi_6,$$

or even shorter we can record this sequence as $T, O, C(2), O, T, C(1)$.

(5) To complete the proof, we need the following two facts:

> *Between any two crossing moves, there is at least one touching move.*
>
> *Between any crossing move of order d and the next touching move, there are at least $d-1$ ordinary moves.*

In fact, after a crossing move of order d the barrier is contained in a symmetric decreasing substring of length $2d$, with d letters on each side of the barrier. For the next crossing move the central barrier must be brought into an increasing substring of length at least 2. But only touching moves affect whether the barrier is in an increasing substring. This yields the first fact.

For the second fact, note that with each ordinary move (reversing some *increasing* substrings) the decreasing $2d$-string can get shortened by only one letter on each side. And, as long as the decreasing string has at least 4 letters, a touching move is impossible. This yields the second fact.

If we construct the sequence of permutations starting with the same initial projection but using a clockwise rotation, then we obtain the reversed sequence of permutations. Thus the sequence that we do have recorded must also satisfy the opposite of our second fact, namely

> *Between a touching move and the next crossing move, of order* d*, there are at least* $d-1$ *ordinary moves.*

(6) The T-O-C-pattern of the infinite sequence of permutations, as derived in **(2)**, is obtained by repeating over and over again the T-O-C-pattern of length t of the sequence $\pi_0 \longrightarrow \ldots \longrightarrow \pi_t$. Thus with the facts of **(5)** we see that in the infinite sequence of moves, each crossing move of order d is embedded into a T-O-C-pattern of the type

$$T, \underbrace{O, O, \ldots, O}_{\geq\, d-1}, C(d), \underbrace{O, O, \ldots, O}_{\geq\, d-1}, \qquad (*)$$

of length $1 + (d-1) + 1 + (d-1) = 2d$.

In the infinite sequence, we may consider a finite segment of length t that starts with a touching move. This segment consists of substrings of the type $(*)$, plus possibly extra inserted T's. This implies that its length t satisfies

$$t \;\geq\; \sum_{i=1}^{c} 2d_i \;\geq\; n,$$

which completes the proof. □

References

[1] J. E. Goodman & R. Pollack: *A combinatorial perspective on some problems in geometry,* Congressus Numerantium **32** (1981), 383-394.

[2] R. E. Jamison & D. Hill: *A catalogue of slope-critical configurations,* Congressus Numerantium **40** (1983), 101-125.

[3] P. R. Scott: *On the sets of directions determined by* n *points,* Amer. Math. Monthly **77** (1970), 502-505.

[4] P. Ungar: $2N$ *noncollinear points determine at least* $2N$ *directions,* J. Combinatorial Theory, Ser. A **33** (1982), 343-347.

Three applications of Euler's formula

Chapter 11

A graph is *planar* if it can be drawn in the plane $\mathbb{R}^2$ without crossing edges (or, equivalently, on the 2-dimensional sphere S^2). We talk of a *plane* graph if such a drawing is already given and fixed. Any such drawing decomposes the plane or sphere into a finite number of connected regions, including the outer (unbounded) region, which are referred to as *faces*. Euler's formula exhibits a beautiful relation between the number of vertices, edges and faces that is valid for any plane graph. Euler mentioned this result for the first time in a letter to his friend Goldbach in 1750, but he did not have a complete proof at the time. Among the many proofs of Euler's formula, we present a pretty and "self-dual" one that gets by without induction. It can be traced back to von Staudt's book "Geometrie der Lage" from 1847.

Leonhard Euler

Euler's formula. *If G is a connected plane graph with n vertices, e edges and f faces, then*

$$n - e + f = 2.$$

■ **Proof.** Let $T \subseteq E$ be the edge set of a spanning tree for G, that is, of a minimal subgraph that connects all the vertices of G. This graph does not contain a cycle because of the minimality assumption.

We now need the *dual graph* G^* of G: to construct it, put a vertex into the interior of each face of G, and connect two such vertices of G^* by edges that correspond to common boundary edges between the corresponding faces. If there are several common boundary edges, then we draw several connecting edges in the dual graph. (Thus G^* may have multiple edges even if the original graph G is simple.)

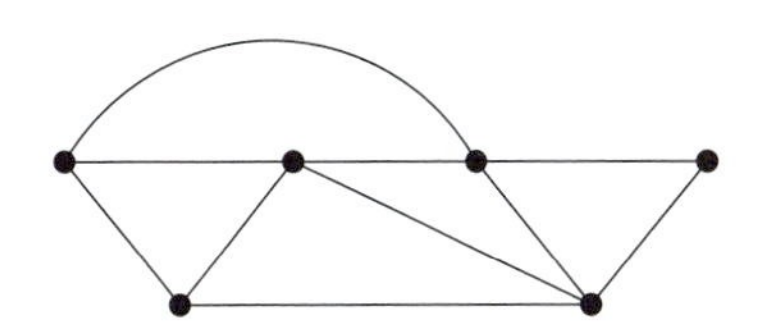
A plane graph G: $n = 6$, $e = 10$, $f = 6$

Consider the collection $T^* \subseteq E^*$ of edges in the dual graph that corresponds to edges in $E \backslash T$. The edges in T^* connect all the faces, since T does not have a cycle; but also T^* does not contain a cycle, since otherwise it would separate some vertices of G inside the cycle from vertices outside (and this cannot be, since T is a spanning subgraph, and the edges of T and of T^* do not intersect). Thus T^* is a spanning tree for G^*.

For every tree the number of vertices is one larger than the number of edges. To see this, choose one vertex as the root, and direct all edges "away from the root": this yields a bijection between the non-root vertices and the edges, by matching each edge with the vertex it points at. Applied to the tree T this yields $n = e_T + 1$, while for the tree T^* it yields $f = e_{T^*} + 1$. Adding both equations we get $n+f = (e_T+1)+(e_{T^*}+1) = e + 2$. □

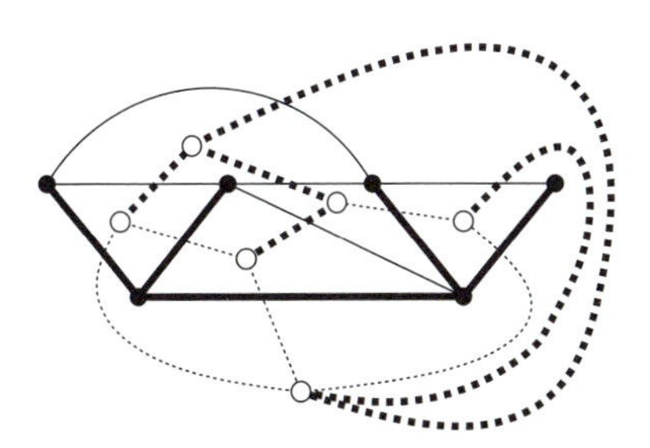
Dual spanning trees in G and in G^*

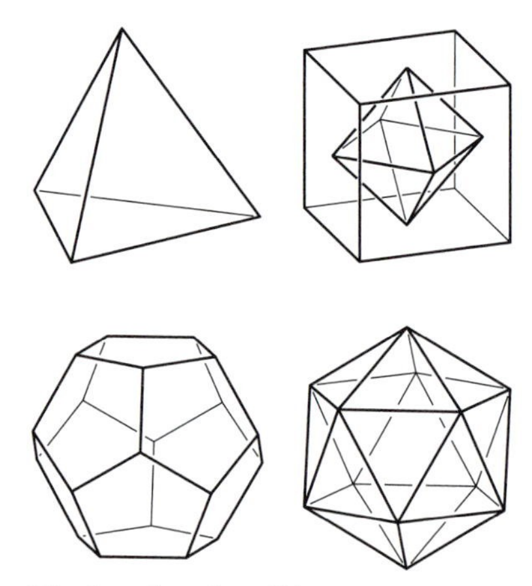

The five platonic solids

Euler's formula thus produces a strong *numerical* conclusion from a *geometric-topological* situation: the numbers of vertices, edges, and faces of a finite graph G satisfy $n - e + f = 2$ whenever the graph is *or can be* drawn in the plane or on a sphere.

Many well-known and classical consequences can be derived from Euler's formula. Among them are the classification of the regular convex polyhedra (the platonic solids), the fact that K_5 and $K_{3,3}$ are not planar (see below), and the five-color theorem that every planar map can be colored with at most five colors such that no two adjacent countries have the same color. But for this we have a much better proof, which does not even need Euler's formula — see Chapter 30.

This chapter collects three other beautiful proofs that have Euler's formula at their core. The first two — a proof of the Sylvester-Gallai theorem, and a theorem on two-colored point configurations — use Euler's formula in clever combination with other arithmetic relationships between basic graph parameters. Let us first look at these parameters.

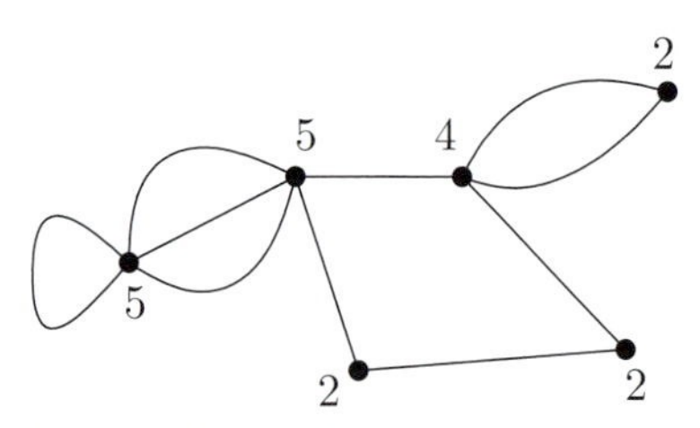

Here the degree is written next to each vertex. Counting the vertices of given degree yields $n_2 = 3$, $n_3 = 0$, $n_4 = 1$, $n_5 = 2$.

The *degree* of a vertex is the number of edges that end in the vertex, where loops count double. Let n_i denote the number of vertices of degree i in G. Counting the vertices according to their degrees, we obtain

$$n = n_0 + n_1 + n_2 + n_3 + \dots \tag{1}$$

On the other hand, every edge has two ends, so it contributes 2 to the sum of all degrees, and we obtain

$$2e = n_1 + 2n_2 + 3n_3 + 4n_4 + \dots \tag{2}$$

You may interpret this identity as counting in two ways the ends of the edges, that is, the edge-vertex incidences. The *average degree* $\overline{d}$ of the vertices is therefore

$$\overline{d} = \frac{2e}{n}.$$

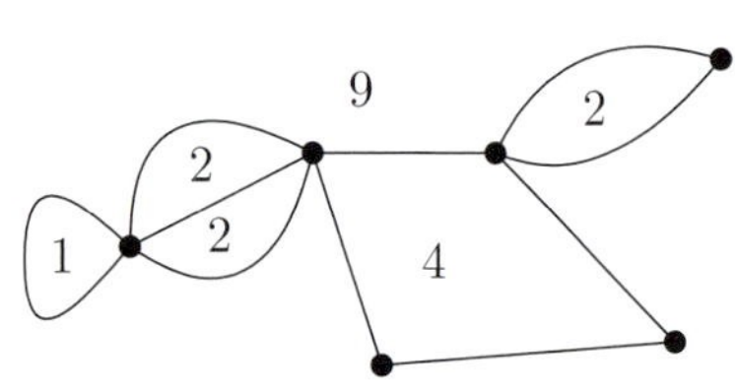

The number of sides is written into each region. Counting the faces with a given number of sides yields $f_1 = 1$, $f_2 = 3$, $f_4 = 1$, $f_9 = 1$, and $f_i = 0$ otherwise.

Next we count the faces of a plane graph according to their number of sides: a *k-face* is a face that is bounded by k edges (where an edge that on both sides borders the same region has to be counted twice!). Let f_k be the number of k-faces. Counting all faces we find

$$f = f_1 + f_2 + f_3 + f_4 + \dots \tag{3}$$

Counting the edges according to the faces of which they are sides, we get

$$2e = f_1 + 2f_2 + 3f_3 + 4f_4 + \dots \tag{4}$$

As before, we can interpret this as double-counting of edge-face incidences. Note that the average number of sides of faces is given by

$$\overline{f} = \frac{2e}{f}.$$

Let us deduce from this — together with Euler's formula — quickly that the complete graph K_5 and the complete bipartite graph $K_{3,3}$ are not planar. For a hypothetical plane drawing of K_5 we calculate $n = 5$, $e = \binom{5}{2} = 10$, thus $f = e + 2 - n = 7$ and $\overline{f} = \frac{2e}{f} = \frac{20}{7} < 3$. But if the average number of sides is smaller than 3, then the embedding would have a face with at most two sides, which cannot be.

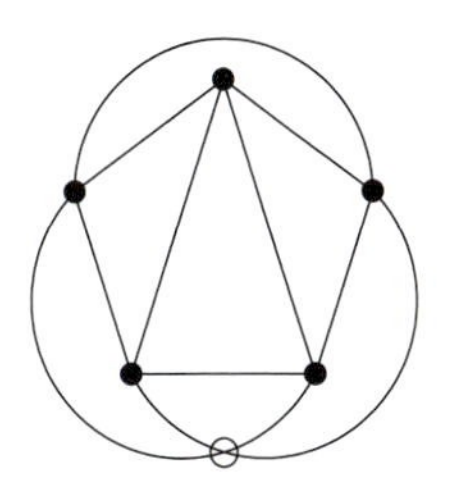
K_5 drawn with one crossing

Similarly for $K_{3,3}$ we get $n = 6$, $e = 9$, and $f = e + 2 - n = 5$, and thus $\overline{f} = \frac{2e}{f} = \frac{18}{5} < 4$, which cannot be since $K_{3,3}$ is simple and bipartite, so all its cycles have length at least 4.

It is no coincidence, of course, that the equations (3) and (4) for the f_i's look so similar to the equations (1) and (2) for the n_i's. They are transformed into each other by the dual graph construction $G \to G^*$ explained above.

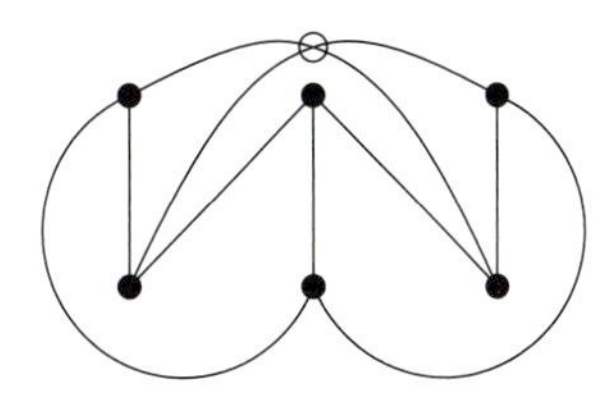
$K_{3,3}$ drawn with one crossing

From the double counting identities, we get the following important "local" consequences of Euler's formula.

Proposition. *Let G be any simple plane graph with $n > 2$ vertices. Then*

(A) *G has a vertex of degree at most 5.*

(B) *G has at most $3n - 6$ edges.*

(C) *If the edges of G are two-colored, then there is a vertex of G with at most two color-changes in the cyclic order of the edges around the vertex.*

■ **Proof.** For each of the three statements, we may assume that G is connected.

(A) Every face has at least 3 sides (since G is simple), so (3) and (4) yield

$$\begin{aligned} f &= f_3 + f_4 + f_5 + \dots \\ 2e &= 3f_3 + 4f_4 + 5f_5 + \dots \end{aligned}$$

and thus $2e - 3f \geq 0$.

Now if each vertex has degree at least 6, then (1) and (2) imply

$$\begin{aligned} n &= n_6 + n_7 + n_8 + \dots \\ 2e &= 6n_6 + 7n_7 + 8n_8 + \dots \end{aligned}$$

and thus $2e - 6n \geq 0$,

Taking both inequalities together, we get

$$6(e - n - f) = (2e - 6n) + 2(2e - 3f) \geq 0$$

and thus $e \geq n + f$, contradicting Euler's formula.

(B) As in the first step of part (A), we obtain $2e - 3f \geq 0$, and thus

$$3n - 6 = 3e - 3f \geq e$$

from Euler's formula.

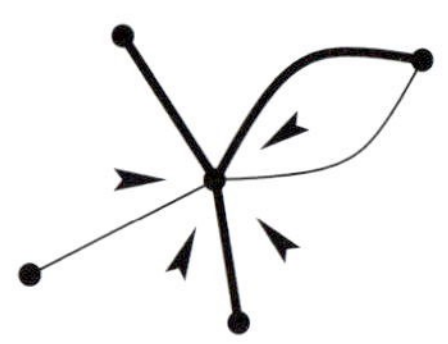

Arrows point to the corners with color changes.

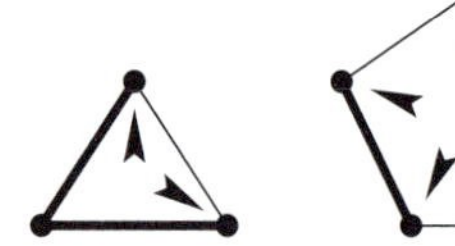

(C) Let c be the number of corners where color changes occur. Suppose the statement is false, then we have $c \geq 4n$ corners with color changes, since at every vertex there is an even number of changes. Now every face with $2k$ or $2k+1$ sides has at most $2k$ such corners, so we conclude that

$$\begin{aligned} 4n \leq c &\leq 2f_3 + 4f_4 + 4f_5 + 6f_6 + 6f_7 + 8f_8 + \dots \\ &\leq 2f_3 + 4f_4 + 6f_5 + 8f_6 + 10f_7 + \dots \\ &= 2(3f_3 + 4f_4 + 5f_5 + 6f_6 + 7f_7 + \dots) \\ &\quad -4(f_3 + f_4 + f_5 + f_6 + f_7 + \dots) \\ &= 4e - 4f \end{aligned}$$

using again (3) and (4). So we have $e \geq n + f$, again contradicting Euler's formula. □

1. The Sylvester-Gallai theorem, revisited

It was first noted by Norman Steenrod, it seems, that part (A) of the proposition yields a strikingly simple proof of the Sylvester-Gallai theorem (see Chapter 9).

The Sylvester-Gallai theorem. *Given any set of $n \geq 3$ points in the plane, not all on one line, there is always a line that contains exactly two of the points.*

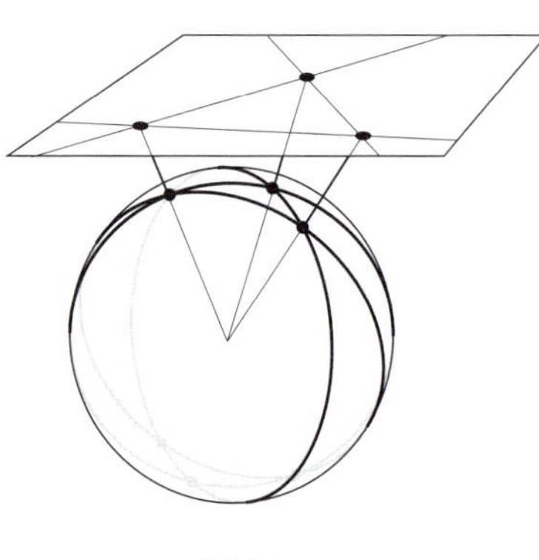

■ **Proof.** (Sylvester-Gallai via Euler)

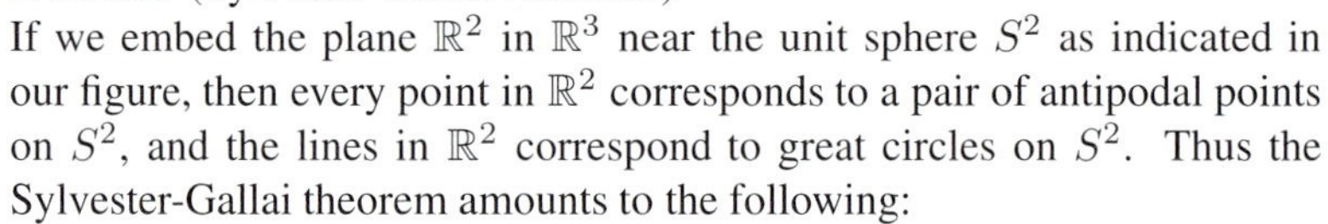

If we embed the plane $\mathbb{R}^2$ in $\mathbb{R}^3$ near the unit sphere S^2 as indicated in our figure, then every point in $\mathbb{R}^2$ corresponds to a pair of antipodal points on S^2, and the lines in $\mathbb{R}^2$ correspond to great circles on S^2. Thus the Sylvester-Gallai theorem amounts to the following:

Given any set of $n \geq 3$ pairs of antipodal points on the sphere, not all on one great circle, there is always a great circle that contains exactly two of the antipodal pairs.

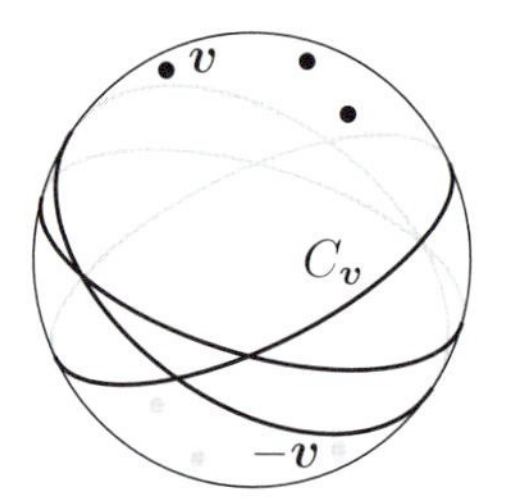

Now we dualize, replacing each pair of antipodal points by the corresponding great circle on the sphere. That is, instead of points $\pm\boldsymbol{v} \in S^2$ we consider the orthogonal circles given by $C_{\boldsymbol{v}} := \{\boldsymbol{x} \in S^2 : \langle \boldsymbol{x}, \boldsymbol{v} \rangle = 0\}$. (This $C_{\boldsymbol{v}}$ is the equator if we consider $\boldsymbol{v}$ as the north pole of the sphere.) Then the Sylvester-Gallai problem asks us to prove:

Given any collection of $n \geq 3$ great circles on S^2, not all of them passing through one point, there is always a point that is on exactly two of the great circles.

But the arrangement of great circles yields a simple plane graph on S^2, whose vertices are the intersection points of two of the great circles, which divide the great circles into edges. All the vertex degrees are even, and they are at least 4 — by construction. Thus part (A) of the proposition yields the existence of a vertex of degree 4. That's it! □

2. Monochromatic lines

The following proof of a "colorful" relative of the Sylvester-Gallai theorem is due to Don Chakerian.

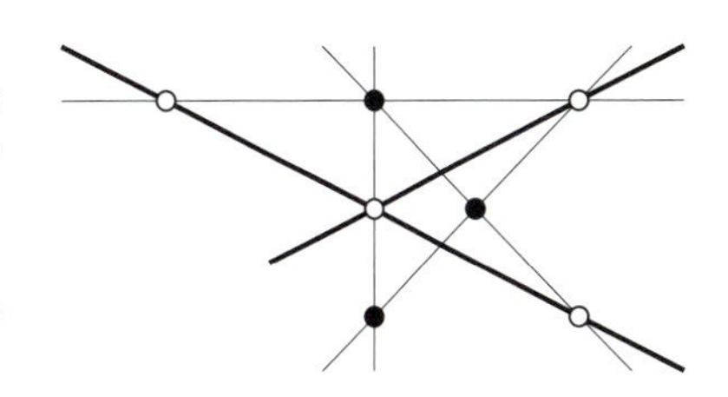

Theorem. *Given any finite configuration of "black" and "white" points in the plane, not all on one line, there is always a "monochromatic" line: a line that contains at least two points of one color and none of the other.*

■ **Proof.** As for the Sylvester-Gallai problem, we transfer the problem to the unit sphere and dualize it there. So we must prove:

> *Given any finite collection of "black" and "white" great circles on the unit sphere, not all passing through one point, there is always an intersection point that lies either only on white great circles, or only on black great circles.*

Now the (positive) answer is clear from part (C) of the proposition, since in every vertex where great circles of different colors intersect, we always have at least 4 corners with sign changes. □

3. Pick's theorem

Pick's theorem from 1899 is a beautiful and surprising result in itself, but it is also a "classical" consequence of Euler's formula. For the following, call a convex polygon $P \subseteq \mathbb{R}^2$ *elementary* if its vertices are integral (that is, they lie in the *lattice* $\mathbb{Z}^2$), but if it does not contain any further lattice points.

Lemma. *Every elementary triangle $\Delta = \text{conv}\{\boldsymbol{p}_0, \boldsymbol{p}_1, \boldsymbol{p}_2\} \subseteq \mathbb{R}^2$ has area $A(\Delta) = \frac{1}{2}$.*

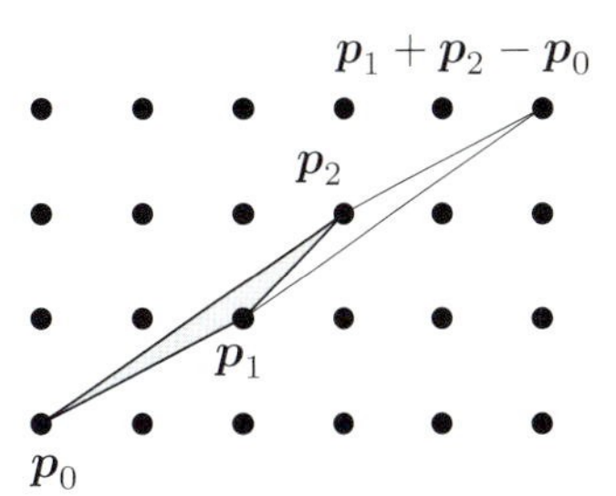

■ **Proof.** Both the parallelogram P with corners $\boldsymbol{p}_0, \boldsymbol{p}_1, \boldsymbol{p}_2, \boldsymbol{p}_1 + \boldsymbol{p}_2 - \boldsymbol{p}_0$ and the lattice $\mathbb{Z}^2$ are symmetric with respect to the map

$$\sigma : \boldsymbol{x} \longmapsto \boldsymbol{p}_1 + \boldsymbol{p}_2 - \boldsymbol{x},$$

which is the reflection with respect to the center of the segment from $\boldsymbol{p}_1$ to $\boldsymbol{p}_2$. Thus the parallelogram $P = \Delta \cup \sigma(\Delta)$ is elementary as well, and its integral translates tile the plane. Hence $\{\boldsymbol{p}_1 - \boldsymbol{p}_0, \boldsymbol{p}_2 - \boldsymbol{p}_0\}$ is a basis of the lattice $\mathbb{Z}^2$, it has determinant ± 1, P is a parallelogram of area 1, and Δ has area $\frac{1}{2}$. (For an explanation of these terms see the box on the next page.) □

Theorem. *The area of any (not necessarily convex) polygon $Q \subseteq \mathbb{R}^2$ with integral vertices is given by*

$$A(Q) = n_{int} + \frac{1}{2} n_{bd} - 1,$$

where n_{int} and n_{bd} are the numbers of integral points in the interior respectively on the boundary of Q.

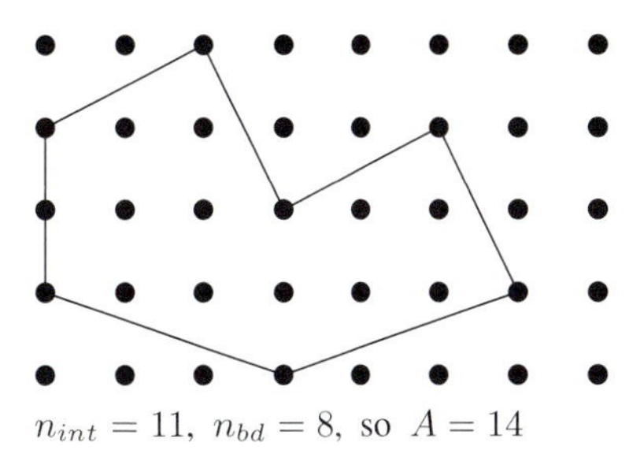

$n_{int} = 11$, $n_{bd} = 8$, so $A = 14$

Lattice bases

A *basis* of $\mathbb{Z}^2$ is a pair of linearly independent vectors $\boldsymbol{e}_1, \boldsymbol{e}_2$ such that

$$\mathbb{Z}^2 = \{\lambda_1\boldsymbol{e}_1 + \lambda_2\boldsymbol{e}_2 : \lambda_1, \lambda_2 \in \mathbb{Z}\}.$$

Let $\boldsymbol{e}_1 = \binom{a}{b}$ and $\boldsymbol{e}_2 = \binom{c}{d}$, then the area of the parallelogram spanned by $\boldsymbol{e}_1$ and $\boldsymbol{e}_2$ is given by $A(\boldsymbol{e}_1, \boldsymbol{e}_2) = |\det(\boldsymbol{e}_1, \boldsymbol{e}_2)| = |\det\left(\begin{smallmatrix} a & c \\ b & d \end{smallmatrix}\right)|$. If $\boldsymbol{f}_1 = \binom{r}{s}$ and $\boldsymbol{f}_2 = \binom{t}{u}$ is another basis, then there exists an invertible $\mathbb{Z}$-matrix Q with $\left(\begin{smallmatrix} r & t \\ s & u \end{smallmatrix}\right) = \left(\begin{smallmatrix} a & c \\ b & d \end{smallmatrix}\right)Q$. Since $QQ^{-1} = \left(\begin{smallmatrix} 1 & 0 \\ 0 & 1 \end{smallmatrix}\right)$, and the determinants are integers, it follows that $|\det Q| = 1$, and hence $|\det(\boldsymbol{f}_1, \boldsymbol{f}_2)| = |\det(\boldsymbol{e}_1, \boldsymbol{e}_2)|$. Therefore all basis parallelograms have the same area 1, since $A(\binom{1}{0}, \binom{0}{1}) = 1$.

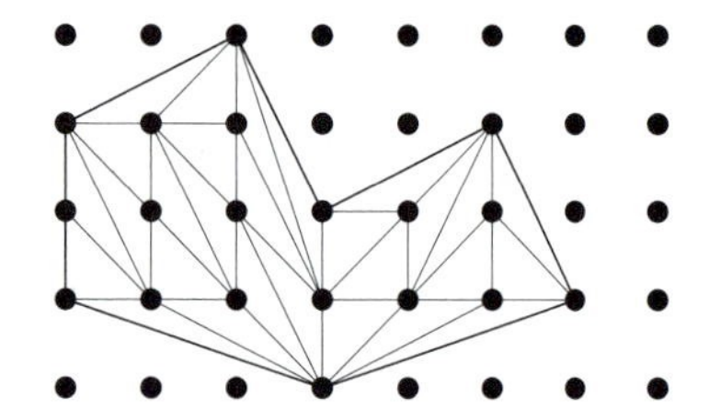

■ **Proof.** Every such polygon can be triangulated using all the n_{int} lattice points in the interior, and all the n_{bd} lattice points on the boundary of Q. (This is not quite obvious, in particular if Q is not required to be convex, but the argument given in Chapter 31 on the art gallery problem proves this.)

Now we interpret the triangulation as a plane graph, which subdivides the plane into one unbounded face plus $f - 1$ triangles of area $\frac{1}{2}$, so

$$A(Q) = \frac{1}{2}(f - 1).$$

Every triangle has three sides, where each of the e_{int} interior edges bounds two triangles, while the e_{bd} boundary edges appear in one single triangle each. So $3(f-1) = 2e_{int} + e_{bd}$ and thus $f = 2(e-f) - e_{bd} + 3$. Also, there is the same number of boundary edges and vertices, $e_{bd} = n_{bd}$. These two facts together with Euler's formula yield

$$\begin{aligned} f &= 2(e - f) - e_{bd} + 3 \\ &= 2(n - 2) - n_{bd} + 3 = 2n_{int} + n_{bd} - 1, \end{aligned}$$

and thus

$$A(Q) = \tfrac{1}{2}(f - 1) = n_{int} + \tfrac{1}{2}n_{bd} - 1. \qquad \square$$

References

[1] G. D. CHAKERIAN: *Sylvester's problem on collinear points and a relative,* Amer. Math. Monthly **77** (1970), 164-167.

[2] G. PICK: *Geometrisches zur Zahlenlehre,* Sitzungsberichte Lotos (Prag), Natur-med. Verein für Böhmen **19** (1899), 311-319.

[3] K. G. C. VON STAUDT: *Geometrie der Lage,* Verlag der Fr. Korn'schen Buchhandlung, Nürnberg 1847.

[4] N. E. STEENROD: *Solution 4065/Editorial Note,* Amer. Math. Monthly **51** (1944), 170-171.

Cauchy's rigidity theorem

Chapter 12

A famous result that depends on Euler's formula (specifically, on part (C) of the proposition in the previous chapter) is Cauchy's rigidity theorem for 3-dimensional polyhedra.

For the notions of congruence and of combinatorial equivalence that are used in the following we refer to the appendix on polytopes and polyhedra in the chapter on Hilbert's third problem, see page 50.

Augustin Cauchy

Theorem. *If two 3-dimensional convex polyhedra P and P' are combinatorially equivalent with corresponding facets being congruent, then also the angles between corresponding pairs of adjacent facets are equal (and thus P is congruent to P').*

The illustration in the margin shows two 3-dimensional polyhedra that are combinatorially equivalent, such that the corresponding faces are congruent. But they are not congruent, and only one of them is convex. Thus the assumption of convexity is essential for Cauchy's theorem!

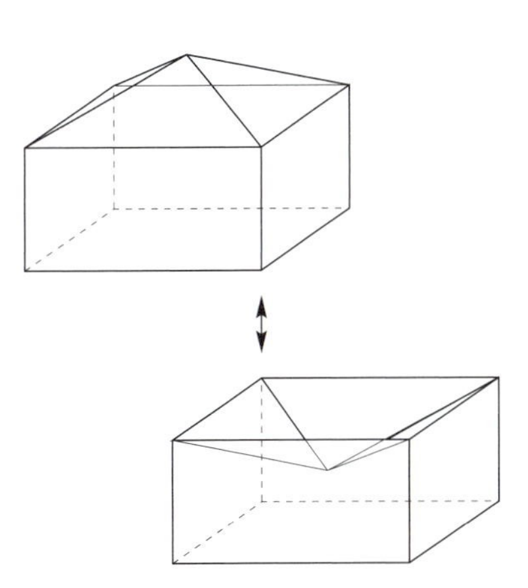

■ **Proof.** The following is essentially Cauchy's original proof. Assume that two convex polyhedra P and P' with congruent faces are given. We color the edges of P as follows: an edge is black (or "positive") if the corresponding interior angle between the two adjacent facets is larger in P' than in P; it is white (or "negative") if the corresponding angle is smaller in P' than in P.

The black and the white edges of P together form a 2-colored plane graph on the surface of P, which by radial projection, assuming that the origin is in the interior of P, we may transfer to the surface of the unit sphere. If P and P' have unequal corresponding facet-angles, then the graph is nonempty. With part (C) of the proposition in the previous chapter we find that there is a vertex p that is adjacent to at least one black or white edge, such that there are at most two changes between black and white edges (in cyclic order).

Now we intersect P with a small sphere S_ε (of radius ε) centered at the vertex p, and we intersect P' with a sphere S'_ε of the same radius ε centered at the corresponding vertex p'. In S_ε and S'_ε we find convex spherical polygons Q and Q' such that corresponding arcs have the same lengths, because of the congruence of the facets of P and P', and since we have chosen the same radius ε.

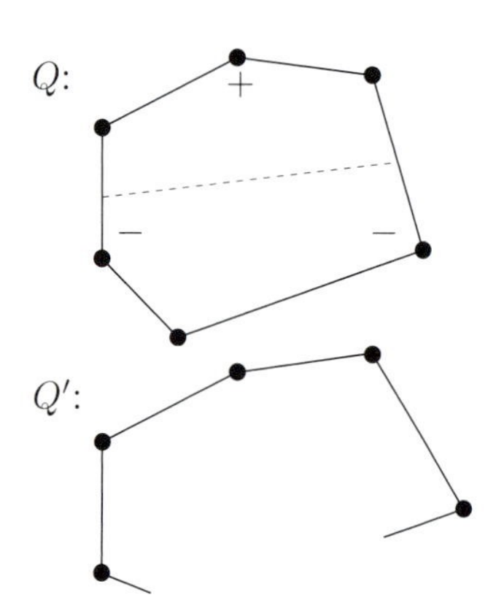

Now we mark by $+$ the angles of Q for which the corresponding angle in Q' is larger, and by $-$ the angles whose corresponding angle of Q' is smaller. That is, when moving from Q to Q' the $+$ angles are "opened," the $-$ angles are "closed," while all side lengths and the unmarked angles stay constant.

From our choice of p we know that *some* $+$ or $-$ sign occurs, and that in cyclic order there are at most two $+/-$ changes. If only one type of signs occurs, then the lemma below directly gives a contradiction, saying that one edge must change its length. If both types of signs occur, then (since there are only two sign changes) there is a "separation line" that connects the midpoints of two edges and separates all the $+$ signs from all the $-$ signs. Again we get a contradiction from the lemma below, since the separation line cannot be both longer and shorter in Q' than in Q. $\square$

Cauchy's arm lemma.
If Q and Q' are convex (planar or spherical) n-gons, labeled as in the figure,

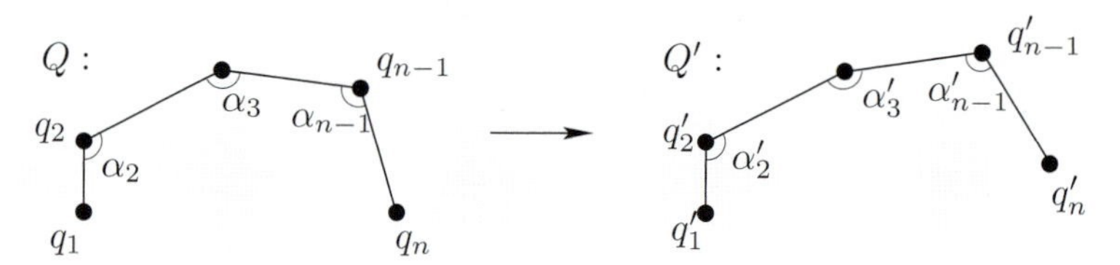

such that $\overline{q_i q_{i+1}} = \overline{q'_i q'_{i+1}}$ holds for the lengths of corresponding edges for $1 \le i \le n-1$, and $\alpha_i \le \alpha'_i$ holds for the sizes of corresponding angles for $2 \le i \le n-1$, then the "missing" edge length satisfies

$$\overline{q_1 q_n} \;\le\; \overline{q'_1 q'_n},$$

with equality if and only if $\alpha_i = \alpha'_i$ holds for all i.

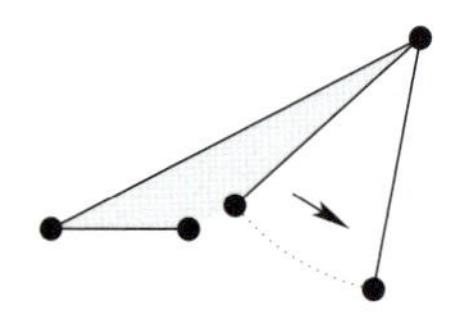

It is interesting that Cauchy's original proof of the lemma was false: a continuous motion that opens angles and keeps side-lengths fixed may destroy convexity — see the figure! On the other hand, both the lemma and its proof given here, from a letter by I. J. Schoenberg to S. K. Zaremba, are valid both for planar and for spherical polygons.

■ **Proof.** We use induction on n. The case $n = 3$ is easy: If in a triangle we increase the angle γ between two sides of fixed lengths a and b, then the length c of the opposite side also increases. Analytically, this follows from the cosine theorem

$$c^2 \;=\; a^2 + b^2 - 2ab\cos\gamma$$

in the planar case, and from the analogous result

$$\cos c \;=\; \cos a \cos b + \sin a \sin b \cos\gamma$$

in spherical trigonometry. Here the lengths a, b, c are measured on the surface of a sphere of radius 1, and thus have values in the interval $[0, \pi]$.

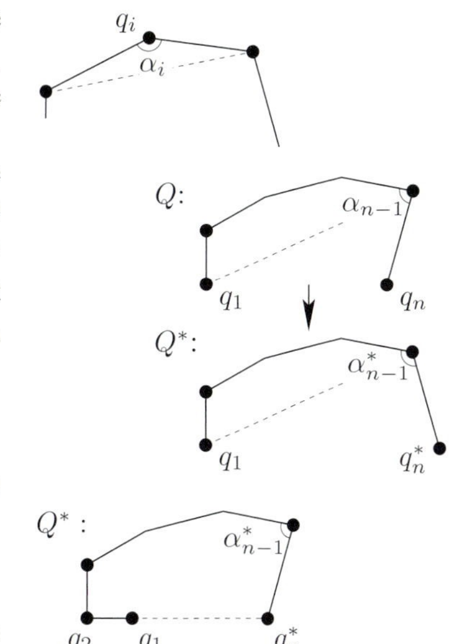

Now let $n \geq 4$. If for any $i \in \{2, \ldots, n-1\}$ we have $\alpha_i = \alpha'_i$, then the corresponding vertex can be cut off by introducing the diagonal from q_{i-1} to q_{i+1} resp. from q'_{i-1} to q'_{i+1}, with $\overline{q_{i-1}q_{i+1}} = \overline{q'_{i-1}q'_{i+1}}$, so we are done by induction. Thus we may assume $\alpha_i < \alpha'_i$ for $2 \leq i \leq n-1$.

Now we produce a new polygon Q^* from Q by replacing α_{n-1} by the largest possible angle $\alpha^*_{n-1} \leq \alpha'_{n-1}$ that keeps Q^* convex. For this we replace q_n by q^*_n, keeping all the other q_i, edge lengths, and angles from Q.

If indeed we can choose $\alpha^*_{n-1} = \alpha'_{n-1}$ keeping Q^* convex, then we get $\overline{q_1q_n} < \overline{q_1q^*_n} \leq \overline{q'_1q'_n}$, using the case $n = 3$ for the first step and induction as above for the second.

Otherwise after a nontrivial move that yields

$$\overline{q_1q^*_n} > \overline{q_1q_n} \tag{1}$$

we "get stuck" in a situation where q_2, q_1 and q^*_n are collinear, with

$$\overline{q_2q_1} + \overline{q_1q^*_n} = \overline{q_2q^*_n}. \tag{2}$$

Now we compare this Q^* with Q' and find

$$\overline{q_2q^*_n} \leq \overline{q'_2q'_n} \tag{3}$$

by induction on n (ignoring the vertex q_1 resp. q'_1). Thus we obtain

$$\overline{q'_1q'_n} \overset{(*)}{\geq} \overline{q'_2q'_n} - \overline{q'_1q'_2} \overset{(3)}{\geq} \overline{q_2q^*_n} - \overline{q_1q_2} \overset{(2)}{=} \overline{q_1q^*_n} \overset{(1)}{>} \overline{q_1q_n},$$

where $(*)$ is just the triangle inequality, and all other relations have already been derived. □

We have seen an example which shows that Cauchy's theorem is not true for *non-convex* polyhedra. The special feature of this example is, of course, that a non-continuous "flip" takes one polyhedron to the other, keeping the facets congruent while the dihedral angles "jump." One can ask for more:

> *Could there be, for some non-convex polyhedron, a* continuous *deformation that would keep the facets flat and congruent?*

It was conjectured that no triangulated surface, convex or not, admits such a motion. So, it was quite a surprise when in 1977 — more than 160 years after Cauchy's work — Robert Connelly presented counterexamples: closed triangulated spheres embedded in $\mathbb{R}^3$ (without self-intersections) that are flexible, with a continuous motion that keeps all the edge lengths constant, and thus keeps the triangular faces congruent.

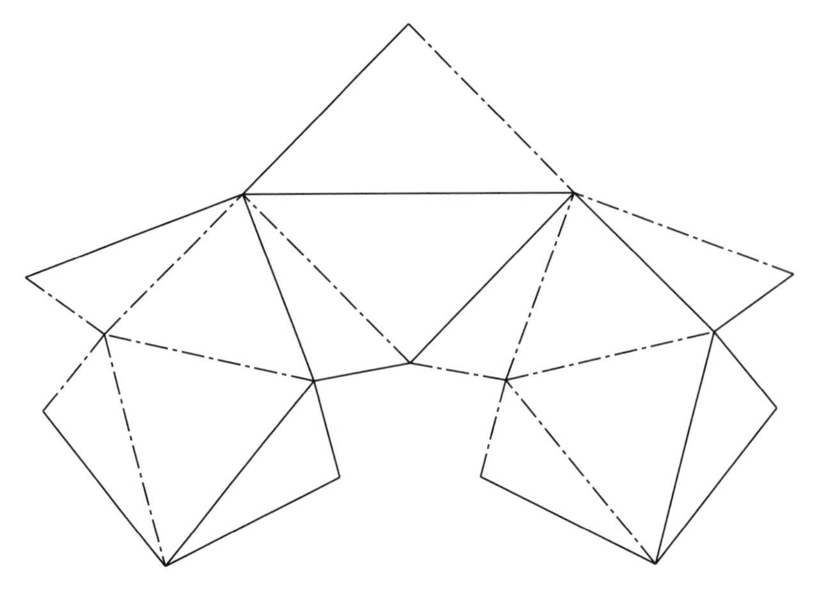

A beautiful example of a flexible surface constructed by Klaus Steffen: The dashed lines represent the non-convex edges in this "cut-out" paper model. Fold the normal lines as "mountains" and the dashed lines as "valleys." The edges in the model have lengths 5, 10, 11, 12 and 17 units.

The rigidity theory of surfaces has even more surprises in store: only very recently Connelly, Sabitov and Walz managed to prove that when any such flexing surface moves, the *volume* it encloses must be constant. Their proof is beautiful also in its use of algebraic machinery (outside the scope of this book).

References

[1] A. CAUCHY: *Sur les polygones et les polyèdres, seconde mémoire,* J. École Polytechnique XVIe Cahier, Tome IX (1813), 87-98; Œuvres Complètes, IIe Série, Vol. 1, Paris 1905, 26-38.

[2] R. CONNELLY: *A counterexample to the rigidity conjecture for polyhedra,* Inst. Haut. Etud. Sci., Publ. Math. **47** (1978), 333-338.

[3] R. CONNELLY: *The rigidity of polyhedral surfaces,* Mathematics Magazine **52** (1979), 275-283.

[4] R. CONNELLY, I. SABITOV & A. WALZ: *The bellows conjecture,* Beiträge zur Algebra und Geometrie/Contributions to Algebra and Geometry **38** (1997), 1-10.

[5] J. SCHOENBERG & S.K. ZAREMBA: *On Cauchy's lemma concerning convex polygons,* Canadian J. Math. **19** (1967), 1062-1071.

Touching simplices

Chapter 13

How many d-dimensional simplices can be positioned in $\mathbb{R}^d$ so that they touch pairwise, that is, so that all their pairwise intersections are $(d-1)$-dimensional?

This is an old and very natural question. We shall call $f(d)$ the answer to this problem, and record $f(1) = 2$, which is trivial. For $d = 2$ the configuration of four triangles in the margin shows $f(2) \geq 4$. There is no similar configuration with five triangles, because from this the dual graph construction, which for our example with four triangles yields a planar drawing of K_4, would give a planar embedding of K_5, which is impossible (see page 67). Thus we have

$$f(2) = 4.$$

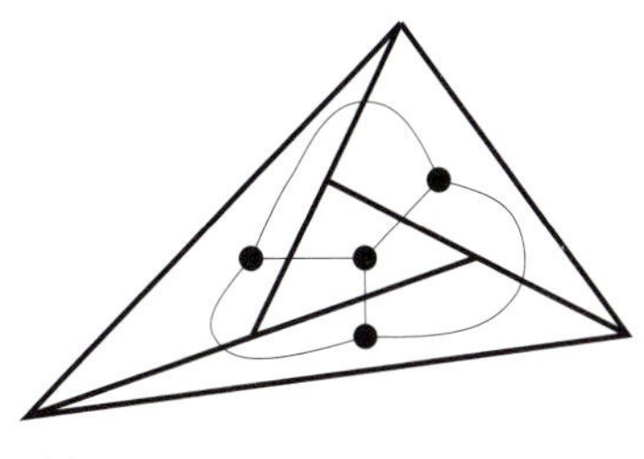

$f(2) \geq 4$

In three dimensions, $f(3) \geq 8$ is quite easy to see. For that we use the configuration of eight triangles depicted on the right. The four shaded triangles are joined to some point x below the "plane of drawing," which yields four tetrahedra that touch the plane from below. Similarly, the four white triangles are joined to some point y above the plane of drawing. So we obtain a configuration of eight touching tetrahedra in $\mathbb{R}^3$, that is, $f(3) \geq 8$.

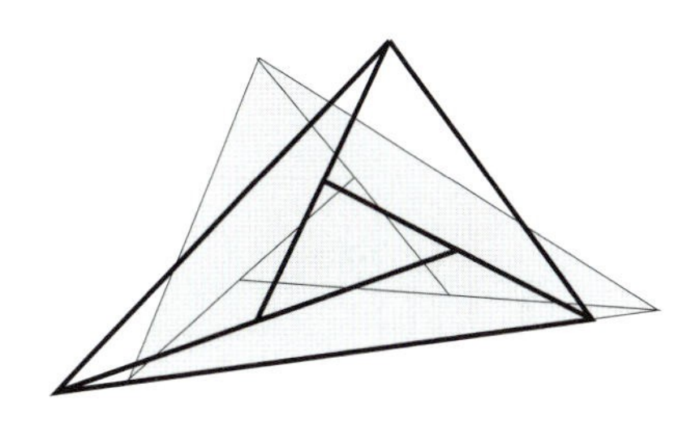

$f(3) \geq 8$

In 1965, Baston wrote a book proving $f(3) \leq 9$, and in 1991 it took Zaks another book to establish

$$f(3) = 8.$$

With $f(1) = 2$, $f(2) = 4$ and $f(3) = 8$, it doesn't take much inspiration to arrive at the following conjecture, first posed by Bagemihl in 1956.

Conjecture. *The maximal number of pairwise touching d-simplices in a configuration in $\mathbb{R}^d$ is*

$$f(d) = 2^d.$$

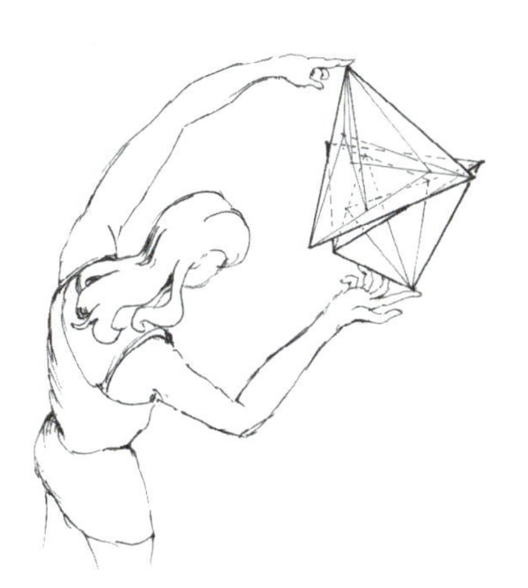

"Touching simplices"

The lower bound, $f(d) \geq 2^d$, is easy to verify "if we do it right." This amounts to a heavy use of affine coordinate tranformations, and to an induction on the dimension that establishes the following stronger result, due to Joseph Zaks [4].

Theorem 1. *For every $d \geq 2$, there is a family of 2^d pairwise touching d-simplices in $\mathbb{R}^d$ together with a transversal line that hits the interior of every single one of them.*

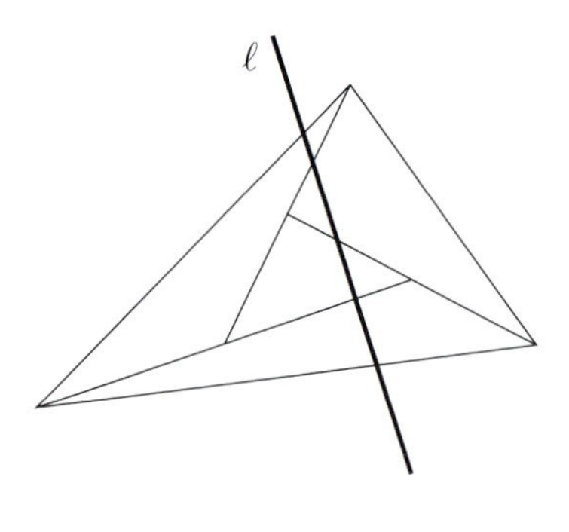

■ **Proof.** For $d = 2$ the family of four triangles that we had considered does have such a transversal line. Now consider any d-dimensional configuration of touching simplices that has a transversal line ℓ. Any nearby parallel line ℓ' is a transversal line as well. If we choose ℓ' and ℓ parallel and close enough, then each of the simplices contains an orthogonal (shortest) connecting interval between the two lines. Only a bounded part of the lines ℓ and ℓ' is contained in the simplices of the configuration, and we may add two connecting segments outside the configuration, such that the rectangle spanned by the two outside connecting lines (that is, their convex hull) contains all the other connecting segments. Thus, we have placed a "ladder" such that each of the simplices of the configuration has one of the ladder's steps in its interior, while the four ends of the ladder are outside the configuration.

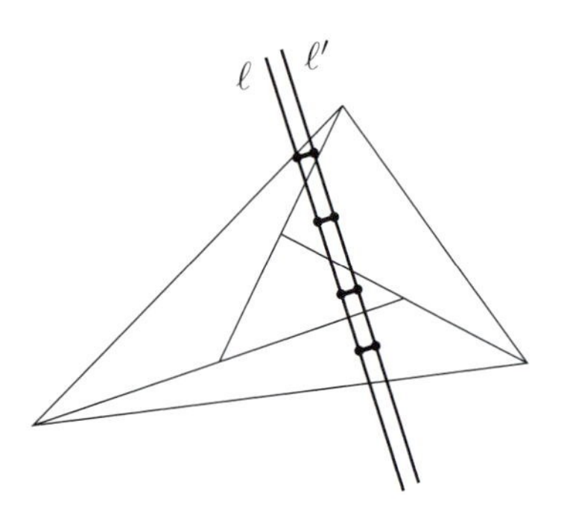

Now the main step is that we perform an (affine) coordinate transformation that maps $\mathbb{R}^d$ to $\mathbb{R}^d$, and takes the rectangle spanned by the ladder to the rectangle (half-square) as shown in the figure below, given by

$$R^1 \;=\; \{(x_1, x_2, 0, \dots, 0)^T : -1 \le x_1 \le 0; -1 \le x_2 \le 1\}.$$

Thus the configuration of touching simplices Σ^1 in $\mathbb{R}^d$ which we obtain has the x_1-axis as a transversal line, and it is placed such that each of the simplices contains a segment

$$S^1(\alpha) \;=\; \{(\alpha, x_2, 0, \dots, 0)^T : -1 \le x_2 \le 1\}$$

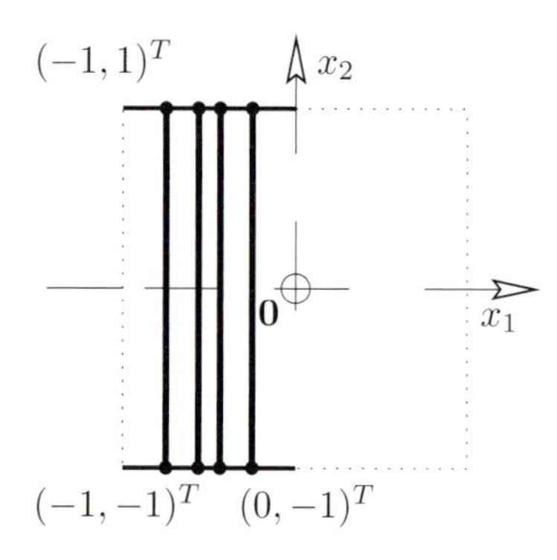

in its interior (for some α with $-1 < \alpha < 0$), while the origin $\mathbf{0}$ is outside all simplices.

Now we produce a second copy Σ^2 of this configuration by reflecting the first one in the hyperplane given by $x_1 = x_2$. This second configuration has the x_2-axis as a transversal line, and each simplex contains a segment

$$S^2(\beta) \;=\; \{(x_1, \beta, 0, \dots, 0)^T : -1 \le x_1 \le 1\}$$

in its interior, with $-1 < \beta < 0$. But each segment $S^1(\alpha)$ intersects each segment $S^2(\beta)$, and thus the interior of each simplex of Σ^1 intersects each simplex of Σ^2 in its interior. Thus if we add a new $(d+1)$-st coordinate x_{d+1}, and take Σ to be

$$\{\text{conv}(P_i \cup \{-\mathbf{e}_{d+1}\}) : P_i \in \Sigma^1\} \;\cup\; \{\text{conv}(P_j \cup \{\mathbf{e}_{d+1}\}) : P_j \in \Sigma^2\},$$

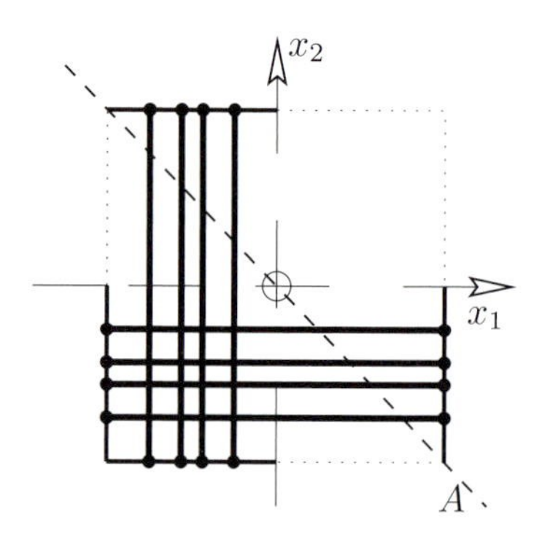

then we get a configuration of touching $(d+1)$-simplices in $\mathbb{R}^{d+1}$. Furthermore, the antidiagonal

$$A \;=\; \{(x, -x, 0, \dots, 0)^T : x \in \mathbb{R}\} \;\subseteq\; \mathbb{R}^d$$

intersects all segments $S^1(\alpha)$ and $S^2(\beta)$. We can "tilt" it a little, and obtain a line

$$L_\varepsilon \;=\; \{(x, -x, 0, \dots, 0, \varepsilon x)^T : x \in \mathbb{R}\} \;\subseteq\; \mathbb{R}^{d+1},$$

which for all small enough $\varepsilon > 0$ intersects all the simplices of Σ. This completes our induction step. □

In contrast to this exponential lower bound, tight upper bounds are harder to get. A naive inductive argument (considering all the facet hyperplanes in a touching configuration separately) yields only

$$f(d) \leq \frac{2}{3}(d+1)!,$$

and this is quite far from the lower bound of Theorem 1. However, Micha Perles found the following "magical" proof for a much better bound.

Theorem 2. *For all $d \geq 1$, we have $f(d) < 2^{d+1}$.*

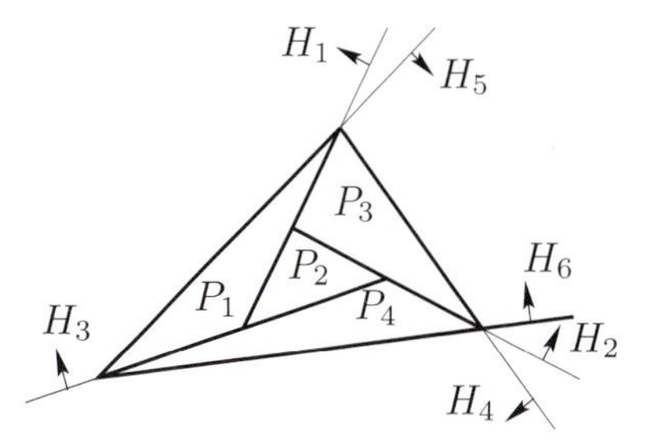

■ **Proof.** Given a configuration of r touching d-simplices $P_1, P_2, \ldots, P_r$ in $\mathbb{R}^d$, first enumerate the different hyperplanes $H_1, H_2, \ldots, H_s$ spanned by facets of the P_i, and for each of them arbitrarily choose a positive side H_i^+, and call the other side H_i^-.

For example, for the 2-dimensional configuration of $r = 4$ triangles depicted on the right we find $s = 6$ hyperplanes (which are lines for $d = 2$).

From these data, we construct the *B-matrix*, an $(r \times s)$-matrix with entries in $\{+1, -1, 0\}$, as follows:

$$B_{ij} := \begin{cases} +1 & \text{if } P_i \text{ has a facet in } H_j, \text{ and } P_i \subseteq H_j^+, \\ -1 & \text{if } P_i \text{ has a facet in } H_j, \text{ and } P_i \subseteq H_j^-, \\ 0 & \text{if } P_i \text{ does not have a facet in } H_j. \end{cases}$$

For example, the 2-dimensional configuration in the margin gives rise to the matrix

$$B = \begin{pmatrix} 1 & 0 & 1 & 0 & 1 & 0 \\ -1 & -1 & 1 & 0 & 0 & 0 \\ -1 & 1 & 0 & 1 & 0 & 0 \\ 0 & -1 & -1 & 0 & 0 & 1 \end{pmatrix}.$$

Three properties of the B-matrix are worth recording. First, since every d-simplex has $d+1$ facets, we find that every row of B has exactly $d+1$ nonzero entries, and thus has exactly $s-(d+1)$ zero entries. Secondly, we are dealing with a configuration of pairwise touching simplices, and thus for every pair of rows we find one column in which one row has a $+1$ entry, while the entry in the other row is -1. That is, the rows are different *even if we disregard their zero entries*. Thirdly, the rows of B "represent" the simplices P_i, via

$$P_i = \bigcap_{j:B_{ij}=1} H_j^+ \cap \bigcap_{j:B_{ij}=-1} H_j^-. \qquad (*)$$

Now we derive from B a new matrix C, in which every row of B is replaced by all the row vectors that one can generate from it by replacing all the zeros by either $+1$ or -1. Since each row of B has $s-d-1$ zeros, and B has r rows, the matrix C has $2^{s-d-1}r$ rows.

For our example, this matrix C is a (32×6)-matrix that starts

$$C = \begin{pmatrix} 1 & 1 & 1 & 1 & 1 & 1 \\ 1 & 1 & 1 & 1 & 1 & -1 \\ 1 & 1 & 1 & -1 & 1 & 1 \\ 1 & 1 & 1 & -1 & 1 & -1 \\ 1 & -1 & 1 & 1 & 1 & 1 \\ 1 & -1 & 1 & 1 & 1 & -1 \\ 1 & -1 & 1 & -1 & 1 & 1 \\ 1 & -1 & 1 & -1 & 1 & -1 \\ \hline -1 & -1 & 1 & 1 & 1 & 1 \\ -1 & -1 & 1 & 1 & 1 & -1 \\ \vdots & \vdots & \vdots & \vdots & \vdots & \vdots \end{pmatrix},$$

where the first eight rows of C are derived from the first row of B, the second eight rows come from the second row of B, etc.

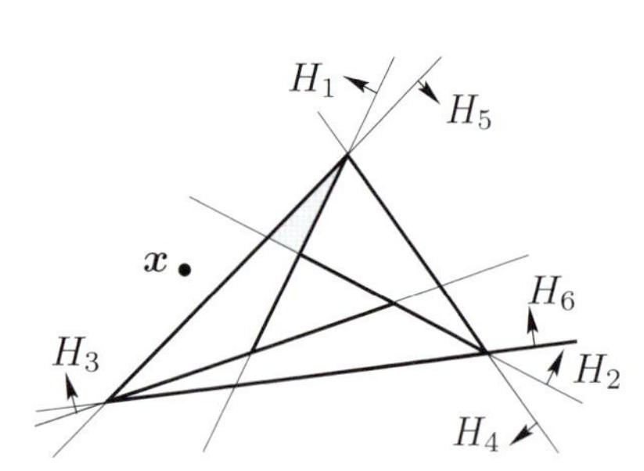

The first row of the C-matrix represents the shaded triangle, while the second row corresponds to an empty intersection of the halfspaces. The point $\boldsymbol{x}$ leads to the vector

$$\begin{pmatrix} 1 & -1 & 1 & 1 & -1 & 1 \end{pmatrix}$$

that does not appear in the C-matrix.

The point now is that all the rows of C are different: If two rows are derived from the same row of B, then they are different since their zeros have been replaced differently; if they are derived from different rows of B, then they differ no matter how the zeros have been replaced. But the rows of C are (± 1)-vectors of length s, and there are only 2^s different such vectors. Thus since the rows of C are distinct, C can have at most 2^s rows, that is,

$$2^{s-d-1}r \;\leq\; 2^s.$$

However, not all possible (± 1)-vectors appear in C, which yields a strict inequality $2^{s-d-1}r < 2^s$, and thus $r < 2^{d+1}$. To see this, we note that every row of C represents an intersection of halfspaces — just as for the rows of B before, via the formula $(*)$. This intersection is a subset of the simplex P_i, which was given by the corresponding row of B. Let us take a point $\boldsymbol{x} \in \mathbb{R}^d$ that does not lie on any of the hyperplanes H_j, and not in any of the simplices P_i. From this $\boldsymbol{x}$ we derive a (± 1)-vector that records for each j whether $\boldsymbol{x} \in H_j^+$ or $\boldsymbol{x} \in H_j^-$. This (± 1)-vector does not occur in C, because its halfspace intersection according to $(*)$ contains $\boldsymbol{x}$ and thus is not contained in any simplex P_i. □

References

[1] F. BAGEMIHL: *A conjecture concerning neighboring tetrahedra,* Amer. Math. Monthly **63** (1956) 328-329.

[2] V. J. D. BASTON: *Some Properties of Polyhedra in Euclidean Space,* Pergamon Press, Oxford 1965.

[3] M. A. PERLES: *At most 2^{d+1} neighborly simplices in E^d*, Annals of Discrete Math. **20** (1984), 253-254.

[4] J. ZAKS: *Neighborly families of 2^d d-simplices in E^d,* Geometriae Dedicata **11** (1981), 279-296.

[5] J. ZAKS: *No Nine Neighborly Tetrahedra Exist,* Memoirs Amer. Math. Soc. No. 447, Vol. 91, 1991.

Every large point set has an obtuse angle

Chapter 14

Around 1950 Paul Erdős conjectured that every set of more than 2^d points in $\mathbb{R}^d$ determines at least one *obtuse angle*, that is, an angle that is strictly greater than $\frac{\pi}{2}$. In other words, any set of points in $\mathbb{R}^d$ which only has acute angles (including right angles) has size at most 2^d. This problem was posed as a "prize question" by the Dutch Mathematical Society — but solutions were received only for $d = 2$ and for $d = 3$.

For $d = 2$ the problem is easy: The five points may determine a convex pentagon, which always has an obtuse angle (in fact, at least one angle of at least $108°$). Otherwise we have one point contained in the convex hull of three others that form a triangle. But this point "sees" the three edges of the triangle in three angles that sum to $360°$, so one of the angles is at least $120°$. (The second case also includes situations where we have three points on a line, and thus a $180°$ angle.)

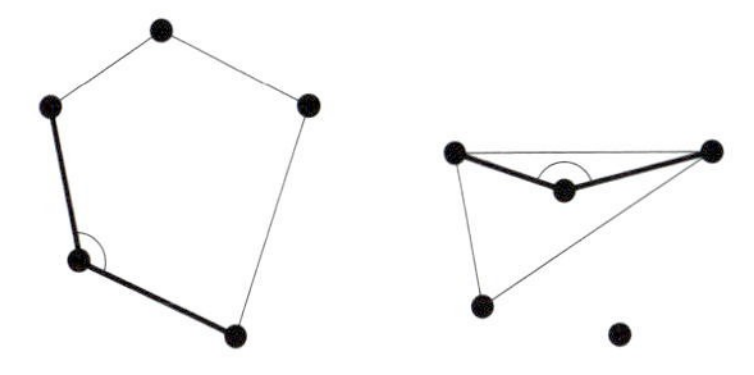

Unrelated to this, Victor Klee asked a few years later — and Erdős spread the question — how large a point set in $\mathbb{R}^d$ could be and still have the following "antipodality property": For *any* two points in the set there is a strip (bounded by two parallel hyperplanes) that contains the point set, and that has the two chosen points on different sides on the boundary.

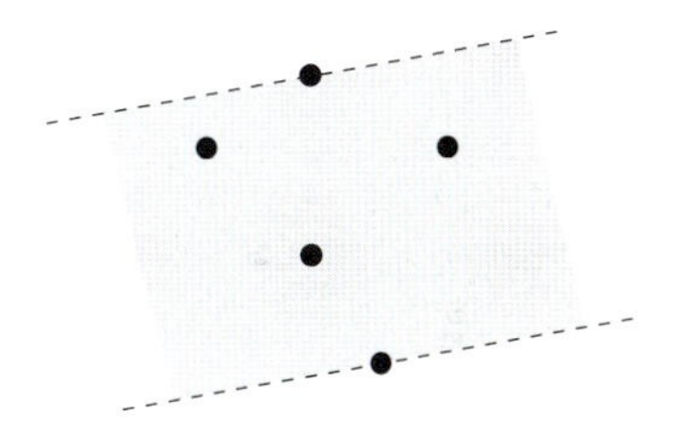

Then, in 1962, Ludwig Danzer and Branko Grünbaum solved both problems in one stroke: They sandwiched both maximal sizes into a chain of inequalities, which starts and ends in 2^d. Thus the answer is 2^d both for Erdős' and for Klee's problem.

In the following, we consider (finite) sets $S \subseteq \mathbb{R}^d$ of points, their convex hulls $\text{conv}(S)$, and general convex polytopes $Q \subseteq \mathbb{R}^d$. (See the appendix on polytopes on page 50 for the basic concepts.) We assume that these sets have the full dimension d, that is, they are not contained in a hyperplane. Two convex sets *touch* if they have at least one boundary point in common, while their interiors do not intersect. For any set $Q \subseteq \mathbb{R}^d$ and any vector $\boldsymbol{s} \in \mathbb{R}^d$ we denote by $Q + \boldsymbol{s}$ the image of Q under the translation that moves $\mathbf{0}$ to $\boldsymbol{s}$. Similarly, $Q - \boldsymbol{s}$ is the translate obtained by the map that moves $\boldsymbol{s}$ to the origin.

Don't be intimidated: This chapter is an excursion into d-dimensional geometry, but the arguments in the following do not require any "high-dimensional intuition," since they all can be followed, visualized (and thus *understood*) in three dimensions, or even in the plane. Hence, our figures will illustrate the proof for $d = 2$ (where a "hyperplane" is just a line), and you could create your own pictures for $d = 3$ (where a "hyperplane" is a plane).

Theorem 1. *For every d, one has the following chain of inequalities:*

$$2^d \overset{(1)}{\leq} \max\left\{\#S \mid S \subseteq \mathbb{R}^d,\ \sphericalangle(\boldsymbol{s}_i,\boldsymbol{s}_j,\boldsymbol{s}_k) \leq \tfrac{\pi}{2} \text{ for every } \{\boldsymbol{s}_i,\boldsymbol{s}_j,\boldsymbol{s}_k\} \subseteq S\right\}$$

$$\overset{(2)}{\leq} \max\left\{\#S \;\middle|\; \begin{array}{l} S \subseteq \mathbb{R}^d \text{ such that for any two points } \{\boldsymbol{s}_i,\boldsymbol{s}_j\} \subseteq S \\ \text{there is a strip } \mathcal{S}(i,j) \text{ that contains } S, \text{ with } \boldsymbol{s}_i \text{ and } \boldsymbol{s}_j \\ \text{lying in the parallel boundary hyperplanes of } \mathcal{S}(i,j) \end{array}\right\}$$

$$\overset{(3)}{=} \max\left\{\#S \;\middle|\; \begin{array}{l} S \subseteq \mathbb{R}^d \text{ such that the translates } P-\boldsymbol{s}_i,\ \boldsymbol{s}_i \in S, \text{ of} \\ \text{the convex hull } P := \mathrm{conv}(S) \text{ intersect in a common} \\ \text{point, but they only touch} \end{array}\right\}$$

$$\overset{(4)}{\leq} \max\left\{\#S \;\middle|\; \begin{array}{l} S \subseteq \mathbb{R}^d \text{ such that the translates } Q+\boldsymbol{s}_i \text{ of some d-} \\ \text{dimensional convex polytope } Q \subseteq \mathbb{R}^d \text{ touch pairwise} \end{array}\right\}$$

$$\overset{(5)}{=} \max\left\{\#S \;\middle|\; \begin{array}{l} S \subseteq \mathbb{R}^d \text{ such that the translates } Q^*+\boldsymbol{s}_i \text{ of some} \\ \text{d-dimensional centrally symmetric convex polytope} \\ Q^* \subseteq \mathbb{R}^d \text{ touch pairwise} \end{array}\right\}$$

$$\overset{(6)}{\leq} 2^d.$$

■ **Proof.** We have six claims (equalities and inequalities) to verify. Let's get going.

(1) Take $S := \{0,1\}^d$ to be the vertex set of the standard unit cube in $\mathbb{R}^d$, and choose $\boldsymbol{s}_i, \boldsymbol{s}_j, \boldsymbol{s}_k \in S$. By symmetry we may assume that $\boldsymbol{s}_j = \boldsymbol{0}$ is the zero vector. Hence the angle can be computed from

$$\cos \sphericalangle(\boldsymbol{s}_i,\boldsymbol{s}_j,\boldsymbol{s}_k) = \frac{\langle \boldsymbol{s}_i, \boldsymbol{s}_k\rangle}{|\boldsymbol{s}_i||\boldsymbol{s}_k|}$$

which is clearly nonnegative. Thus S is a set with $|S| = 2^d$ that has no obtuse angles.

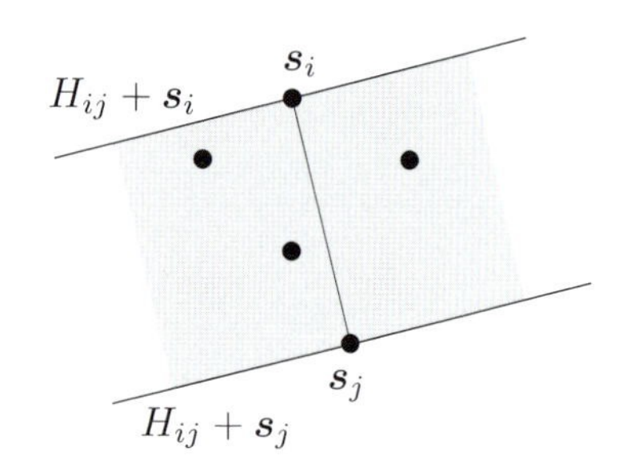

(2) If S contains no obtuse angles, then for any $\boldsymbol{s}_i, \boldsymbol{s}_j \in S$ we may define $H_{ij}+\boldsymbol{s}_i$ and $H_{ij}+\boldsymbol{s}_j$ to be the parallel hyperplanes through $\boldsymbol{s}_i$ resp. $\boldsymbol{s}_j$ that are orthogonal to the edge $[\boldsymbol{s}_i, \boldsymbol{s}_j]$. Here $H_{ij} = \{\boldsymbol{x} \in \mathbb{R}^d : \langle \boldsymbol{x}, \boldsymbol{s}_i - \boldsymbol{s}_j\rangle = 0\}$ is the hyperplane *through the origin* that is orthogonal to the line through $\boldsymbol{s}_i$ and $\boldsymbol{s}_j$, and $H_{ij} + \boldsymbol{s}_j = \{\boldsymbol{x} + \boldsymbol{s}_j : \boldsymbol{x} \in H_{ij}\}$ is the translate of H_{ij} that passes through $\boldsymbol{s}_j$, etc. Hence the strip between $H_{ij}+\boldsymbol{s}_i$ and $H_{ij}+\boldsymbol{s}_j$ consists, besides $\boldsymbol{s}_i$ and $\boldsymbol{s}_j$, exactly of all the points $\boldsymbol{x} \in \mathbb{R}^d$ such that the angles $\sphericalangle(\boldsymbol{s}_i,\boldsymbol{s}_j,\boldsymbol{x})$ and $\sphericalangle(\boldsymbol{s}_j,\boldsymbol{s}_i,\boldsymbol{x})$ are non-obtuse. Thus the strip contains all of S.

(3) P is contained in the halfspace of $H_{ij}+\boldsymbol{s}_j$ that contains $\boldsymbol{s}_i$ if and only if $P - \boldsymbol{s}_j$ is contained in the halfspace of H_{ij} that contains $\boldsymbol{s}_i - \boldsymbol{s}_j$: A property "an object is contained in a halfspace" is not destroyed if we translate both the object and the halfspace by the same amount (namely by $-\boldsymbol{s}_j$). Similarly, P is contained in the halfspace of $H_{ij}+\boldsymbol{s}_i$ that contains $\boldsymbol{s}_j$ if and only if $P - \boldsymbol{s}_i$ is contained in the halfspace of H_{ij} that contains $\boldsymbol{s}_j - \boldsymbol{s}_i$.

Putting both statements together, we find that the polytope P is contained in the strip between $H_{ij}+\boldsymbol{s}_i$ and $H_{ij}+\boldsymbol{s}_j$ if and only if $P-\boldsymbol{s}_i$ and $P-\boldsymbol{s}_j$ lie in different halfspaces with respect to the hyperplane H_{ij}.

This correspondence is illustrated by the sketch in the margin.

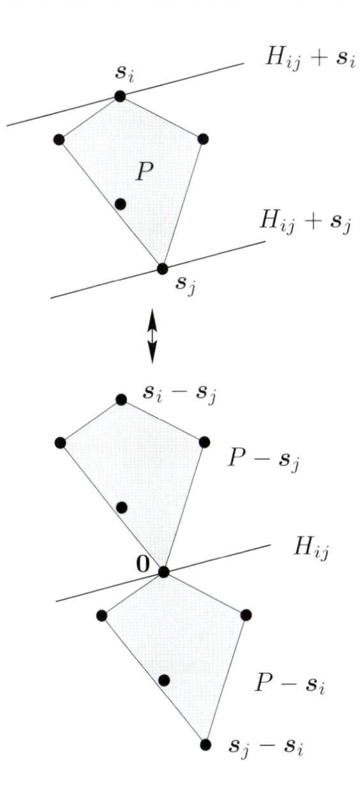

Furthermore, from $s_i \in P = \mathrm{conv}(S)$ we get that the origin $\mathbf{0}$ is contained in all the translates $P - s_i$ $(s_i \in S)$. Thus we see that the sets $P - s_i$ all intersect in $\mathbf{0}$, but they only touch: their interiors are pairwise disjoint, since they lie on opposite sides of the corresponding hyperplanes H_{ij}.

(4) This we get for free: "the translates must touch pairwise" is a weaker condition than "they intersect in a common point, but only touch." Similarly, we can relax the conditions by letting P be an arbitrary convex d-polytope in $\mathbb{R}^d$. Furthermore, we may replace S by $-S$.

(5) Here "$\geq$" is trivial, but that is not the interesting direction for us. We have to start with a configuration $S \subseteq \mathbb{R}^d$ and an arbitrary d-polytope $Q \subseteq \mathbb{R}^d$ such that the translates $Q + s_i$ $(s_i \in S)$ touch pairwise. The claim is that in this situation we can use

$$Q^* := \left\{ \tfrac{1}{2}(x - y) \in \mathbb{R}^d : x, y \in Q \right\}$$

instead of Q. But this is not hard to see: First, Q^* is d-dimensional, convex, and centrally symmetric. One can check that Q^* is a polytope (its vertices are of the form $\frac{1}{2}(q_i - q_j)$, for vertices q_i, q_j of Q), but this is not important for us.

Now we will show that $Q + s_i$ and $Q + s_j$ touch *if and only if* $Q^* + s_i$ and $Q^* + s_j$ touch. For this we note, in the footsteps of Minkowski, that

$$\begin{aligned}
(Q^* + s_i) \cap (Q^* + s_j) \neq \varnothing & \\
\iff & \exists q_i', q_i'', q_j', q_j'' \in Q : \tfrac{1}{2}(q_i' - q_i'') + s_i = \tfrac{1}{2}(q_j' - q_j'') + s_j \\
\iff & \exists q_i', q_i'', q_j', q_j'' \in Q : \tfrac{1}{2}(q_i' + q_j'') + s_i = \tfrac{1}{2}(q_j' + q_i'') + s_j \\
\iff & \exists q_i, q_j \in Q : q_i + s_i = q_j + s_j \\
\iff & (Q + s_i) \cap (Q + s_j) \neq \varnothing,
\end{aligned}$$

where in the third (and crucial) equivalence "$\iff$" we use that every $q \in Q$ can be written as $q = \frac{1}{2}(q + q)$ to get "$\Leftarrow$", and that Q is convex and thus $\frac{1}{2}(q_i' + q_j''), \frac{1}{2}(q_j' + q_i'') \in Q$ to see "$\Rightarrow$".

Thus the passage from Q to Q^* (known as *Minkowski symmetrization*) preserves the property that two translates $Q + s_i$ and $Q + s_j$ intersect. That is, we have shown that for any convex set Q, two translates $Q + s_i$ and $Q + s_j$ intersect if and only if the translates $Q^* + s_i$ and $Q^* + s_j$ intersect.

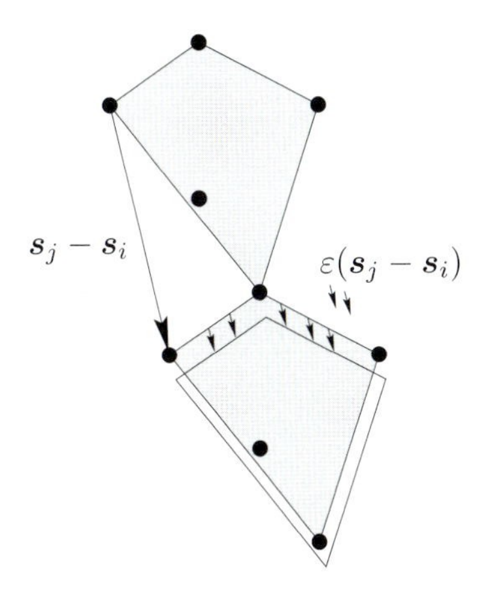

The following characterization shows that Minkowski symmetrization also preserves the property that two translates touch:

> *$Q + s_i$ and $Q + s_j$ touch if and only if they intersect, while $Q + s_i$ and $Q + s_j + \varepsilon(s_j - s_i)$ do not intersect for any $\varepsilon > 0$.*

(6) Assume that $Q^* + s_i$ and $Q^* + s_j$ touch. For every intersection point

$$x \in (Q^* + s_i) \cap (Q^* + s_j)$$

we have

$$\boldsymbol{x} - \boldsymbol{s}_i \in Q^* \quad \text{and} \quad \boldsymbol{x} - \boldsymbol{s}_j \in Q^*,$$

thus, since Q^* is centrally symmetric,

$$\boldsymbol{s}_i - \boldsymbol{x} = -(\boldsymbol{x} - \boldsymbol{s}_i) \in Q^*,$$

and hence, since Q^* is convex,

$$\tfrac{1}{2}(\boldsymbol{s}_i - \boldsymbol{s}_j) = \tfrac{1}{2}\left((\boldsymbol{x} - \boldsymbol{s}_j) + (\boldsymbol{s}_i - \boldsymbol{x})\right) \in Q^*.$$

We conclude that $\frac{1}{2}(\boldsymbol{s}_i + \boldsymbol{s}_j)$ is contained in $Q^* + \boldsymbol{s}_j$ for all i. Consequently, for $P := \text{conv}(S)$ we get

$$P_j := \tfrac{1}{2}(P + \boldsymbol{s}_j) = \text{conv}\left\{\tfrac{1}{2}(\boldsymbol{s}_i + \boldsymbol{s}_j) : \boldsymbol{s}_i \in S\right\} \subseteq Q^* + \boldsymbol{s}_j,$$

which implies that the sets $P_j = \frac{1}{2}(P + \boldsymbol{s}_j)$ can only touch.

Finally, the sets P_j are contained in P, because all the points $\boldsymbol{s}_i$, $\boldsymbol{s}_j$ and $\frac{1}{2}(\boldsymbol{s}_i + \boldsymbol{s}_j)$ are in P, since P is convex. But the P_j are just smaller, scaled, translates of P, contained in P. The scaling factor is $\frac{1}{2}$, which implies that

$$\text{vol}(P_j) \;=\; \frac{1}{2^d}\text{vol}(P),$$

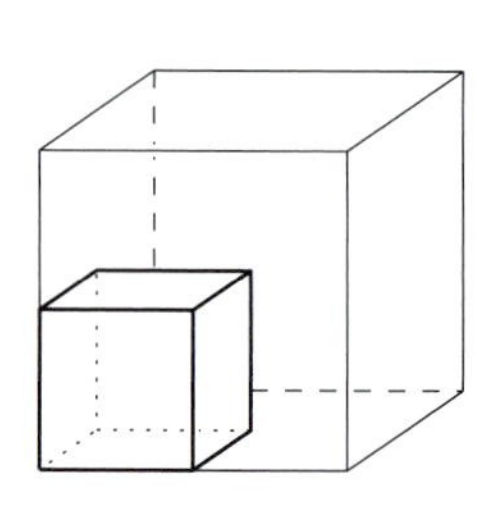
Scaling factor $\frac{1}{2}$, $\text{vol}(P_j) = \frac{1}{8}\text{vol}(P)$

since we are dealing with d-dimensional sets. This means that at most 2^d sets P_j fit into P, and hence $|S| \leq 2^d$.

This completes our proof: the chain of inequalities is closed. □

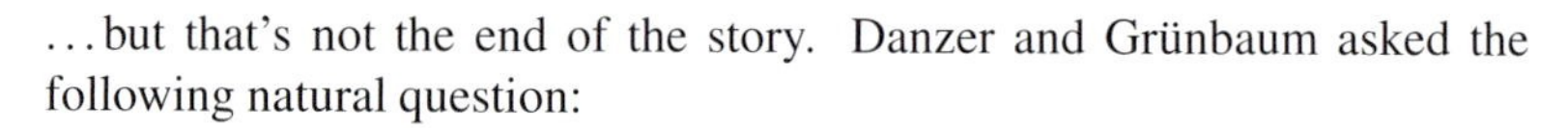
. . . but that's not the end of the story. Danzer and Grünbaum asked the following natural question:

> *What happens if one requires all angles to be* ***acute*** *rather than just non-obtuse, that is, if right angles are forbidden?*

They constructed configurations of $2d - 1$ points in $\mathbb{R}^d$ with only acute angles, conjecturing that this may be best possible. Grünbaum proved that this is indeed true for $d \leq 3$. But twenty-one years later, in 1983, Paul Erdős and Zoltan Füredi showed that the conjecture is false — quite dramatically, if the dimension is high! Their proof is a great example for the power of probabilistic arguments; see Chapter 35 for an introduction to the "probabilistic method." Our version of the proof uses a slight improvement in the choice of the parameters due to our reader David Bevan.

Theorem 2. *For every $d \geq 2$, there is a set $S \subseteq \{0,1\}^d$ of $2\lfloor\frac{\sqrt{6}}{9}\left(\frac{2}{\sqrt{3}}\right)^d\rfloor$ points in $\mathbb{R}^d$ (vertices of the unit d-cube) that determine only acute angles. In particular, in dimension $d = 34$ there is a set of $72 > 2{\cdot}34 - 1$ points with only acute angles.*

■ **Proof.** Set $m := \lfloor\frac{\sqrt{6}}{9}\left(\frac{2}{\sqrt{3}}\right)^d\rfloor$, and pick $3m$ vectors

$$\boldsymbol{x}(1), \boldsymbol{x}(2), \ldots, \boldsymbol{x}(3m) \;\in\; \{0,1\}^d$$

by choosing all their coordinates independently and randomly, to be either 0 or 1, with probability $\frac{1}{2}$ for each alternative. (You may toss a perfect coin $3md$ times for this; however, if d is large you may get bored by this soon.)

We have seen above that all angles determined by $0/1$-vectors are non-obtuse. Three vectors $\boldsymbol{x}(i), \boldsymbol{x}(j), \boldsymbol{x}(k)$ determine a right angle with apex $\boldsymbol{x}(j)$ if and only if the scalar product $\langle \boldsymbol{x}(i) - \boldsymbol{x}(j), \boldsymbol{x}(k) - \boldsymbol{x}(j) \rangle$ vanishes, that is, if we have

$$x(i)_\ell - x(j)_\ell = 0 \quad \text{or} \quad x(k)_\ell - x(j)_\ell = 0 \quad \text{for each coordinate } \ell.$$

We call (i, j, k) a *bad triple* if this happens. (If $\boldsymbol{x}(i) = \boldsymbol{x}(j)$ or $\boldsymbol{x}(j) = \boldsymbol{x}(k)$, then the angle is not defined, but also then the triple (i, j, k) is certainly bad.)

The probability that one specific triple is bad is exactly $\left(\frac{3}{4}\right)^d$: Indeed, it will be good if and only if, for one of the d coordinates ℓ, we get

$$\begin{array}{lll} \text{either} & x(i)_\ell = x(k)_\ell = 0, & x(j)_\ell = 1, \\ \text{or} & x(i)_\ell = x(k)_\ell = 1, & x(j)_\ell = 0. \end{array}$$

This leaves us with six bad options out of eight equally likely ones, and a triple will be bad if and only if one of the bad options (with probability $\frac{3}{4}$) happens for each of the d coordinates.

The number of triples we have to consider is $3\binom{3m}{3}$, since there are $\binom{3m}{3}$ sets of three vectors, and for each of them there are three choices for the apex. Of course the probabilities that the various triples are bad are not independent: but *linearity of expectation* (which is what you get by averaging over all possible selections; see the appendix) yields that the *expected* number of bad triples is exactly $3\binom{3m}{3}\left(\frac{3}{4}\right)^d$. This means — and this is the point where the probabilistic method shows its power — that there is *some* choice of the $3m$ vectors such that there are at most $3\binom{3m}{3}\left(\frac{3}{4}\right)^d$ bad triples, where

$$3\binom{3m}{3}\left(\tfrac{3}{4}\right)^d \;<\; 3\frac{(3m)^3}{6}\left(\tfrac{3}{4}\right)^d \;=\; m^3\left(\tfrac{9}{\sqrt{6}}\right)^2\left(\tfrac{3}{4}\right)^d \;\le\; m,$$

by the choice of m.

But if there are not more than m bad triples, then we can remove m of the $3m$ vectors $\boldsymbol{x}(i)$ in such a way that the remaining $2m$ vectors don't contain a bad triple, that is, they determine acute angles only. □

The "probabilistic construction" of a large set of $0/1$-points without right angles can be easily implemented, using a random number generator to "flip the coin." David Bevan has thus constructed a set of 31 points in dimension $d = 15$ that determines only acute angles.

Appendix: Three tools from probability

Here we gather three basic tools from discrete probability theory which will come up several times: random variables, linearity of expectation and Markov's inequality.

Let (Ω, p) be a finite *probability space*, that is, Ω is a finite set and $p = \text{Prob}$ is a map from Ω into the interval $[0, 1]$ with $\sum_{\omega \in \Omega} p(\omega) = 1$. A *random variable* X on Ω is a mapping $X : \Omega \longrightarrow \mathbb{R}$. We define a probability space on the image set $X(\Omega)$ by setting $p(X = x) := \sum_{X(\omega)=x} p(\omega)$. A simple example is an unbiased dice (all $p(\omega) = \frac{1}{6}$) with $X =$ "the number on top when the dice is thrown."

The *expectation* EX of X is the average to be expected, that is,

$$EX = \sum_{\omega \in \Omega} p(\omega) X(\omega).$$

Now suppose X and Y are two random variables on Ω, then the sum $X + Y$ is again a random variable, and we obtain

$$\begin{aligned} E(X+Y) &= \sum_{\omega} p(\omega)(X(\omega) + Y(\omega)) \\ &= \sum_{\omega} p(\omega) X(\omega) + \sum_{\omega} p(\omega) Y(\omega) \;=\; EX + EY. \end{aligned}$$

Clearly, this can be extended to any finite linear combination of random variables — this is what is called the *linearity of expectation*. Note that it needs no assumption that the random variables have to be "independent" in any sense!

Our third tool concerns random variables X which take only nonnegative values, shortly denoted $X \geq 0$. Let

$$\text{Prob}(X \geq a) = \sum_{\omega : X(\omega) \geq a} p(\omega)$$

be the probability that X is at least as large as some $a > 0$. Then

$$EX = \sum_{\omega : X(\omega) \geq a} p(\omega) X(\omega) + \sum_{\omega : X(\omega) < a} p(\omega) X(\omega) \geq a \sum_{\omega : X(\omega) \geq a} p(\omega),$$

and we have proved *Markov's inequality*

$$\text{Prob}(X \geq a) \leq \frac{EX}{a}.$$

References

[1] L. DANZER & B. GRÜNBAUM: *Über zwei Probleme bezüglich konvexer Körper von P. Erdös und von V. L. Klee,* Math. Zeitschrift **79** (1962), 95-99.

[2] P. ERDŐS & Z. FÜREDI: *The greatest angle among n points in the d-dimensional Euclidean space,* Annals of Discrete Math. **17** (1983), 275-283.

[3] H. MINKOWSKI: *Dichteste gitterförmige Lagerung kongruenter Körper,* Nachrichten Ges. Wiss. Göttingen, Math.-Phys. Klasse 1904, 311-355.

Borsuk's conjecture

Chapter 15

Karol Borsuk

Karol Borsuk's paper "Three theorems on the n-dimensional euclidean sphere" from 1933 is famous because it contained an important result (conjectured by Stanisław Ulam) that is now known as the Borsuk-Ulam theorem:

Every continuous map $f : S^d \to \mathbb{R}^d$ maps two antipodal points of the sphere S^d to the same point in $\mathbb{R}^d$.

The same paper is famous also because of a problem posed at its end, which became known as Borsuk's Conjecture:

> *Can every set $S \subseteq \mathbb{R}^d$ of bounded diameter diam$(S) > 0$ be partitioned into at most $d + 1$ sets of smaller diameter?*

The bound of $d+1$ is best possible: if S is a regular d-dimensional simplex, or just the set of its $d + 1$ vertices, then no part of a diameter-reducing partition can contain more than one of the simplex vertices. If $f(d)$ denotes the smallest number such that every bounded set $S \subseteq \mathbb{R}^d$ has a diameter-reducing partition into $f(d)$ parts, then the example of a regular simplex establishes $f(d) \geq d + 1$.

Borsuk's conjecture was proved for the case when S is a sphere (by Borsuk himself), for smooth bodies S (using the Borsuk-Ulam theorem), for $d \leq 3$, ... but the general conjecture remained open. The best available upper bound for $f(d)$ was established by Oded Schramm, who showed that

$$f(d) \leq (1.23)^d$$

for all large enough d. This bound looks quite weak compared with the conjecture "$f(d) = d + 1$", but it suddenly seemed reasonable when Jeff Kahn and Gil Kalai dramatically disproved Borsuk's conjecture in 1993. Sixty years after Borsuk's paper, Kahn and Kalai proved that $f(d) \geq (1.2)^{\sqrt{d}}$ holds for large enough d.

A Book version of the Kahn-Kalai proof was provided by A. Nilli: brief and self-contained, it yields an explicit counterexample to Borsuk's conjecture in dimension $d = 946$. We present here a modification of this proof, due to Andrei M. Raigorodskii and to Bernulf Weißbach, which reduces the dimension to $d = 561$, and even to $d = 560$. The current "record" is $d = 298$, achieved by Aicke Hinrichs and Christian Richter in 2002.

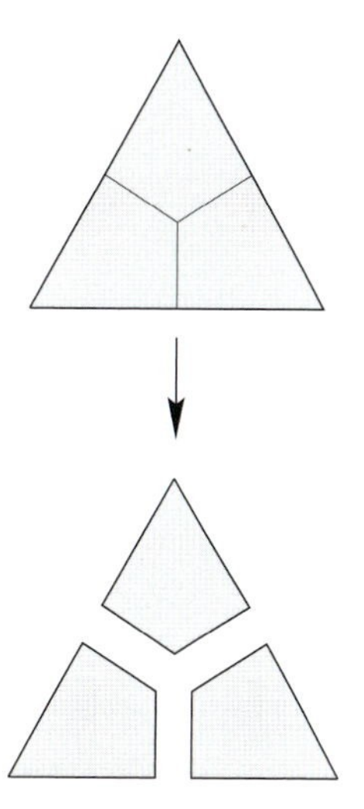

Any d-simplex *can* be split into $d + 1$ pieces, each of smaller diameter.

A. Nilli

Theorem. *Let $q = p^m$ be a prime power, $n := 4q-2$, and $d := \binom{n}{2} = (2q-1)(4q-3)$. Then there is a set $S \subseteq \{+1,-1\}^d$ of 2^{n-2} points in $\mathbb{R}^d$ such that every partition of S, whose parts have smaller diameter than S, has at least*

$$\frac{2^{n-2}}{\sum_{i=0}^{q-2}\binom{n-1}{i}}$$

parts. For $q = 9$ this implies that the Borsuk conjecture is false in dimension $d = 561$. Furthermore, $f(d) > (1.2)^{\sqrt{d}}$ holds for all large enough d.

■ **Proof.** The construction of the set S proceeds in four steps.

(1) Let q be a prime power, set $n = 4q-2$, and let

$$Q \;:=\; \Big\{\boldsymbol{x} \in \{+1,-1\}^n : x_1 = 1,\ \#\{i : x_i = -1\} \text{ is even}\Big\}.$$

This Q is a set of 2^{n-2} vectors in $\mathbb{R}^n$. We will see that $\langle \boldsymbol{x}, \boldsymbol{y}\rangle \equiv 2 \pmod 4$ holds for all vectors $\boldsymbol{x}, \boldsymbol{y} \in Q$. We will call $\boldsymbol{x}, \boldsymbol{y}$ *nearly-orthogonal* if $|\langle \boldsymbol{x}, \boldsymbol{y}\rangle| = 2$. We will prove that any subset $Q' \subseteq Q$ which contains no nearly-orthogonal vectors must be "small": $|Q'| \le \sum_{i=0}^{q-2}\binom{n-1}{i}$.

(2) From Q, we construct the set

$$R \;:=\; \{\boldsymbol{x}\boldsymbol{x}^T : \boldsymbol{x} \in Q\}$$

of 2^{n-2} symmetric $(n\times n)$-matrices of rank 1. We interpret them as vectors with n^2 components, $R \subseteq \mathbb{R}^{n^2}$. We will show that there are only acute angles between these vectors: they have positive scalar products, which are at least 4. Furthermore, if $R' \subseteq R$ contains no two vectors with minimal scalar product 4, then $|R'|$ is "small": $|R'| \le \sum_{i=0}^{q-2}\binom{n-1}{i}$.

$$\boldsymbol{x} = \begin{pmatrix} 1 \\ -1 \\ -1 \\ 1 \\ -1 \end{pmatrix} \Longrightarrow$$

$$\boldsymbol{x}^T = \begin{pmatrix} 1 & -1 & -1 & 1 & -1 \end{pmatrix}$$

$$\boldsymbol{x}\boldsymbol{x}^T = \begin{pmatrix} 1 & -1 & -1 & 1 & -1 \\ -1 & 1 & 1 & -1 & 1 \\ -1 & 1 & 1 & -1 & 1 \\ 1 & -1 & -1 & 1 & -1 \\ -1 & 1 & 1 & -1 & 1 \end{pmatrix}$$

Vectors, matrices, and scalar products

In our notation all vectors $\boldsymbol{x}, \boldsymbol{y}, \ldots$ are column vectors; the transposed vectors $\boldsymbol{x}^T, \boldsymbol{y}^T, \ldots$ are thus row vectors. The matrix product $\boldsymbol{x}\boldsymbol{x}^T$ is a matrix of rank 1, with $(\boldsymbol{x}\boldsymbol{x}^T)_{ij} = x_i x_j$.
If $\boldsymbol{x}, \boldsymbol{y}$ are column vectors, then their *scalar product* is

$$\langle \boldsymbol{x}, \boldsymbol{y}\rangle \;=\; \sum_i x_i y_i \;=\; \boldsymbol{x}^T\boldsymbol{y}.$$

We will also need scalar products for matrices $X, Y \in \mathbb{R}^{n\times n}$ which can be interpreted as vectors of length n^2, and thus their scalar product is

$$\langle X, Y\rangle \;:=\; \sum_{i,j} x_{ij} y_{ij}.$$

(3) From R, we obtain the set of points in $\mathbb{R}^{\binom{n}{2}}$ whose coordinates are the subdiagonal entries of the corresponding matrices:

$$S \;:=\; \{(\boldsymbol{x}\boldsymbol{x}^T)_{i>j} : \boldsymbol{x}\boldsymbol{x}^T \in R\}.$$

Again, S consists of 2^{n-2} points. The maximal distance between these points is precisely obtained for the nearly-orthogonal vectors $\boldsymbol{x}, \boldsymbol{y} \in Q$. We conclude that a subset $S' \subseteq S$ of smaller diameter than S must be "small": $|S'| \leq \sum_{i=0}^{q-2} \binom{n-1}{i}$.

(4) Estimates: From **(3)** we see that one needs at least

$$g(q) \;:=\; \frac{2^{4q-4}}{\sum_{i=0}^{q-2}\binom{4q-3}{i}}$$

parts in every diameter-reducing partition of S. Thus

$$f(d) \;\geq\; \max\{g(q), d+1\} \qquad \text{for } d \;=\; (2q-1)(4q-3).$$

Therefore, whenever we have $g(q) > (2q-1)(4q-3)+1$, then we have a counterexample to Borsuk's conjecture in dimension $d = (2q-1)(4q-3)$. We will calculate below that $g(9) > 562$, which yields the counterexample in dimension $d = 561$, and that

$$g(q) \;>\; \frac{e}{64\,q^2}\left(\frac{27}{16}\right)^q,$$

which yields the asymptotic bound $f(d) > (1.2)^{\sqrt{d}}$ for d large enough.

Details for (1): We start with some harmless divisibility considerations.

Lemma. *The function $P(z) := \binom{z-2}{q-2}$ is a polynomial of degree $q-2$. It yields integer values for all integers z. The integer $P(z)$ is divisible by p if and only if z is not congruent to 0 or 1 modulo q.*

■ **Proof.** For this we write the binomial coefficient as

$$P(z) \;=\; \binom{z-2}{q-2} \;=\; \frac{(z-2)(z-3)\cdot\ldots\cdot(z-q+1)}{(q-2)(q-3)\cdot\;\ldots\;\ldots\;\cdot 2\cdot 1} \qquad (*)$$

and compare the number of p-factors in the denominator and in the numerator. The denominator has the same number of p-factors as $(q-2)!$, or as $(q-1)!$, since $q-1$ is not divisible by p. Indeed, by the claim in the margin we get an integer with the same number of p-factors if we take *any* product of $q-1$ integers, one from each non-zero residue class modulo q.

Now if z is congruent to 0 or $1 \pmod q$, then the numerator is also of this type: All factors in the product are from different residue classes, and the only classes that do not occur are the zero class (the multiples of q), and the class either of -1 or of $+1$, but neither $+1$ nor -1 is divisible by p. Thus denominator and numerator have the same number of p-factors, and hence the quotient is not divisible by p.

Claim. *If $a \equiv b \not\equiv 0 \pmod q$, then a and b have the same number of p-factors.*

■ **Proof.** We have $a = b + sp^m$, where b is not divisible by $p^m = q$. So every power p^k that divides b satisfies $k < m$, and thus it also divides a. The statement is symmetric in a and b. □

On the other hand, if $z \not\equiv 0, 1 \pmod q$, then the numerator of $(*)$ contains one factor that is divisible by $q = p^m$. At the same time, the product has no factors from two adjacent nonzero residue classes: one of them represents numbers that have no p-factors at all, the other one has fewer p-factors than $q = p^m$. Hence there are more p-factors in the numerator than in the denominator, and the quotient is divisible by p. $\square$

Now we consider an arbitrary subset $Q' \subseteq Q$ that does not contain any nearly-orthogonal vectors. We want to establish that Q' must be "small."

> ***Claim 1.*** *If $\boldsymbol{x}, \boldsymbol{y}$ are distinct vectors from Q, then $\frac{1}{4}(\langle \boldsymbol{x}, \boldsymbol{y}\rangle + 2)$ is an integer in the range*
>
> $$-(q-2) \;\leq\; \tfrac{1}{4}(\langle \boldsymbol{x}, \boldsymbol{y}\rangle + 2) \;\leq\; q-1.$$

Both $\boldsymbol{x}$ and $\boldsymbol{y}$ have an even number of (-1)-components, so the number of components in which $\boldsymbol{x}$ and $\boldsymbol{y}$ differ is even, too. Thus

$$\langle \boldsymbol{x}, \boldsymbol{y}\rangle \;=\; (4q-2) \;-\; 2\#\{i : x_i \neq y_i\} \;\equiv\; -2 \pmod 4$$

for all $\boldsymbol{x}, \boldsymbol{y} \in Q$, that is, $\frac{1}{4}(\langle \boldsymbol{x}, \boldsymbol{y}\rangle + 2)$ is an integer.
From $\boldsymbol{x}, \boldsymbol{y} \in \{+1, -1\}^{4q-2}$ we see that $-(4q-2) \leq \langle \boldsymbol{x}, \boldsymbol{y}\rangle \leq 4q-2$, that is, $-(q-1) \leq \frac{1}{4}(\langle \boldsymbol{x}, \boldsymbol{y}\rangle + 2) \leq q$. The lower bound never holds with equality, since $x_1 = y_1 = 1$ implies that $\boldsymbol{x} \neq -\boldsymbol{y}$. The upper bound holds with equality only if $\boldsymbol{x} = \boldsymbol{y}$.

> ***Claim 2.*** *For any $\boldsymbol{y} \in Q'$, the polynomial in n variables $x_1, \ldots, x_n$ of degree $q-2$ given by*
>
> $$F_{\boldsymbol{y}}(\boldsymbol{x}) \;:=\; P\big(\tfrac{1}{4}(\langle \boldsymbol{x}, \boldsymbol{y}\rangle + 2)\big) \;=\; \binom{\frac{1}{4}(\langle \boldsymbol{x}, \boldsymbol{y}\rangle + 2) - 2}{q-2}$$
>
> *satisfies that $F_{\boldsymbol{y}}(\boldsymbol{x})$ is divisible by p for every $\boldsymbol{x} \in Q' \backslash \{\boldsymbol{y}\}$, but not for $\boldsymbol{x} = \boldsymbol{y}$.*

The representation by a binomial coefficient shows that $F_{\boldsymbol{y}}(\boldsymbol{x})$ is an integer-valued polynomial. For $\boldsymbol{x} = \boldsymbol{y}$, we get $F_{\boldsymbol{y}}(\boldsymbol{y}) = 1$. For $\boldsymbol{x} \neq \boldsymbol{y}$, the Lemma yields that $F_{\boldsymbol{y}}(\boldsymbol{x})$ is not divisible by p if and only if $\frac{1}{4}(\langle \boldsymbol{x}, \boldsymbol{y}\rangle + 2)$ is congruent to 0 or 1 $(\bmod q)$. By Claim 1, this happens only if $\frac{1}{4}(\langle \boldsymbol{x}, \boldsymbol{y}\rangle + 2)$ is either 0 or 1, that is, if $\langle \boldsymbol{x}, \boldsymbol{y}\rangle \in \{-2, +2\}$. So $\boldsymbol{x}$ and $\boldsymbol{y}$ must be nearly-orthogonal for this, which contradicts the definition of Q'.

> ***Claim 3.*** *The same is true for the polynomials $\overline{F}_{\boldsymbol{y}}(\boldsymbol{x})$ in the $n-1$ variables $x_2, \ldots, x_n$ that are obtained as follows: Expand $F_{\boldsymbol{y}}(\boldsymbol{x})$ into monomials and remove the variable x_1, and reduce all higher powers of other variables, by substituting $x_1 = 1$, and $x_i^2 = 1$ for $i > 1$. The polynomials $\overline{F}_{\boldsymbol{y}}(\boldsymbol{x})$ have degree at most $q-2$.*

The vectors $\boldsymbol{x} \in Q \subseteq \{+1, -1\}^n$ all satisfy $x_1 = 1$ and $x_i^2 = 1$. Thus the substitutions do not change the values of the polynomials on the set Q. They also do not increase the degree, so $\overline{F}_{\boldsymbol{y}}(\boldsymbol{x})$ has degree at most $q-2$.

> ***Claim 4.*** *There is no linear relation (with rational coefficients) between the polynomials $\overline{F}_{\boldsymbol{y}}(\boldsymbol{x})$, that is, the polynomials $\overline{F}_{\boldsymbol{y}}(\boldsymbol{x})$, $\boldsymbol{y} \in Q'$, are linearly independent over $\mathbb{Q}$. In particular, they are distinct.*

Assume that there is a relation of the form $\sum_{\boldsymbol{y}\in Q'} \alpha_{\boldsymbol{y}} \overline{F}_{\boldsymbol{y}}(\boldsymbol{x}) = 0$ such that not all coefficients $\alpha_{\boldsymbol{y}}$ are zero. After multiplication with a suitable scalar we may assume that all the coefficients are integers, but not all of them are divisible by p. But then for every $\boldsymbol{y} \in Q'$ the evaluation at $\boldsymbol{x} := \boldsymbol{y}$ yields that $\alpha_{\boldsymbol{y}} \overline{F}_{\boldsymbol{y}}(\boldsymbol{y})$ is divisible by p, and hence so is $\alpha_{\boldsymbol{y}}$, since $\overline{F}_{\boldsymbol{y}}(\boldsymbol{y})$ is not.

> ***Claim 5.*** *$|Q'|$ is bounded by the number of squarefree monomials of degree at most $q-2$ in $n-1$ variables, which is $\sum_{i=0}^{q-2} \binom{n-1}{i}$.*

By construction the polynomials $\overline{F}_{\boldsymbol{y}}$ are squarefree: none of their monomials contains a variable with higher degree than 1. Thus each $\overline{F}_{\boldsymbol{y}}(\boldsymbol{x})$ is a linear combination of the squarefree monomials of degree at most $q-2$ in the $n-1$ variables $x_2, \ldots, x_n$. Since the polynomials $\overline{F}_{\boldsymbol{y}}(\boldsymbol{x})$ are linearly independent, their number (which is $|Q'|$) cannot be larger than the number of monomials in question.

Details for (2): The first column of $\boldsymbol{x}\boldsymbol{x}^T$ is $\boldsymbol{x}$. Thus for distinct $\boldsymbol{x} \in Q$ we obtain distinct matrices $M(\boldsymbol{x}) := \boldsymbol{x}\boldsymbol{x}^T$. We interpret these matrices as vectors of length n^2 with components $x_i x_j$. A simple computation

$$\begin{aligned} \langle M(\boldsymbol{x}), M(\boldsymbol{y}) \rangle &= \sum_{i=1}^{n}\sum_{j=1}^{n} (x_i x_j)(y_i y_j) \\ &= \Big(\sum_{i=1}^{n} x_i y_i\Big)\Big(\sum_{j=1}^{n} x_j y_j\Big) = \langle \boldsymbol{x}, \boldsymbol{y} \rangle^2 \geq 4 \end{aligned}$$

shows that the scalar product of $M(\boldsymbol{x})$ and $M(\boldsymbol{y})$ is minimized if and only if $\boldsymbol{x}, \boldsymbol{y} \in Q$ are nearly-orthogonal.

Details for (3): Let $U(\boldsymbol{x}) \in \{+1, -1\}^d$ denote the vector of all subdiagonal entries of $M(\boldsymbol{x})$. Since $M(\boldsymbol{x}) = \boldsymbol{x}\boldsymbol{x}^T$ is symmetric with diagonal values $+1$, we see that $M(\boldsymbol{x}) \neq M(\boldsymbol{y})$ implies $U(\boldsymbol{x}) \neq U(\boldsymbol{y})$. Furthermore,

$$4 \leq \langle M(\boldsymbol{x}), M(\boldsymbol{y}) \rangle = 2\langle U(\boldsymbol{x}), U(\boldsymbol{y}) \rangle + n,$$

that is,

$$\langle U(\boldsymbol{x}), U(\boldsymbol{y}) \rangle \geq -\frac{n}{2} + 2,$$

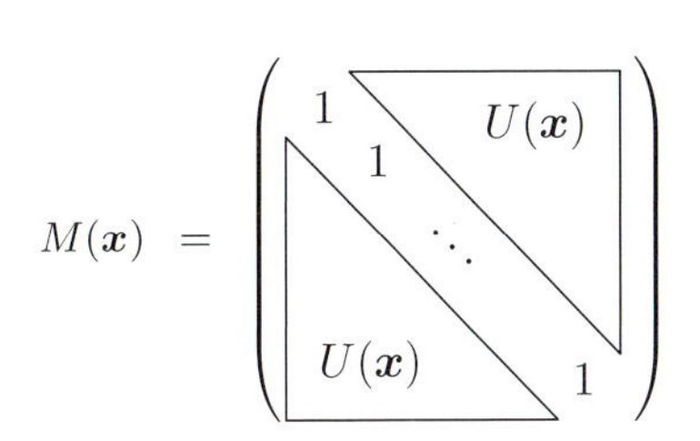

with equality if and only if $\boldsymbol{x}$ and $\boldsymbol{y}$ are nearly-orthogonal. Since all the vectors $U(\boldsymbol{x}) \in S$ have the same length $\sqrt{\langle U(\boldsymbol{x}), U(\boldsymbol{x}) \rangle} = \sqrt{\binom{n}{2}}$, this means that the maximal distance between points $U(\boldsymbol{x}), U(\boldsymbol{y}) \in S$ is achieved exactly when $\boldsymbol{x}$ and $\boldsymbol{y}$ are nearly-orthogonal.

Details for (4): For $q = 9$ we have $g(9) \approx 758.31$, which is greater than $d + 1 = \binom{34}{2} + 1 = 562$.

To obtain a general bound for large d, we use monotonicity and unimodality of the binomial coefficients and the estimates $n! > e(\frac{n}{e})^n$ and $n! < en(\frac{n}{e})^n$ (see the appendix to Chapter 2) and derive

$$\sum_{i=0}^{q-2} \binom{4q-3}{i} < q\binom{4q}{q} = q\,\frac{(4q)!}{q!(3q)!} < q\,\frac{e\,4q\left(\frac{4q}{e}\right)^{4q}}{e\left(\frac{q}{e}\right)^{q} e\left(\frac{3q}{e}\right)^{3q}} = \frac{4q^2}{e}\left(\frac{256}{27}\right)^q.$$

Thus we conclude

$$f(d) \geq g(q) = \frac{2^{4q-4}}{\sum\limits_{i=0}^{q-2}\binom{4q-3}{i}} > \frac{e}{64q^2}\left(\frac{27}{16}\right)^q.$$

From this, with

$$d = (2q-1)(4q-3) = 5q^2 + (q-3)(3q-1) \geq 5q^2 \quad \text{for } q \geq 3,$$

$$q = \tfrac{5}{8} + \sqrt{\tfrac{d}{8} + \tfrac{1}{64}} > \sqrt{\tfrac{d}{8}}, \qquad \text{and} \qquad \left(\tfrac{27}{16}\right)^{\sqrt{\frac{1}{8}}} > 1.2032,$$

we get

$$f(d) > \frac{e}{13d}(1.2032)^{\sqrt{d}} > (1.2)^{\sqrt{d}} \quad \text{for all large enough } d. \qquad \square$$

A counterexample of dimension 560 is obtained by noting that for $q = 9$ the quotient $g(q) \approx 758$ is *much* larger than the dimension $d(q) = 561$. Thus one gets a counterexample for $d = 560$ by taking only the "three fourths" of the points in S that satisfy $x_{21} + x_{31} + x_{32} = -1$.

Borsuk's conjecture is known to be true for $d \leq 3$, but it has not been verified for any larger dimension. In contrast to this, it *is* true up to $d = 8$ if we restrict ourselves to subsets $S \subseteq \{1, -1\}^d$, as constructed above (see [8]). In either case it is quite possible that counterexamples can be found in reasonably small dimensions.

References

[1] K. BORSUK: *Drei Sätze über die n-dimensionale euklidische Sphäre,* Fundamenta Math. **20** (1933), 177-190.

[2] A. HINRICHS & C. RICHTER: *New sets with large Borsuk numbers,* Preprint, February 2002, 10 pages; Discrete Math., to appear.

[3] J. KAHN & G. KALAI: *A counterexample to Borsuk's conjecture,* Bulletin Amer. Math. Soc. **29** (1993), 60-62.

[4] A. NILLI: *On Borsuk's problem,* in: "Jerusalem Combinatorics '93" (H. Barcelo and G. Kalai, eds.), Contemporary Mathematics **178**, Amer. Math. Soc. 1994, 209-210.

[5] A. M. RAIGORODSKII: *On the dimension in Borsuk's problem,* Russian Math. Surveys (6) **52** (1997), 1324-1325.

[6] O. SCHRAMM: *Illuminating sets of constant width,* Mathematika **35** (1988), 180-199.

[7] B. WEISSBACH: *Sets with large Borsuk number,* Beiträge zur Algebra und Geometrie/Contributions to Algebra and Geometry **41** (2000), 417-423.

[8] G. M. ZIEGLER: *Coloring Hamming graphs, optimal binary codes, and the 0/1-Borsuk problem in low dimensions,* Lecture Notes in Computer Science **2122**, Springer-Verlag 2001, 164-175.

Analysis

"Hilbert's seaside resort hotel"

Sets, functions, and the continuum hypothesis

Chapter 16

Set theory, founded by Georg Cantor in the second half of the 19th century, has profoundly transformed mathematics. Modern day mathematics is unthinkable without the concept of a set, or as David Hilbert put it: "Nobody will drive us from the paradise (of set theory) that Cantor has created for us."

Georg Cantor

One of Cantor's basic concepts was the notion of the *size* or *cardinality* of a set M, denoted by $|M|$. For finite sets, this presents no difficulties: we just count the number of elements and say that M is an n-set or has size n, if M contains precisely n elements. Thus two finite sets M and N have equal size, $|M| = |N|$, if they contain the same number of elements.

To carry this notion of *equal* size over to infinite sets, we use the following suggestive thought experiment for finite sets. Suppose a number of people board a bus. When will we say that the number of people is the same as the number of available seats? Simple enough, we let all people sit down. If everyone finds a seat, and no seat remains empty, then and only then do the two sets (of the people and of the seats) agree in number. In other words, the two sizes are the same if there is a *bijection* of one set onto the other.

This is then our definition: Two arbitrary sets M and N (finite or infinite) are said to be of *equal size* or *cardinality*, if and only if there exists a bijection from M onto N. Clearly, this notion of equal size is an equivalence relation, and we can thus associate a number, called *cardinal number*, to every class of equal-sized sets. For example, we obtain for finite sets the cardinal numbers $0, 1, 2, \ldots, n, \ldots$ where n stands for the class of n-sets, and, in particular, 0 for the *empty set* $\varnothing$. We further observe the obvious fact that a proper subset of a finite set M invariably has smaller size than M.

The theory becomes very interesting (and highly non-intuitive) when we turn to infinite sets. Consider the set $\mathbb{N} = \{1, 2, 3, \ldots\}$ of natural numbers. We call a set M *countable* if it can be put in one-to-one correspondence with $\mathbb{N}$. In other words, M is countable if we can list the elements of M as $m_1, m_2, m_3, \ldots$. But now a strange phenomenon occurs. Suppose we add to $\mathbb{N}$ a new element x. Then $\mathbb{N} \cup \{x\}$ is still countable, and hence has equal size with $\mathbb{N}$!

This fact is delightfully illustrated by "Hilbert's hotel." Suppose a hotel has countably many rooms, numbered $1, 2, 3, \ldots$ with guest g_i occupying room i; so the hotel is fully booked. Now a new guest x arrives asking for a room, whereupon the hotel manager tells him: Sorry, all rooms are taken. No problem, says the new arrival, just move guest g_1 to room 2, g_2 to room 3, g_3 to room 4, and so on, and I will then take room 1. To the

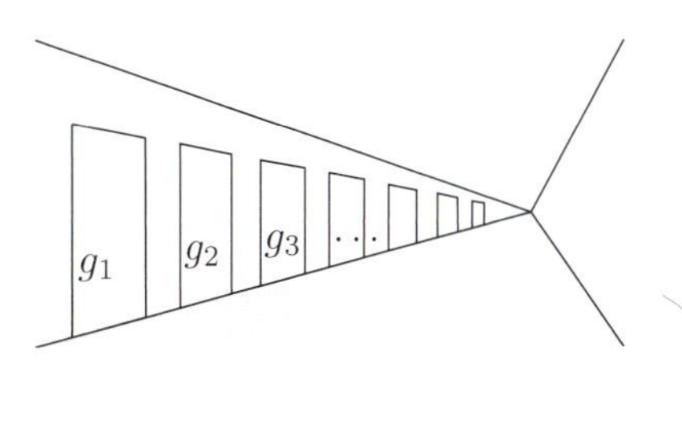

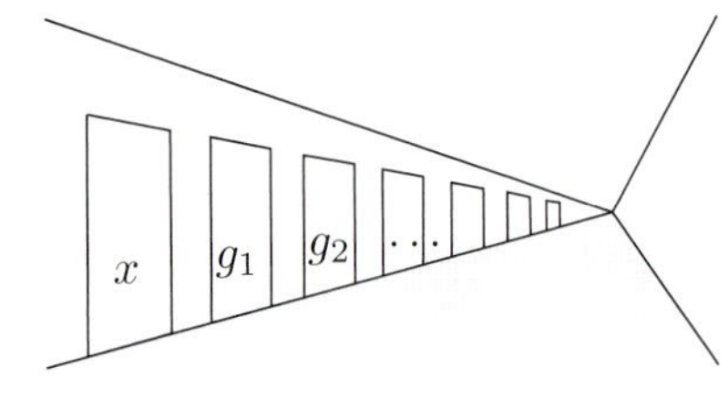

manager's surprise (he is not a mathematician) this works; he can still put up all guests plus the new arrival x!

Now it is clear that he can also put up another guest y, and another one z, and so on. In particular, we note that, in contrast to finite sets, it may well happen that a proper subset of an *infinite* set M has the same size as M. In fact, as we will see, this is a characterization of infinity: A set is infinite if and only if it has the same size as some proper subset.

Let us leave Hilbert's hotel and look at our familiar number sets. The set $\mathbb{Z}$ of integers is again countable, since we may enumerate $\mathbb{Z}$ in the form $\mathbb{Z} = \{0, 1, -1, 2, -2, 3, -3, \dots\}$. It may come more as a surprise that the rationals can be enumerated in a similar way.

Theorem 1. *The set $\mathbb{Q}$ of rational numbers is countable.*

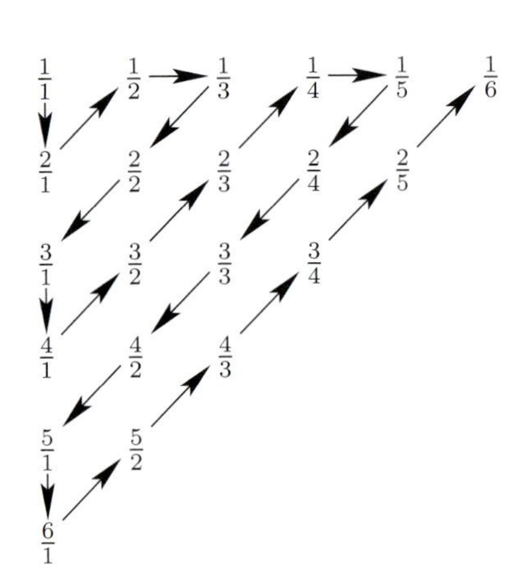

■ **Proof.** By listing the set $\mathbb{Q}^+$ of positive rationals as suggested in the figure in the margin, but leaving out numbers already encountered, we see that $\mathbb{Q}^+$ is countable, and hence so is $\mathbb{Q}$ by listing 0 at the beginning and $-\frac{p}{q}$ right after $\frac{p}{q}$. With this listing

$$\mathbb{Q} = \{0, 1, -1, 2, -2, \tfrac{1}{2}, -\tfrac{1}{2}, \tfrac{1}{3}, -\tfrac{1}{3}, 3, -3, 4, -4, \tfrac{3}{2}, -\tfrac{3}{2}, \dots\}. \qquad \square$$

Another way to interpret the figure is the following statement:

The union of countably many countable sets M_n is again countable.

Indeed, set $M_n = \{a_{n1}, a_{n2}, a_{n3}, \dots\}$ and list

$$\bigcup_{n=1}^{\infty} M_n = \{a_{11}, a_{21}, a_{12}, a_{13}, a_{22}, a_{31}, a_{41}, a_{32}, a_{23}, a_{14}, \dots\}$$

precisely as before.

Let us contemplate Cantor's enumeration of the positive rationals a bit more. Looking at the figure we obtained the sequence

$$\tfrac{1}{1}, \tfrac{2}{1}, \tfrac{1}{2}, \tfrac{1}{3}, \tfrac{2}{2}, \tfrac{3}{1}, \tfrac{4}{1}, \tfrac{3}{2}, \tfrac{2}{3}, \tfrac{1}{4}, \tfrac{1}{5}, \tfrac{2}{4}, \tfrac{3}{3}, \tfrac{4}{2}, \tfrac{5}{1}, \dots$$

and then had to strike out the duplicates such as $\frac{2}{2} = \frac{1}{1}$ or $\frac{2}{4} = \frac{1}{2}$.

But there is a listing that is even more elegant and systematic, and which contains no duplicates — found only quite recently by Neil Calkin and Herbert Wilf. Their new list starts as follows:

$$\tfrac{1}{1}, \tfrac{1}{2}, \tfrac{2}{1}, \tfrac{1}{3}, \tfrac{3}{2}, \tfrac{2}{3}, \tfrac{3}{1}, \tfrac{1}{4}, \tfrac{4}{3}, \tfrac{3}{5}, \tfrac{5}{2}, \tfrac{2}{5}, \tfrac{5}{3}, \tfrac{3}{4}, \tfrac{4}{1}, \dots.$$

Here the denominator of the n-th rational number equals the numerator of the $(n+1)$-st number. In other words, the n-th fraction is $b(n)/b(n+1)$, where $\big(b(n)\big)_{n\geq 0}$ is a sequence that starts with

$$(1, 1, 2, 1, 3, 2, 3, 1, 4, 3, 5, 2, 5, 3, 4, 1, 5, \dots).$$

This sequence has first been studied by a German mathematician, Moritz Abraham Stern, in a paper from 1858, and is has become known as "Stern's diatomic series."

How do we obtain this sequence, and hence the Calkin-Wilf listing of the positive fractions? Consider the infinite binary tree in the margin. We immediately note its recursive rule:

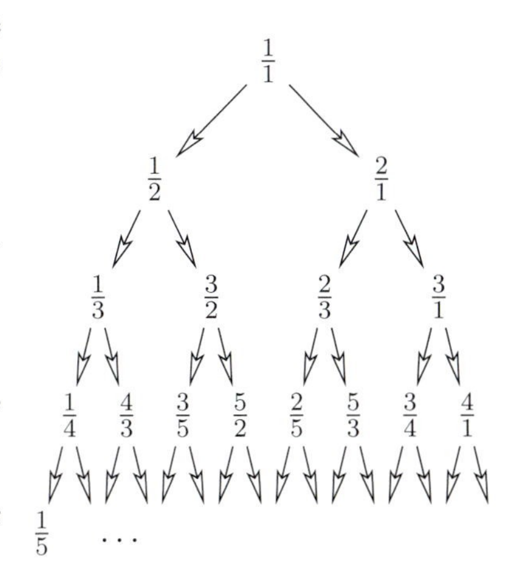

- $\frac{1}{1}$ is on top of the tree, and
- every node $\frac{i}{j}$ has two sons: the left son is $\frac{i}{i+j}$ and the right son is $\frac{i+j}{j}$.

We can easily check the following four properties:

(1) All fractions in the tree are reduced, that is, if $\frac{r}{s}$ appears in the tree, then r and s are relatively prime.

This holds for the top $\frac{1}{1}$, and then we use induction downward. If r and s are relatively prime, then so are r and $r+s$, as well as s and $r+s$.

(2) Every reduced fraction $\frac{r}{s} > 0$ appears in the tree.

We use induction on the sum $r+s$. The smallest value is $r+s=2$, that is $\frac{r}{s}=\frac{1}{1}$, and this appears at the top. If $r>s$, then $\frac{r-s}{s}$ appears in the tree by induction, and so we get $\frac{r}{s}$ as its right son. Similarly, if $r<s$, then $\frac{r}{s-r}$ appears, which has $\frac{r}{s}$ as its left son.

(3) Every reduced fraction appears exactly once.

The argument is similar. If $\frac{r}{s}$ appears more than once, then $r \neq s$, since any node in the tree except the top is of the form $\frac{i}{i+j} < 1$ or $\frac{i+j}{j} > 1$. But if $r>s$ or $r<s$, then we argue by induction as before.

Every positive rational appears therefore exactly once in our tree, and we may write them down listing the numbers level-by-level from left to right. This yields precisely the initial segment shown above.

(4) The denominator of the n-th fraction in our list equals the numerator of the $(n+1)$-st.

This is certainly true for $n=0$, or when the n-th fraction is a left son. Suppose the n-th number $\frac{r}{s}$ is a right son. If $\frac{r}{s}$ is at the right boundary, then $s=1$, and the successor lies at the left boundary and has numerator 1. Finally, if $\frac{r}{s}$ is in the interior, and $\frac{r'}{s'}$ is the next fraction in our sequence, then $\frac{r}{s}$ is the right son of $\frac{r-s}{s}$, $\frac{r'}{s'}$ is the left son of $\frac{r'}{s'-r'}$, and by induction the denominator of $\frac{r-s}{s}$ is the numerator of $\frac{r'}{s'-r'}$, so we get $s=r'$.

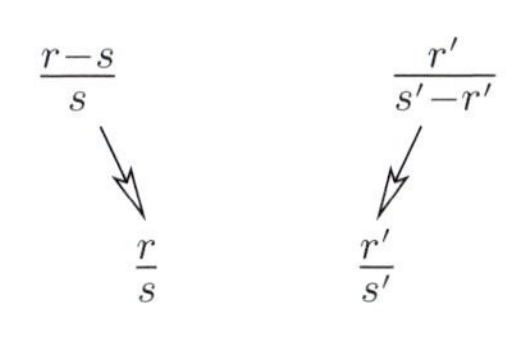

Well, this is nice, but there is even more to come. There are two natural questions:

– Does the sequence $(b(n))_{n\geq 0}$ have a "meaning"? That is, does $b(n)$ count anything simple?

– Given $\frac{r}{s}$, is there an easy way to determine the successor in the listing?

To answer the first question, we work out that the node $b(n)/b(n+1)$ has the two sons $b(2n+1)/b(2n+2)$ and $b(2n+2)/b(2n+3)$. By the set-up of the tree we obtain the recursions

$$b(2n+1) = b(n) \quad \text{and} \quad b(2n+2) = b(n) + b(n+1). \tag{1}$$

With $b(0) = 1$ the sequence $(b(n))_{n\geq 0}$ is completely determined by (1).

So, is there a "nice" "known" sequence which obeys the same recursion? Yes, there is. We know that any number n can be uniquely written as a sum of distinct powers of 2 — this is the usual binary representation of n. A *hyper-binary* representation of n is a representation of n a sum of powers of 2, where every power 2^k appears at most *twice*. Let $h(n)$ be the number of such representations for n. You are invited to check that the sequence $h(n)$ obeys the recursion (1), and this gives $b(n) = h(n)$ for all n.

For example, $h(6) = 3$, with the hyper-binary representations
$6 = 4 + 2$
$6 = 4 + 1 + 1$
$6 = 2 + 2 + 1 + 1$.

Incidentally, we have proved a surprising fact: Let $\frac{r}{s}$ be a reduced fraction, there exists precisely one integer n with $r = h(n)$ and $s = h(n+1)$.

Let us look at the second question. We have in our tree

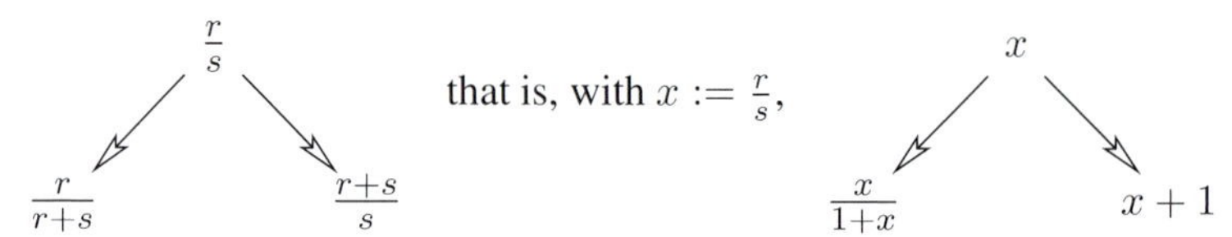

We now use this to generate an even larger infinite binary tree (without a root) as follows:

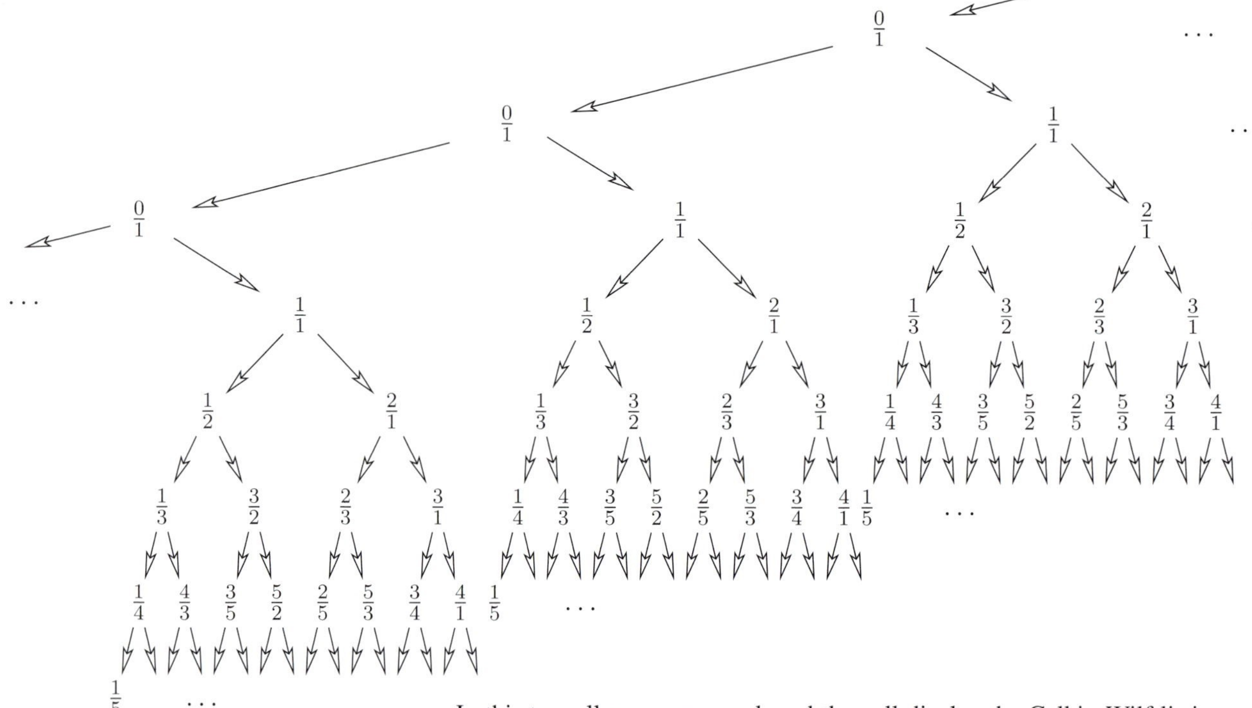

In this tree all rows are equal, and they all display the Calkin-Wilf listing of the positive rationals (starting with an additional $\frac{0}{1}$).

So how does one get from one rational to the next? To answer this, we first record that for every rational x its right son is $x+1$, the right grand-son is $x+2$, so the k-fold right son is $x+k$. Similarly, the left son of x is $\frac{x}{1+x}$, whose left son is $\frac{x}{1+2x}$, and so on: The k-fold left son of x is $\frac{x}{1+kx}$.
Now to find how to get from $\frac{r}{s} = x$ to the "next" rational $f(x)$ in the listing, we have to analyze the situation depicted in the margin. In fact, if we consider any nonnegative rational number x in our infinite binary tree, then it is the k-fold right son of the left son of some rational $y \geq 0$ (for some $k \geq 0$), while $f(x)$ is given as the k-fold left son of the right son of the same y. Thus with the formulas for k-fold left sons and k-fold right sons, we get

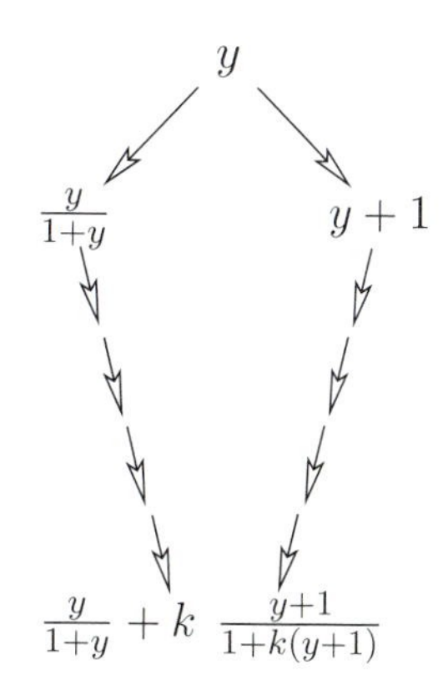

$$x = \frac{y}{1+y} + k,$$

as claimed in the figure in the margin. Here $k = \lfloor x \rfloor$ is the integral part of x, while $\frac{y}{1+y} = \{x\}$ is the fractional part. And from this we obtain

$$f(x) = \frac{y+1}{1+k(y+1)} = \frac{1}{\frac{1}{y+1}+k} = \frac{1}{k+1-\frac{y}{y+1}} = \frac{1}{\lfloor x \rfloor + 1 - \{x\}}.$$

Thus we have obtained a beautiful formula for the successor $f(x)$ of x, found very recently by Moshe Newman:

The function

$$x \longmapsto f(x) = \frac{1}{\lfloor x \rfloor + 1 - \{x\}}$$

generates the Calkin-Wilf sequence

$$\tfrac{1}{1} \mapsto \tfrac{1}{2} \mapsto \tfrac{2}{1} \mapsto \tfrac{1}{3} \mapsto \tfrac{3}{2} \mapsto \tfrac{2}{3} \mapsto \tfrac{3}{1} \mapsto \tfrac{1}{4} \mapsto \tfrac{4}{3} \mapsto \ldots$$

which contains every positive rational number exactly once.

The Calkin-Wilf-Newman way to enumerate the positive rationals has a number of additional remarkable properties. For example, one may ask for a fast way to determine the n-th fraction in the sequence, say for $n = 10^6$. Here it is:

> To find the n-th fraction in the Calkin-Wilf sequence, express n as a binary number $n = (b_k b_{k-1} \ldots b_1 b_0)_2$, and then follow the path in the Calkin-Wilf tree that is determined by its digits, starting at $\frac{s}{t} = \frac{0}{1}$. Here $b_i = 1$ means "take the right son," that is, "add the denominator to the numerator," while $b_i = 0$ means "take the left son," that is, "add the numerator to the denominator."

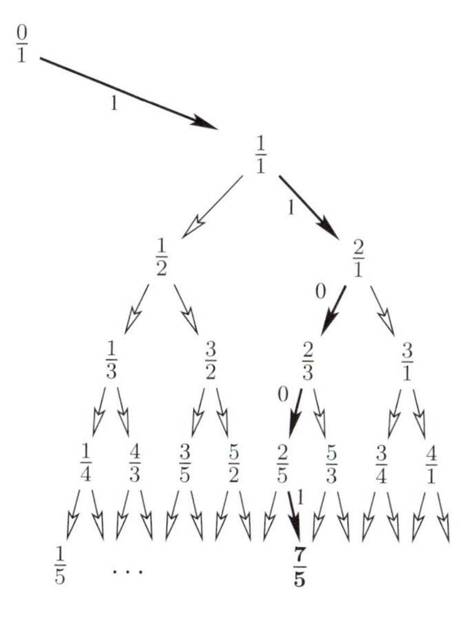

The figure in the margin shows the resulting path for $n = 25 = (11001)_2$: So the 25th number in the Calkin-Wilf sequence is $\frac{7}{5}$. The reader could easily work out a similar scheme that computes for a given fraction $\frac{s}{t}$ (the binary representation of) its position n in the Calkin-Wilf sequence.

Let us move on to the real numbers $\mathbb{R}$. Are they still countable? No, they are not, and the means by which this is shown — Cantor's *diagonalization method* — is not only of fundamental importance for all of set theory, but certainly belongs into The Book as a rare stroke of genius.

Theorem 2. *The set $\mathbb{R}$ of real numbers is **not** countable.*

■ **Proof.** Any subset N of a countable set $M = \{m_1, m_2, m_3, \ldots\}$ is *at most countable* (that is, finite or countable). In fact, just list the elements of N as they appear in M. Accordingly, if we can find a subset of $\mathbb{R}$ which is not countable, then a fortiori $\mathbb{R}$ cannot be countable. The subset M of $\mathbb{R}$ we want to look at is the interval $(0, 1]$ of all positive real numbers r with $0 < r \leq 1$. Suppose, to the contrary, that M is countable, and let $M = \{r_1, r_2, r_3, \ldots\}$ be a listing of M. We write r_n as its unique *infinite* decimal expansion without an infinite sequence of zeros at the end:

$$r_n = 0.a_{n1}a_{n2}a_{n3}\ldots$$

where $a_{ni} \in \{0, 1, \ldots, 9\}$ for all n and i. For example, $0.7 = 0.6999\ldots$ Consider now the doubly infinite array

$$\begin{array}{ccl} r_1 & = & 0.a_{11}a_{12}a_{13}\ldots \\ r_2 & = & 0.a_{21}a_{22}a_{23}\ldots \\ \vdots & & \vdots \\ r_n & = & 0.a_{n1}a_{n2}a_{n3}\ldots \\ \vdots & & \vdots \end{array}$$

For every n, choose $b_n \in \{1, \ldots, 8\}$ different from a_{nn}; clearly this can be done. Then $b = 0.b_1b_2b_3\ldots b_n\ldots$ is a real number in our set M and hence must have an index, say $b = r_k$. But this cannot be, since b_k is different from a_{kk}. And this is the whole proof! □

Let us stay with the real numbers for a moment. We note that all four types of intervals $(0, 1)$, $(0, 1]$, $[0, 1)$ and $[0, 1]$ have the same size. As an example, we verify that $(0, 1]$ and $(0, 1)$ have equal cardinality. The map $f : (0, 1] \longrightarrow (0, 1)$, $x \longmapsto y$ defined by

$$y := \begin{cases} \frac{3}{2} - x & \text{for } \frac{1}{2} < x \leq 1, \\ \frac{3}{4} - x & \text{for } \frac{1}{4} < x \leq \frac{1}{2}, \\ \frac{3}{8} - x & \text{for } \frac{1}{8} < x \leq \frac{1}{4}, \\ \quad \vdots & \end{cases}$$

does the job. Indeed, the map is bijective, since the range of y in the first line is $\frac{1}{2} \leq y < 1$, in the second line $\frac{1}{4} \leq y < \frac{1}{2}$, in the third line $\frac{1}{8} \leq y < \frac{1}{4}$, and so on.

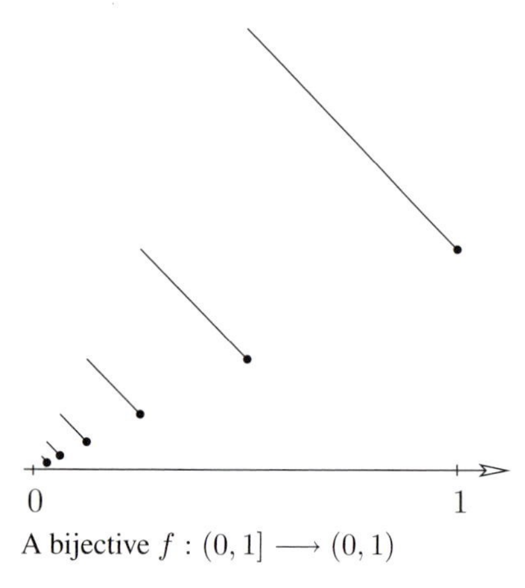

A bijective $f : (0, 1] \longrightarrow (0, 1)$

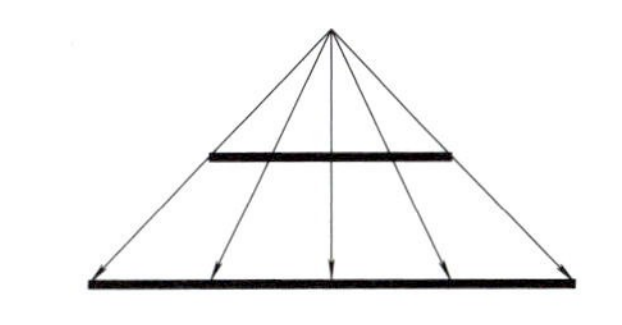

Next we find that *any* two intervals (of finite length > 0) have equal size by considering the central projection as in the figure. Even more is true: Every interval (of length > 0) has the same size as the whole real line $\mathbb{R}$. To see this, look at the bent open interval $(0, 1)$ and project it onto $\mathbb{R}$ from the center S.

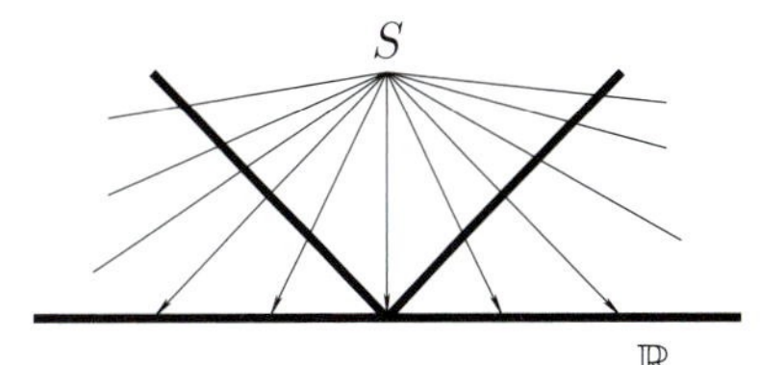

So, in conclusion, any open, half-open, closed (finite or infinite) interval of length > 0 has the same size, and we denote this size by c, where c stands for *continuum* (a name sometimes used for the interval [0,1]).

That finite and infinite intervals have the same size may come expected on second thought, but here is a fact that is downright counter-intuitive.

Theorem 3. *The set $\mathbb{R}^2$ of all ordered pairs of real numbers (that is, the real plane) has the same size as $\mathbb{R}$.*

■ **Proof.** To see this, it suffices to prove that the set of all pairs (x, y), $0 < x, y \leq 1$, can be mapped bijectively onto $(0, 1]$. The proof is again from The Book. Consider the pair (x, y) and write x, y in their unique non-terminating decimal expansion as in the following example:

$$\begin{array}{rcllllll} x & = & 0.3 & 01 & 2 & 007 & 08 & \dots \\ y & = & 0.009 & 2 & 05 & 1 & 0008 & \dots \end{array}$$

Note that we have separated the digits of x and y into groups by always going to the next nonzero digit, inclusive. Now we associate to (x, y) the number $z \in (0, 1]$ by writing down the first x-group, after that the first y-group, then the second x-group, and so on. Thus, in our example, we obtain

$$z = 0.3\,009\,01\,2\,2\,05\,007\,1\,08\,0008\,\dots$$

Since neither x nor y exhibits only zeros from a certain point on, we find that the expression for z is again a non-terminating decimal expansion. Conversely, from the expansion of z we can immediately read off the preimage (x, y), and the map is bijective — end of proof. $\square$

As $(x, y) \longmapsto x + iy$ is a bijection from $\mathbb{R}^2$ onto the complex numbers $\mathbb{C}$, we conclude that $|\mathbb{C}| = |\mathbb{R}| = c$. Why is the result $|\mathbb{R}^2| = |\mathbb{R}|$ so unexpected? Because it goes against our intuition of *dimension*. It says that the 2-dimensional plane $\mathbb{R}^2$ (and, in general, by induction, the n-dimensional space $\mathbb{R}^n$) can be mapped bijectively onto the 1-dimensional line $\mathbb{R}$. Thus dimension is not generally preserved by bijective maps. If, however, we require the map and its inverse to be continuous, then the dimension is preserved, as was first shown by Luitzen Brouwer.

Let us go a little further. So far, we have the notion of equal size. When will we say that M is at most as large as N? Mappings provide again the key. We say that the cardinal number $\mathfrak{m}$ is *less than or equal to* $\mathfrak{n}$, if for sets M and N with $|M| = \mathfrak{m}$, $|N| = \mathfrak{n}$, there exists an *injection* from M into N. Clearly, the relation $\mathfrak{m} \leq \mathfrak{n}$ is independent of the representative sets M and N chosen. For finite sets this corresponds again to our intuitive notion: An m-set is at most as large as an n-set if and only if $m \leq n$.

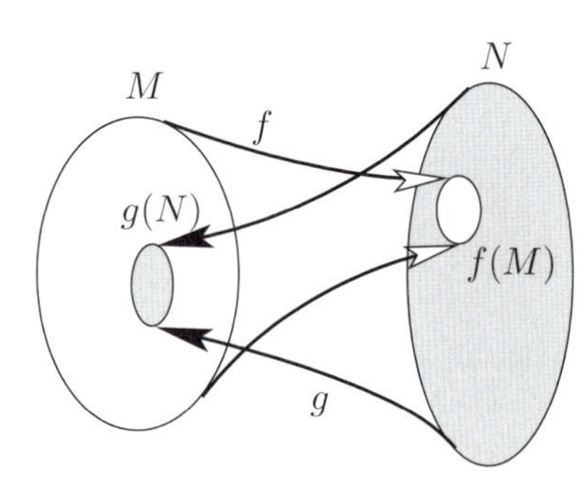

Now we are faced with a basic problem. We would certainly like to have that the usual laws concerning inequalities also hold for cardinal numbers. But is this true for infinite cardinals? In particular, is it true that $\mathfrak{m} \leq \mathfrak{n}$, $\mathfrak{n} \leq \mathfrak{m}$ imply $\mathfrak{m} = \mathfrak{n}$? This is not at all obvious: We are given infinite sets M and N as well as maps $f : M \longrightarrow N$ and $g : N \longrightarrow M$ that are injective but not necessarily surjective. This suggests to construct a bijection by relating some elements $m \in M$ to $f(m) \in N$, and some elements $n \in N$ to $g(n) \in M$. But it is not clear whether the many possible choices can be made to "fit together."

The affirmative answer is provided by the famous Schröder-Bernstein theorem, which Cantor announced in 1883. The first proofs were given by Friedrich Schröder and Felix Bernstein quite some time later. The following proof appears in a little book by one of the twentieth century giants of set theory, Paul Cohen, who is famous for resolving the continuum hypothesis (which we will discuss below).

"Schröder and Bernstein painting"

Theorem 4. *If each of two sets M and N can be mapped injectively into the other, then there is a bijection from M to N, that is, $|M| = |N|$.*

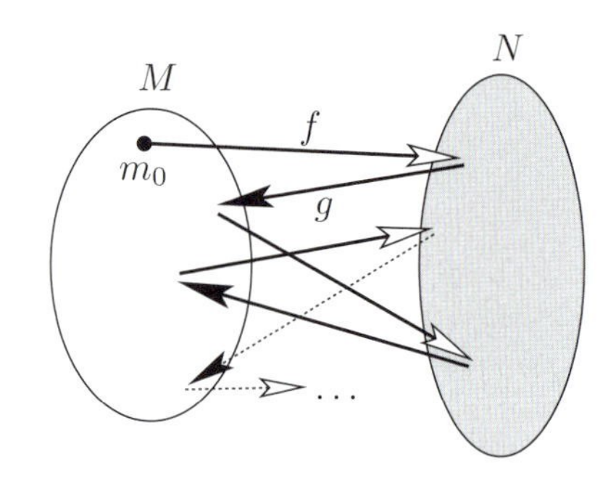

■ **Proof.** We may certainly assume that M and N are disjoint — if not, then we just replace N by a new copy.

Now f and g map back and forth between the elements of M and those of N. One way to bring this potentially confusing situation into perfect clarity and order is to align $M \cup N$ into chains of elements: Take an arbitrary element $m_0 \in M$, say, and from this generate a chain of elements by applying f, then g, then f again, then g, and so on. The chain may close up (this is Case 1) if we reach m_0 again in this process, or it may continue with distinct elements indefinitely. (The first "duplicate" in the chain cannot be an element different from m_0, by injectivity.)

If the chain continues indefinitely, then we try to follow it backwards: From m_0 to $g^{-1}(m_0)$ if m_0 is in the image of g, then to $f^{-1}(g^{-1}(m_0))$ if $g^{-1}(m_0)$ is in the image of f, and so on. Three more cases may arise here: The process of following the chain backwards may go on indefinitely

(Case 2), it may stop in an element of M that does not lie in the image of g (Case 3), or it may stop in an element of N that does not lie in the image of f (Case 4).

Thus $M \cup N$ splits perfectly into four types of chains, whose elements we may label in such a way that a bijection is simply given by putting $F : m_i \longmapsto n_i$. We verify this in the four cases separately:

Case 1. Finite cycles on $2k+2$ distinct elements ($k \geq 0$)

$$m_0 \xrightarrow{f} n_0 \xrightarrow{g} m_1 \xrightarrow{f} \cdots \quad m_k \xrightarrow{f} n_k \xrightarrow{g} m_0$$

Case 2. Two-way infinite chains of distinct elements

$$\cdots \longrightarrow m_0 \xrightarrow{f} n_0 \xrightarrow{g} m_1 \xrightarrow{f} n_1 \xrightarrow{g} m_2 \xrightarrow{f} \cdots$$

Case 3. The one-way infinite chains of distinct elements that start at the elements $m_0 \in M \backslash g(N)$

$$m_0 \xrightarrow{f} n_0 \xrightarrow{g} m_1 \xrightarrow{f} n_1 \xrightarrow{g} m_2 \xrightarrow{f} \cdots$$

Case 4. The one-way infinite chains of distinct elements that start at the elements $n_0 \in N \backslash f(M)$

$$n_0 \xrightarrow{g} m_0 \xrightarrow{f} n_1 \xrightarrow{g} m_1 \xrightarrow{f} \cdots \quad \square$$

What about the other relations governing inequalities? As usual, we set $\mathfrak{m} < \mathfrak{n}$ if $\mathfrak{m} \leq \mathfrak{n}$, but $\mathfrak{m} \neq \mathfrak{n}$. We have just seen that for any two cardinals $\mathfrak{m}$ and $\mathfrak{n}$ at most one of the three possibilities

$$\mathfrak{m} < \mathfrak{n}, \ \mathfrak{m} = \mathfrak{n}, \ \mathfrak{m} > \mathfrak{n}$$

holds, and it follows from the theory of cardinal numbers that, in fact, precisely one relation is true. (See the appendix to this chapter, Proposition 2.)

Furthermore, the Schröder-Bernstein Theorem tells us that the relation $<$ is transitive, that is, $\mathfrak{m} < \mathfrak{n}$ and $\mathfrak{n} < \mathfrak{p}$ imply $\mathfrak{m} < \mathfrak{p}$. Thus the cardinalities are arranged in linear order starting with the finite cardinals $0, 1, 2, 3, \ldots$. Invoking the usual Zermelo-Fraenkel axiom system (in particular, the axiom of choice) we easily find that any infinite set M contains a countable subset. In fact, M contains an element, say m_1. The set $M \setminus \{m_1\}$ is not empty (since it is infinite) and hence contains an element m_2. Considering $M \setminus \{m_1, m_2\}$ we infer the existence of m_3, and so on. So, the size of a countable set is the *smallest infinite* cardinal, usually denoted by $\aleph_0$ (pronounced "aleph zero").

"The smallest infinite cardinal"

With this we have also proved a result announced earlier:
Every infinite set has the same size as some proper subset.

As a corollary to $\aleph_0 \leq \mathfrak{m}$ for any infinite cardinal $\mathfrak{m}$, we can immediately prove "Hilbert's hotel" for any infinite cardinal number $\mathfrak{m}$, that is, we have $|M \cup \{x\}| = |M|$ for any infinite set M. Indeed, M contains a subset $N = \{m_1, m_2, m_3, \ldots\}$. Now map x onto m_1, m_1 onto m_2, and so on, keeping the elements of $M \backslash N$ fixed. This gives the desired bijection.

As another consequence of the Schröder-Bernstein theorem we may prove that the set $\mathcal{P}(\mathbb{N})$ of all subsets of $\mathbb{N}$ has cardinality c. As noted above, it suffices to show that $|\mathcal{P}(\mathbb{N}) \backslash \{\varnothing\}| = |(0,1]|$. An example of an injective map is

$$f : \mathcal{P}(\mathbb{N}) \setminus \{\varnothing\} \longrightarrow (0,1], \qquad A \longmapsto \sum_{i \in A} 10^{-i},$$

while

$$g : (0,1] \longrightarrow \mathcal{P}(\mathbb{N}) \setminus \{\varnothing\}, \qquad 0.b_1 b_2 b_3 \ldots \longmapsto \{b_i 10^i : i \in \mathbb{N}\}$$

defines an injection in the other direction.

Up to now we know the cardinal numbers $0, 1, 2, \ldots, \aleph_0$, and further that the cardinality c of $\mathbb{R}$ is bigger than $\aleph_0$. The passage from $\mathbb{Q}$ with $|\mathbb{Q}| = \aleph_0$ to $\mathbb{R}$ with $|\mathbb{R}| = c$ immediately suggests the next question:

Is $c = |\mathbb{R}|$ the next infinite cardinal number after $\aleph_0$?

Now, of course, we have the problem whether there *is* a next larger cardinal number, or in other words, whether $\aleph_1$ has a meaning at all. It does — the proof for this is outlined in the appendix to this chapter.

The statement $c = \aleph_1$ became known as the *continuum hypothesis*. The question whether the continuum hypothesis is true presented for many decades one of the supreme challenges in all of mathematics. The answer, finally given by Kurt Gödel and Paul Cohen, takes us to the limit of logical thought. They showed that the statement $c = \aleph_1$ is *independent* of the Zermelo-Fraenkel axiom system, in the same way as the parallel axiom is independent of the other axioms of Euclidian geometry. There are models where $c = \aleph_1$ holds, and there are other models of set theory where $c \neq \aleph_1$ holds.

In the light of this fact it is quite interesting to ask whether there are other conditions (from analysis, say) which are equivalent to the continuum hypothesis. Indeed, it is natural to ask for an analysis example, since historically the first substantial applications of Cantor's set theory occurred in analysis, specifically in complex function theory. In the following we want to present one such instance and its extremely elegant and simple solution by Paul Erdős. In 1962, Wetzel asked the following question:

> *Let $\{f_\alpha\}$ be a family of pairwise distinct analytic functions on the complex numbers such that for each $z \in \mathbb{C}$ the set of values $\{f_\alpha(z)\}$ is at most countable (that is, it is either finite or countable); let us call this property (P_0).*
> *Does it then follow that the family itself is at most countable?*

Very shortly afterwards Erdős showed that, surprisingly, the answer depends on the continuum hypothesis.

Theorem 5. *If $c > \aleph_1$, then every family $\{f_\alpha\}$ satisfying (P_0) is countable. If, on the other hand, $c = \aleph_1$, then there exists some family $\{f_\alpha\}$ with property (P_0) which has size c.*

For the proof we need some basic facts on cardinal and ordinal numbers. For readers who are unfamiliar with these concepts, this chapter has an appendix where all the necessary results are collected.

■ **Proof of Theorem 5.** Assume first $c > \aleph_1$. We shall show that for any family $\{f_\alpha\}$ of size $\aleph_1$ of analytic functions there exists a complex number z_0 such that *all* $\aleph_1$ values $f_\alpha(z_0)$ are distinct. Consequently, if a family of functions satisfies (P_0), then it must be countable.

To see this, we make use of our knowledge of ordinal numbers. First, we well-order the family $\{f_\alpha\}$ according to the initial ordinal number ω_1 of $\aleph_1$. This means by Proposition 1 of the appendix that the index set runs through all ordinal numbers α which are smaller than ω_1. Next we show that the set of pairs (α, β), $\alpha < \beta < \omega_1$, has size $\aleph_1$. Since any $\beta < \omega_1$ is a countable ordinal, the set of pairs (α, β), $\alpha < \beta$, is countable for every fixed β. Taking the union over all $\aleph_1$-many β, we find from Proposition 6 of the appendix that the set of all pairs (α, β), $\alpha < \beta$, has size $\aleph_1$.

Consider now for any pair $\alpha < \beta$ the set

$$S(\alpha, \beta) = \{z \in \mathbb{C} : f_\alpha(z) = f_\beta(z)\}.$$

We claim that each set $S(\alpha, \beta)$ is countable. To verify this, consider the disks C_k of radius $k = 1, 2, 3, \ldots$ around the origin in the complex plane. If f_α and f_β agree on infinitely many points in some C_k, then f_α and f_β are identical by a well-known result on analytic functions. Hence f_α and f_β agree only in finitely many points in each C_k, and hence in at most countably many points altogether. Now we set $S := \bigcup_{\alpha<\beta} S(\alpha, \beta)$. Again by Proposition 6, we find that S has size $\aleph_1$, as each set $S(\alpha, \beta)$ is countable. And here is the punch line: Because, as we know, $\mathbb{C}$ has size c, and c is larger than $\aleph_1$ by assumption, there exists a complex number z_0 not in S, and for this z_0 all $\aleph_1$ values $f_\alpha(z_0)$ are distinct.

Next we assume $c = \aleph_1$. Consider the set $D \subseteq \mathbb{C}$ of complex numbers $p + iq$ with rational real and imaginary part. Since for each p the set $\{p + iq : q \in \mathbb{Q}\}$ is countable, we find that D is countable. Furthermore, D is a *dense* set in $\mathbb{C}$: Every open disk in the complex plane contains some point of D. Let $\{z_\alpha : 0 \leq \alpha < \omega_1\}$ be a well-ordering of $\mathbb{C}$. We shall now construct a family $\{f_\beta : 0 \leq \beta < \omega_1\}$ of $\aleph_1$-many distinct analytic functions such that

$$f_\beta(z_\alpha) \in D \text{ whenever } \alpha < \beta. \tag{1}$$

Any such family satisfies the condition (P_0). Indeed, each point $z \in \mathbb{C}$ has some index, say $z = z_\alpha$. Now, for all $\beta > \alpha$, the values $\{f_\beta(z_\alpha)\}$ lie in the *countable* set D. Since α is a countable ordinal number, the functions f_β with $\beta \le \alpha$ will contribute at most countably further values $f_\beta(z_\alpha)$, so that the set of all values $\{f_\beta(z_\alpha)\}$ is likewise at most countable. Hence, if we can construct a family $\{f_\beta\}$ satisfying (1), then the second part of the theorem is proved.

The construction of $\{f_\beta\}$ is by transfinite induction. For f_0 we may take any analytic function, for example $f_0 =$ constant. Suppose f_β has already been constructed for all $\beta < \gamma$. Since γ is a countable ordinal, we may reorder $\{f_\beta : 0 \le \beta < \gamma\}$ into a sequence $g_1, g_2, g_3, \ldots$. The same reordering of $\{z_\alpha : 0 \le \alpha < \gamma\}$ yields a sequence $w_1, w_2, w_3, \ldots$. We shall now construct a function f_γ satisfying for each n the conditions

$$f_\gamma(w_n) \in D \qquad \text{and} \qquad f_\gamma(w_n) \ne g_n(w_n). \tag{2}$$

The second condition will ensure that all functions f_γ $(0 \le \gamma < \omega_1)$ are distinct, and the first condition is just (1), implying (P_0) by our previous argument. Notice that the condition $f_\gamma(w_n) \ne g_n(w_n)$ is once more a diagonalization argument.

To construct f_γ, we write

$$\begin{aligned} f_\gamma(z) \;:=\; & \varepsilon_0 + \varepsilon_1(z-w_1) + \varepsilon_2(z-w_1)(z-w_2) \\ & + \varepsilon_3(z-w_1)(z-w_2)(z-w_3) + \ldots. \end{aligned}$$

If γ is a finite ordinal, then f_γ is a polynomial and hence analytic, and we can certainly choose numbers ε_i such that (2) is satisfied. Now suppose γ is a countable ordinal, then

$$f_\gamma(z) \;=\; \sum_{n=0}^{\infty} \varepsilon_n(z-w_1)\cdots(z-w_n). \tag{3}$$

Note that the values of ε_m $(m \ge n)$ have no influence on the value $f_\gamma(w_n)$, hence we may choose the ε_n step by step. If the sequence (ε_n) converges to 0 sufficiently fast, then (3) defines an analytic function. Finally, since D is a dense set, we may choose this sequence (ε_n) so that f_γ meets the requirements of (2), and the proof is complete. □

Appendix: On cardinal and ordinal numbers

Let us first discuss the question whether to each cardinal number there exists a next larger one. As a start we show that to every cardinal number $\mathfrak{m}$ there always is a cardinal number $\mathfrak{n}$ larger than $\mathfrak{m}$. To do this we employ again a version of Cantor's diagonalization method.

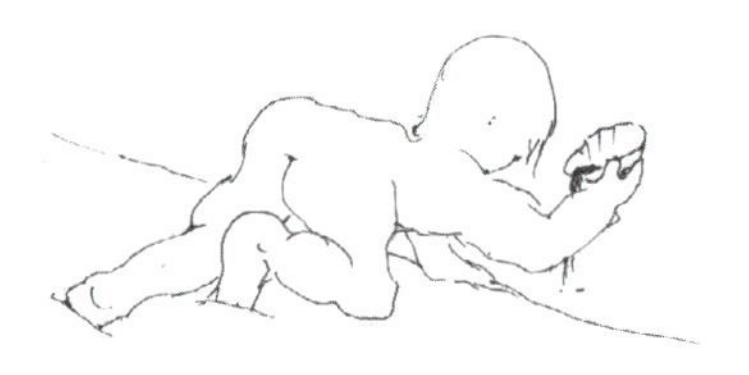

"A legend talks about St. Augustin who, walking along the seashore and contemplating infinity, saw a child trying to empty the ocean with a small shell..."

Let M be a set, then we claim that the set $\mathcal{P}(M)$ *of all subsets* of M has larger size than M. By letting $m \in M$ correspond to $\{m\} \in \mathcal{P}(M)$, we see that M can be mapped bijectively onto a subset of $\mathcal{P}(M)$, which implies $|M| \le |\mathcal{P}(M)|$ by definition. It remains to show that $\mathcal{P}(M)$ *can not* be mapped bijectively onto a subset of M. Suppose, on the contrary,

$\varphi : N \longrightarrow \mathcal{P}(M)$ is a bijection of $N \subseteq M$ onto $\mathcal{P}(M)$. Consider the subset $U \subseteq N$ of all elements of N which are *not* contained in their image under φ, that is, $U = \{m \in N : m \notin \varphi(m)\}$. Since φ is a bijection, there exists $u \in N$ with $\varphi(u) = U$. Now, either $u \in U$ or $u \notin U$, but both alternatives are impossible! Indeed, if $u \in U$, then $u \notin \varphi(u) = U$ by the definition of U, and if $u \notin U = \varphi(u)$, then $u \in U$, contradiction.

Most likely, the reader has seen this argument before. It is the old barber riddle: "A barber is the man who shaves all men who do not shave themselves. Does the barber shave himself?"

To get further in the theory we introduce another great concept of Cantor's, ordered sets and ordinal numbers. A set M is *ordered* by $<$ if the relation $<$ is transitive, and if for any two distinct elements a and b of M we either have $a < b$ or $b < a$. For example, we can order $\mathbb{N}$ in the usual way according to magnitude, $\mathbb{N} = \{1, 2, 3, 4, \ldots\}$, but, of course, we can also order $\mathbb{N}$ the other way round, $\mathbb{N} = \{\ldots, 4, 3, 2, 1\}$, or $\mathbb{N} = \{1, 3, 5, \ldots, 2, 4, 6, \ldots\}$ by listing first the odd numbers and then the even numbers.

Here is the seminal concept. An ordered set M is called *well-ordered* if every nonempty subset of M has a first element. Thus the first and third orderings of $\mathbb{N}$ above are well-orderings, but not the second ordering. The fundamental *well-ordering theorem*, implied by the axioms (including the axiom of choice), now states that *every* set M admits a well-ordering. From now on, we only consider sets endowed with a well-ordering.

Let us say that two well-ordered sets M and N are *similar* (or of the *same order-type*) if there exists a bijection φ from M on N which respects the ordering, that is, $m <_M n$ implies $\varphi(m) <_N \varphi(n)$. Note that any ordered set which is similar to a well-ordered set is itself well-ordered.

The well-ordered sets $\mathbb{N} = \{1, 2, 3, \ldots\}$ and $\mathbb{N} = \{1, 3, 5, \ldots, 2, 4, 6, \ldots\}$ are *not* similar: the first ordering has only one element without an immediate predecessor, while the second one has two.

Similarity is obviously an equivalence relation, and we can thus speak of an *ordinal number* α belonging to a class of similar sets. For finite sets, any two orderings are similar well-orderings, and we use again the ordinal number n for the class of n-sets. Note that, by definition, two similar sets have the same cardinality. Hence it makes sense to speak of the *cardinality* $|\alpha|$ of an ordinal number α. Note further that any subset of a well-ordered set is also well-ordered under the induced ordering.

As we did for cardinal numbers, we now compare ordinal numbers. Let M be a well-ordered set, $m \in M$, then $M_m = \{x \in M : x < m\}$ is called the *(initial) segment* of M determined by m; N is a segment of M if $N = M_m$ for some m. Thus, in particular, M_m is the empty set when m is the first element of M. Now let μ and ν be the ordinal numbers of the well-ordered sets M and N. We say that μ is *smaller* than ν, $\mu < \nu$, if M is similar to a segment of N. Again, we have the transitive law that $\mu < \nu$, $\nu < \pi$ implies $\mu < \pi$, since under a similarity mapping a segment is mapped onto a segment.

The ordinal number of $\{1, 2, 3, \ldots\}$ is smaller than the ordinal number of $\{1, 3, 5, \ldots, 2, 4, 6, \ldots\}$.

Clearly, for finite sets, $m < n$ corresponds to the usual meaning. Let us denote by ω the ordinal number of $\mathbb{N} = \{1, 2, 3, 4, \ldots\}$ ordered according to magnitude. By considering the segment $\mathbb{N}_{n+1}$ we find $n < \omega$ for any finite n. Next we see that $\omega \leq \alpha$ holds for any infinite ordinal number α. Indeed, if the infinite well-ordered set M has ordinal number α, then M contains a first element m_1, the set $M\backslash\{m_1\}$ contains a first element m_2, $M\backslash\{m_1, m_2\}$ contains a first element m_3. Continuing in this way, we produce the sequence $m_1 < m_2 < m_3 < \ldots$ in M. If $M = \{m_1, m_2, m_3, \ldots\}$, then M is similar to $\mathbb{N}$, and hence $\alpha = \omega$. If, on the other hand, $M\backslash\{m_1, m_2, \ldots\}$ is nonempty, then it contains a first element m, and we conclude that $\mathbb{N}$ is similar to the segment M_m, that is, $\omega < \alpha$ by definition.

We now state (without the proofs, which are not difficult) three basic results on ordinal numbers. The first says that any ordinal number μ has a "standard" representative well-ordered set W_μ.

Proposition 1. *Let μ be an ordinal number and denote by W_μ the set of ordinal numbers smaller than μ. Then the following holds:*

(i) *The elements of W_μ are pairwise comparable.*

(ii) *If we order W_μ according to magnitude, then W_μ is well-ordered and has ordinal number μ.*

Proposition 2. *Any two ordinal numbers μ and ν satisfy precisely one of the relations $\mu < \nu$, $\mu = \nu$, or $\mu > \nu$.*

Proposition 3. *Every set of ordinal numbers (ordered according to magnitude) is well-ordered.*

After this excursion to ordinal numbers we come back to cardinal numbers. Let $\mathfrak{m}$ be a cardinal number, and denote by $O_\mathfrak{m}$ the set of all ordinal numbers μ with $|\mu| = \mathfrak{m}$. By Proposition 3 there is a *smallest* ordinal number $\omega_\mathfrak{m}$ in $O_\mathfrak{m}$, which we call the *initial ordinal number* of $\mathfrak{m}$. As an example, ω is the initial ordinal number of $\aleph_0$.

With these preparations we can now prove a basic result for this chapter.

Proposition 4. *For every cardinal number $\mathfrak{m}$ there is a definite next larger cardinal number.*

■ **Proof.** We already know that there is some larger cardinal number $\mathfrak{n}$. Consider now the set $\mathcal{K}$ of all cardinal numbers larger than $\mathfrak{m}$ and at most as large as $\mathfrak{n}$. We associate to each $\mathfrak{p} \in \mathcal{K}$ its initial ordinal number $\omega_\mathfrak{p}$. Among these initial numbers there is a smallest (Proposition 3), and the corresponding cardinal number is then the smallest in $\mathcal{K}$, and thus is the desired next larger cardinal number to $\mathfrak{m}$. □

Proposition 5. *Let the infinite set M have cardinality $\mathfrak{m}$, and let M be well-ordered according to the initial ordinal number $\omega_\mathfrak{m}$. Then M has no last element.*

■ **Proof.** Indeed, if M had a last element m, then the segment M_m would have an ordinal number $\mu < \omega_\mathfrak{m}$ with $|\mu| = \mathfrak{m}$, contradicting the definition of $\omega_\mathfrak{m}$. □

What we finally need is a considerable strenghthening of the result that the union of countably many countable sets is again countable. In the following result we consider *arbitrary* families of countable sets.

Proposition 6. *Suppose $\{A_\alpha\}$ is a family of size $\mathfrak{m}$ of countable sets A_α, where $\mathfrak{m}$ is an infinite cardinal. Then the union $\bigcup_\alpha A_\alpha$ has size at most $\mathfrak{m}$.*

■ **Proof.** We may assume that the sets A_α are pairwise disjoint, since this can only increase the size of the union. Let M with $|M| = \mathfrak{m}$ be the index set, and well-order it according to the initial ordinal number $\omega_{\mathfrak{m}}$. We now replace each $\alpha \in M$ by a countable set $B_\alpha = \{b_{\alpha 1} = \alpha, b_{\alpha 2}, b_{\alpha 3}, \ldots\}$, ordered according to ω, and call the new set $\widetilde{M}$. Then $\widetilde{M}$ is again well-ordered by setting $b_{\alpha i} < b_{\beta j}$ for $\alpha < \beta$ and $b_{\alpha i} < b_{\alpha j}$ for $i < j$. Let $\widetilde{\mu}$ be the ordinal number of $\widetilde{M}$. Since M is a subset of $\widetilde{M}$, we have $\mu \leq \widetilde{\mu}$ by an earlier argument. If $\mu = \widetilde{\mu}$, then M is similar to $\widetilde{M}$, and if $\mu < \widetilde{\mu}$, then M is similar to a segment of $\widetilde{M}$. Now, since the ordering $\omega_{\mathfrak{m}}$ of M has no last element (Proposition 5), we see that M is in both cases similar to the union of countable sets B_β, and hence of the same cardinality.

The rest is easy. Let $\varphi : \bigcup B_\beta \longrightarrow M$ be a bijection, and suppose that $\varphi(B_\beta) = \{\alpha_1, \alpha_2, \alpha_3, \ldots\}$. Replace each α_i by A_{α_i} and consider the union $\bigcup A_{\alpha_i}$. Since $\bigcup A_{\alpha_i}$ is the union of *countably* many countable sets (and hence countable), we see that B_β has the same size as $\bigcup A_{\alpha_i}$. In other words, there is a bijection from B_β to $\bigcup A_{\alpha_i}$ for all β, and hence a bijection ψ from $\bigcup B_\beta$ to $\bigcup A_\alpha$. But now $\psi\varphi^{-1}$ gives the desired bijection from M to $\bigcup A_\alpha$, and thus $|\bigcup A_\alpha| = \mathfrak{m}$. □

References

[1] L. E. J. BROUWER: *Beweis der Invarianz der Dimensionszahl,* Math. Annalen **70** (1911), 161-165.

[2] N. CALKIN & H. WILF: *Recounting the rationals,* Amer. Math. Monthly **107** (2000), 360-363.

[3] P. COHEN: *Set Theory and the Continuum Hypothesis,* W. A. Benjamin, New York 1966.

[4] P. ERDŐS: *An interpolation problem associated with the continuum hypothesis,* Michigan Math. J. **11** (1964), 9-10.

[5] E. KAMKE: *Theory of Sets,* Dover Books 1950.

[6] M. A. STERN: *Ueber eine zahlentheoretische Funktion,* Journal für die reine und angewandte Mathematik **55** (1858), 193-220.

"Infinitely many more cardinals"

In praise of inequalities

Chapter 17

Analysis abounds with inequalities, as witnessed for example by the famous book "Inequalities" by Hardy, Littlewood and Pólya. Let us single out two of the most basic inequalities with two applications each, and let us listen in to George Pólya, who was himself a champion of the Book Proof, about what he considers the most appropriate proofs.

Our first inequality is variously attributed to Cauchy, Schwarz and/or to Buniakowski:

Theorem I (Cauchy-Schwarz inequality)
Let $\langle \boldsymbol{a}, \boldsymbol{b}\rangle$ be an inner product on a real vector space V (with the norm $|\boldsymbol{a}|^2 := \langle \boldsymbol{a}, \boldsymbol{a}\rangle$). Then

$$\langle \boldsymbol{a}, \boldsymbol{b}\rangle^2 \leq |\boldsymbol{a}|^2|\boldsymbol{b}|^2$$

holds for all vectors $\boldsymbol{a}, \boldsymbol{b} \in V$, with equality if and only if $\boldsymbol{a}$ and $\boldsymbol{b}$ are linearly dependent.

■ **Proof.** The following (folklore) proof is probably the shortest. Consider the quadratic function

$$|x\boldsymbol{a} + \boldsymbol{b}|^2 = x^2|\boldsymbol{a}|^2 + 2x\langle \boldsymbol{a}, \boldsymbol{b}\rangle + |\boldsymbol{b}|^2$$

in the variable x. We may assume $\boldsymbol{a} \neq \boldsymbol{0}$. If $\boldsymbol{b} = \lambda\boldsymbol{a}$, then clearly $\langle \boldsymbol{a}, \boldsymbol{b}\rangle^2 = |\boldsymbol{a}|^2|\boldsymbol{b}|^2$. If, on the other hand, $\boldsymbol{a}$ and $\boldsymbol{b}$ are linearly independent, then $|x\boldsymbol{a} + \boldsymbol{b}|^2 > 0$ for all x, and thus the discriminant $\langle \boldsymbol{a}, \boldsymbol{b}\rangle^2 - |\boldsymbol{a}|^2|\boldsymbol{b}|^2$ is less than 0. □

Our second example is the *inequality of the harmonic, geometric and arithmetic mean*:

Theorem II (Harmonic, geometric and arithmetic mean)
Let $a_1, \ldots, a_n$ be positive real numbers, then

$$\frac{n}{\frac{1}{a_1} + \ldots + \frac{1}{a_n}} \leq \sqrt[n]{a_1 a_2 \ldots a_n} \leq \frac{a_1 + \ldots + a_n}{n}$$

with equality in both cases if and only if all a_i's are equal.

■ **Proof.** The following beautiful non-standard induction proof is attributed to Cauchy (see [7]). Let $P(n)$ be the statement of the second inequality, written in the form

$$a_1 a_2 \ldots a_n \leq \left(\frac{a_1 + \ldots + a_n}{n}\right)^n.$$

For $n = 2$, we have $a_1 a_2 \le (\frac{a_1+a_2}{2})^2 \iff (a_1 - a_2)^2 \ge 0$, which is true. Now we proceed in the following two steps:

(A) $P(n) \Longrightarrow P(n-1)$

(B) $P(n)$ and $P(2) \Longrightarrow P(2n)$

which will clearly imply the full result.

To prove **(A)**, set $A := \sum_{k=1}^{n-1} \frac{a_k}{n-1}$, then

$$\Big(\prod_{k=1}^{n-1} a_k\Big) A \overset{P(n)}{\le} \left(\frac{\sum_{k=1}^{n-1} a_k + A}{n}\right)^n = \Big(\frac{(n-1)A + A}{n}\Big)^n = A^n$$

and hence $\prod_{k=1}^{n-1} a_k \le A^{n-1} = \left(\frac{\sum_{k=1}^{n-1} a_k}{n-1}\right)^{n-1}$.

For **(B)**, we see

$$\prod_{k=1}^{2n} a_k = \Big(\prod_{k=1}^{n} a_k\Big)\Big(\prod_{k=n+1}^{2n} a_k\Big) \overset{P(n)}{\le} \Big(\sum_{k=1}^{n} \frac{a_k}{n}\Big)^n \Big(\sum_{k=n+1}^{2n} \frac{a_k}{n}\Big)^n$$

$$\overset{P(2)}{\le} \left(\frac{\sum_{k=1}^{2n} \frac{a_k}{n}}{2}\right)^{2n} = \left(\frac{\sum_{k=1}^{2n} a_k}{2n}\right)^{2n}.$$

The condition for equality is derived just as easily.

The left-hand inequality, between the harmonic and the geometric mean, follows now by considering $\frac{1}{a_1}, \dots, \frac{1}{a_n}$. □

■ **Another Proof.** Of the many other proofs of the arithmetic-geometric mean inequality (the monograph [2] lists more than 50), let us single out a particularly striking one by Alzer which is of recent date. As a matter of fact, this proof yields the stronger inequality

$$a_1^{p_1} a_2^{p_2} \dots a_n^{p_n} \le p_1 a_1 + p_2 a_2 + \dots + p_n a_n$$

for any positive numbers $a_1, \dots, a_n, p_1, \dots, p_n$ with $\sum_{i=1}^n p_i = 1$. Let us denote the expression on the left side by G, and on the right side by A. We may assume $a_1 \le \dots \le a_n$. Clearly $a_1 \le G \le a_n$, so there must exist some k with $a_k \le G \le a_{k+1}$. It follows that

$$\sum_{i=1}^{k} p_i \int_{a_i}^{G} \Big(\frac{1}{t} - \frac{1}{G}\Big)\, dt + \sum_{i=k+1}^{n} p_i \int_{G}^{a_i} \Big(\frac{1}{G} - \frac{1}{t}\Big)\, dt \ge 0 \qquad (1)$$

since all integrands are ≥ 0. Rewriting (1) we obtain

$$\sum_{i=1}^{n} p_i \int_{G}^{a_i} \frac{1}{G}\, dt \ge \sum_{i=1}^{n} p_i \int_{G}^{a_i} \frac{1}{t}\, dt$$

where the left-hand side equals

$$\sum_{i=1}^{n} p_i \frac{a_i - G}{G} = \frac{A}{G} - 1,$$

while the right-hand side is

$$\sum_{i=1}^{n} p_i(\log a_i - \log G) = \log \prod_{i=1}^{n} a_i^{p_i} - \log G = 0.$$

We conclude $\frac{A}{G} - 1 \geq 0$, which is $A \geq G$. In the case of equality, all integrals in (1) must be 0, which implies $a_1 = \ldots = a_n = G$. □

Our first application is a beautiful result of Laguerre (see [7]) concerning the location of roots of polynomials.

Theorem 1. *Suppose all roots of the polynomial $x^n + a_{n-1}x^{n-1} + \ldots + a_0$ are real. Then the roots are contained in the interval with the endpoints*

$$-\frac{a_{n-1}}{n} \pm \frac{n-1}{n}\sqrt{a_{n-1}^2 - \frac{2n}{n-1}a_{n-2}}\,.$$

■ **Proof.** Let y be one of the roots and $y_1, \ldots, y_{n-1}$ the others. Then the polynomial is $(x-y)(x-y_1)\cdots(x-y_{n-1})$. Thus by comparing coefficients

$$\begin{aligned} -a_{n-1} &= y + y_1 + \ldots + y_{n-1}, \\ a_{n-2} &= y(y_1 + \ldots + y_{n-1}) + \sum_{i<j} y_i y_j, \end{aligned}$$

and so

$$a_{n-1}^2 - 2a_{n-2} - y^2 = \sum_{i=1}^{n-1} y_i^2.$$

By Cauchy's inequality applied to $(y_1, \ldots, y_{n-1})$ and $(1, \ldots, 1)$,

$$\begin{aligned} (a_{n-1} + y)^2 &= (y_1 + y_2 + \ldots + y_{n-1})^2 \\ &\leq (n-1)\sum_{i=1}^{n-1} y_i^2 = (n-1)(a_{n-1}^2 - 2a_{n-2} - y^2), \end{aligned}$$

or

$$y^2 + \frac{2a_{n-1}}{n}y + \frac{2(n-1)}{n}a_{n-2} - \frac{n-2}{n}a_{n-1}^2 \leq 0.$$

Thus y (and hence all y_i) lie between the two roots of the quadratic function, and these roots are our bounds. □

For our second application we start from a well-known elementary property of a parabola. Consider the parabola described by $f(x) = 1 - x^2$ between $x = -1$ and $x = 1$. We associate to $f(x)$ the *tangential triangle* and the *tangential rectangle* as in the figure.

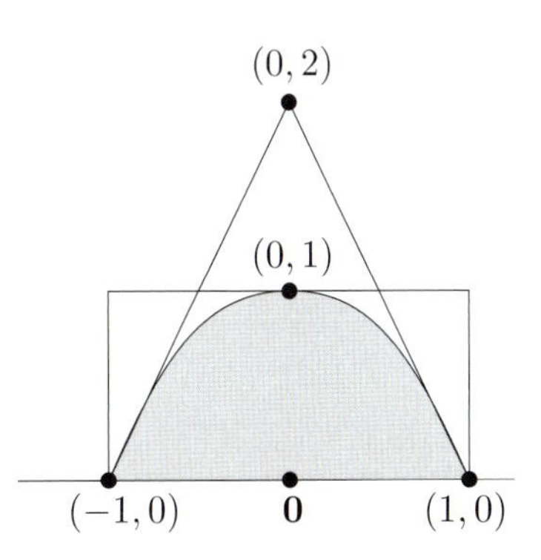

We find that the shaded area $A = \int_{-1}^{1}(1-x^2)dx$ is equal to $\frac{4}{3}$, and the areas T and R of the triangle and rectangle are both equal to 2. Thus $\frac{T}{A} = \frac{3}{2}$ and $\frac{R}{A} = \frac{3}{2}$.

In a beautiful paper, Paul Erdős and Tibor Gallai asked what happens when $f(x)$ is an arbitrary n-th degree real polynomial with $f(x) > 0$ for $-1 < x < 1$, and $f(-1) = f(1) = 0$. The area A is then $\int_{-1}^{1} f(x)dx$. Suppose that $f(x)$ assumes in $(-1,1)$ its maximum value at b, then $R = 2f(b)$. Computing the tangents at -1 and at 1, it is readily seen (see the box) that

$$T = \frac{2f'(1)f'(-1)}{f'(1) - f'(-1)}, \tag{2}$$

respectively $T = 0$ for $f'(1) = f'(-1) = 0$.

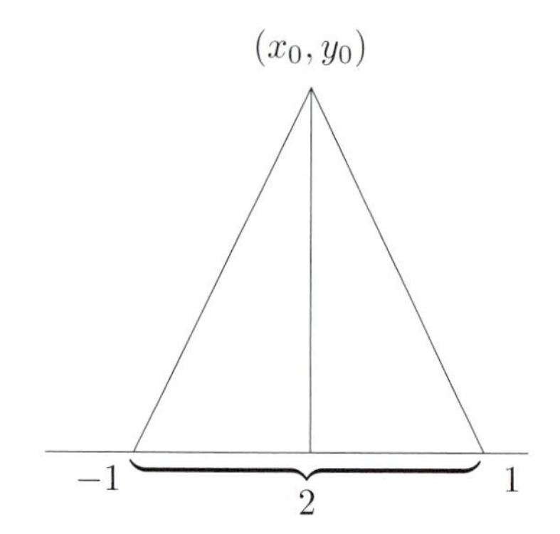

The tangential triangle

The area T of the tangential triangle is precisely y_0, where (x_0, y_0) is the point of intersection of the two tangents. The equation of these tangents are $y = f'(-1)(x+1)$ and $y = f'(1)(x-1)$, hence

$$x_0 = \frac{f'(1) + f'(-1)}{f'(1) - f'(-1)},$$

and thus

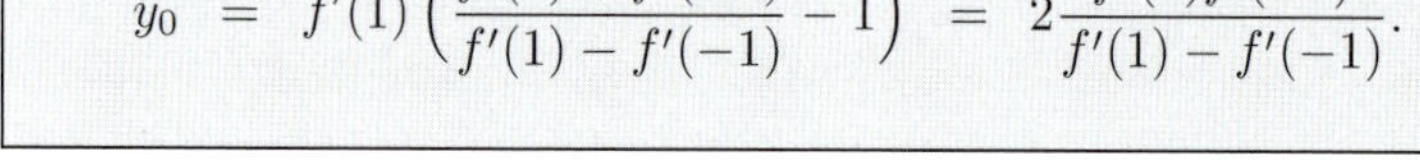

$$y_0 = f'(1)\left(\frac{f'(1) + f'(-1)}{f'(1) - f'(-1)} - 1\right) = 2\frac{f'(1)f'(-1)}{f'(1) - f'(-1)}.$$

In general, there are no nontrivial bounds for $\frac{T}{A}$ and $\frac{R}{A}$. To see this, take $f(x) = 1 - x^{2n}$. Then $T = 2n$, $A = \frac{4n}{2n+1}$, and thus $\frac{T}{A} > n$. Similarly, $R = 2$ and $\frac{R}{A} = \frac{2n+1}{2n}$, which approaches 1 with n to infinity.

But, as Erdős and Gallai showed, for polynomials which have only real roots such bounds do indeed exist.

Theorem 2. *Let $f(x)$ be a real polynomial of degree $n \geq 2$ with only real roots, such that $f(x) > 0$ for $-1 < x < 1$ and $f(-1) = f(1) = 0$. Then*

$$\frac{2}{3}T \leq A \leq \frac{2}{3}R,$$

and equality holds in both cases only for $n = 2$.

Erdős and Gallai established this result with an intricate induction proof. In the review of their paper, which appeared on the first page of the first issue of the Mathematical Reviews in 1940, George Pólya explained how the first inequality can also be proved by the inequality of the arithmetic and geometric mean — a beautiful example of a conscientious review and a Book Proof at the same time.

Mathematical Reviews

Vol. 1, No. 1 — JANUARY, 1940 — Pages 1-32

Erdös, P. and Grünwald, T. On polynomials with only real roots. Ann. of Math. **40**, 537–548 (1939). [MF 93]
Es sei $f(x)$ ein Polynom mit nur reellen Wurzeln,

$$f(-1) = f(1) = 0, \quad 0 < f(x) \leqq f(\mu) \quad \text{für } -1 < x < 1,$$

wobei $-1 < \mu < 1$, so dass μ die Stelle des Maximums von $f(x)$ im Intervall $(-1, 1)$ bedeutet. Dann ist

$$\frac{2}{3}\frac{2f'(1)f'(-1)}{f'(1) - f'(-1)} \leqq \int_{-1}^{1} f(x)dx \leqq \frac{2}{3} \cdot 2f(\mu),$$

■ **Proof of $\frac{2}{3}T \le A$.** Since $f(x)$ has only real roots, and none of them in the open interval $(-1,1)$, it can be written — apart from a constant positive factor which cancels out in the end — in the form

$$f(x) = (1-x^2)\prod_i(\alpha_i - x)\prod_j(\beta_j + x) \tag{3}$$

with $\alpha_i \ge 1$, $\beta_j \ge 1$. Hence

$$A = \int_{-1}^{1}(1-x^2)\prod_i(\alpha_i - x)\prod_j(\beta_j + x)dx.$$

By making the substitution $x \longmapsto -x$, we find that also

$$A = \int_{-1}^{1}(1-x^2)\prod_i(\alpha_i + x)\prod_j(\beta_j - x)dx,$$

and hence by the inequality of the arithmetic and the geometric mean (note that all factors are ≥ 0)

$$\begin{aligned} A &= \int_{-1}^{1}\frac{1}{2}\Big[\ (1-x^2)\prod_i(\alpha_i - x)\prod_j(\beta_j + x)\ + \\ &\qquad\qquad (1-x^2)\prod_i(\alpha_i + x)\prod_j(\beta_j - x)\ \Big]dx \\ &\ge \int_{-1}^{1}(1-x^2)\Big(\prod_i(\alpha_i^2 - x^2)\prod_j(\beta_j^2 - x^2)\Big)^{1/2}dx \\ &\ge \int_{-1}^{1}(1-x^2)\Big(\prod_i(\alpha_i^2 - 1)\prod_j(\beta_j^2 - 1)\Big)^{1/2}dx \\ &= \frac{4}{3}\Big(\prod_i(\alpha_i^2 - 1)\prod_j(\beta_j^2 - 1)\Big)^{1/2}. \end{aligned}$$

Let us compute $f'(1)$ and $f'(-1)$. (We may assume $f'(-1), f'(1) \ne 0$, since otherwise $T = 0$ and the inequality $\frac{2}{3}T \le A$ becomes trivial.) By (3) we see

$$f'(1) = -2\prod_i(\alpha_i - 1)\prod_j(\beta_j + 1),$$

and similarly

$$f'(-1) = 2\prod_i(\alpha_i + 1)\prod_j(\beta_j - 1).$$

Hence we conclude

$$A \ge \frac{2}{3}(-f'(1)f'(-1))^{1/2}.$$

Applying now the inequality of the harmonic and the geometric mean to $-f'(1)$ and $f'(1)$, we arrive by (2) at the conclusion

$$A \;\geq\; \frac{2}{3}\frac{2}{\frac{1}{-f'(1)}+\frac{1}{f'(-1)}} \;=\; \frac{4}{3}\frac{f'(1)f'(-1)}{f'(1)-f'(-1)} \;=\; \frac{2}{3}T,$$

which is what we wanted to show. By analyzing the case of equality in all our inequalities the reader can easily supply the last statement of the theorem. □

The reader is invited to search for an equally inspired proof of the second inequality in Theorem 2.

Well, analysis is inequalities after all, but here is an example from graph theory where the use of inequalities comes in quite unexpected. In Chapter 32 we will discuss Turán's theorem. In the simplest case it takes on the following form:

Theorem 3. *Suppose G is a graph on n vertices without triangles. Then G has at most $\frac{n^2}{4}$ edges, and equality holds only when n is even and G is the complete bipartite graph $K_{n/2,n/2}$.*

■ **First proof.** This proof, using Cauchy's inequality, is due to Mantel. Let $V=\{1,\dots,n\}$ be the vertex set and E the edge set of G. By d_i we denote the degree of i, hence $\sum_{i\in V} d_i = 2|E|$ (see page 143 in the chapter on double counting). Suppose ij is an edge. Since G has no triangles, we find $d_i+d_j\leq n$ since no vertex is a neighbor of both i and j.

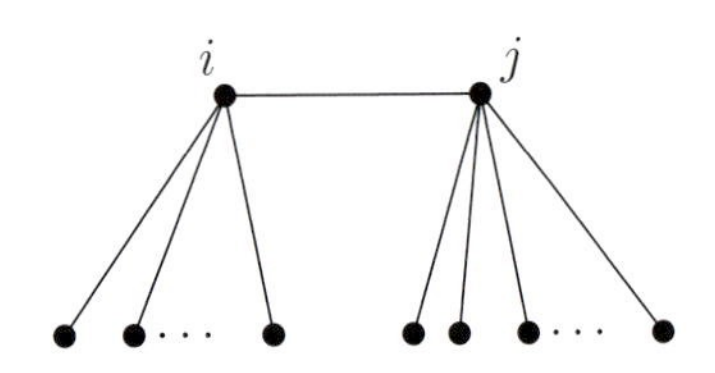

It follows that

$$\sum_{ij\in E}(d_i+d_j) \;\leq\; n|E|.$$

Note that d_i appears exactly d_i times in the sum, so we get

$$n|E| \;\geq\; \sum_{ij\in E}(d_i+d_j) \;=\; \sum_{i\in V} d_i^2,$$

and hence with Cauchy's inequality applied to the vectors $(d_1,\dots,d_n)$ and $(1,\dots,1)$,

$$n|E| \;\geq\; \sum_{i\in V} d_i^2 \;\geq\; \frac{(\sum d_i)^2}{n} \;=\; \frac{4|E|^2}{n},$$

and the result follows. In the case of equality we find $d_i = d_j$ for all i,j, and further $d_i = \frac{n}{2}$ (since $d_i+d_j = n$). Since G is triangle-free, $G=K_{n/2,n/2}$ is immediately seen from this. □

■ **Second proof.** The following proof of Theorem 3, using the inequality of the arithmetic and the geometric mean, is a folklore Book Proof. Let α be the size of a largest independent set A, and set $\beta = n - \alpha$. Since G is triangle-free, the neighbors of a vertex i form an independent set, and we infer $d_i \leq \alpha$ for all i.

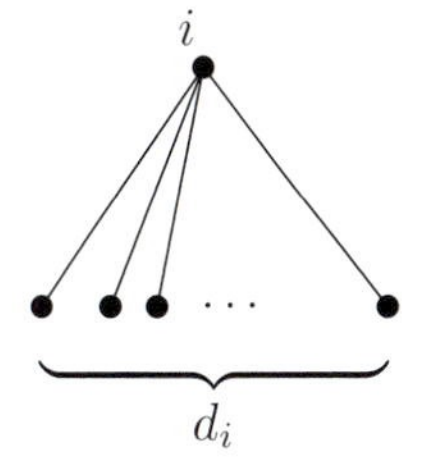

The set $B = V \backslash A$ of size β meets every edge of G. Counting the edges of G according to their endvertices in B, we obtain $|E| \leq \sum_{i \in B} d_i$. The inequality of the arithmetic and geometric mean now yields

$$|E| \leq \sum_{i \in B} d_i \leq \alpha\beta \leq \left(\frac{\alpha+\beta}{2}\right)^2 = \frac{n^2}{4},$$

and again the case of equality is easily dealt with. □

References

[1] H. ALZER: *A proof of the arithmetic mean-geometric mean inequality,* Amer. Math. Monthly **103** (1996), 585.

[2] P. S. BULLEN, D. S. MITRINOVICS & P. M. VASIĆ: *Means and their Inequalities,* Reidel, Dordrecht 1988.

[3] P. ERDŐS & T. GRÜNWALD: *On polynomials with only real roots,* Annals Math. **40** (1939), 537-548.

[4] G. H. HARDY, J. E. LITTLEWOOD & G. PÓLYA: *Inequalities,* Cambridge University Press, Cambridge 1952.

[5] W. MANTEL: *Problem 28,* Wiskundige Opgaven **10** (1906), 60-61.

[6] G. PÓLYA: *Review of* [3], Mathematical Reviews **1** (1940), 1.

[7] G. PÓLYA & G. SZEGŐ: *Problems and Theorems in Analysis, Vol. I,* Springer-Verlag, Berlin Heidelberg New York 1972/78; Reprint 1998.

A theorem of Pólya on polynomials

Chapter 18

Among the many contributions of George Pólya to analysis, the following has always been Erdős' favorite, both for the surprising result and for the beauty of its proof. Suppose that

$$f(z) = z^n + b_{n-1}z^{n-1} + \ldots + b_0$$

is a complex polynomial of degree $n \geq 1$ with leading coefficient 1. Associate with $f(z)$ the set

$$\mathcal{C} := \{z \in \mathbb{C} : |f(z)| \leq 2\},$$

that is, $\mathcal{C}$ is the set of points which are mapped under f into the circle of radius 2 around the origin in the complex plane. So for $n = 1$ the domain $\mathcal{C}$ is just a circular disk of diameter 4.

George Pólya

By an astoundingly simple argument, Pólya revealed the following beautiful property of this set $\mathcal{C}$:

> *Take any line L in the complex plane and consider the orthogonal projection $\mathcal{C}_L$ of the set $\mathcal{C}$ onto L. Then the total length of any such projection never exceeds 4.*

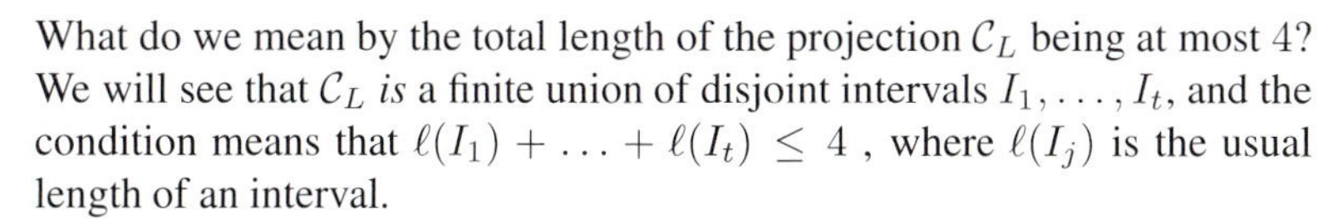

What do we mean by the total length of the projection $\mathcal{C}_L$ being at most 4? We will see that $\mathcal{C}_L$ *is* a finite union of disjoint intervals $I_1, \ldots, I_t$, and the condition means that $\ell(I_1) + \ldots + \ell(I_t) \leq 4$, where $\ell(I_j)$ is the usual length of an interval.

By rotating the plane we see that it suffices to consider the case when L is the real axis of the complex plane. With these comments in mind, let us state Pólya's result.

Theorem 1. *Let $f(z)$ be a complex polynomial of degree at least 1 and leading coefficient 1. Set $\mathcal{C} = \{z \in \mathbb{C} : |f(z)| \leq 2\}$ and let $\mathcal{R}$ be the orthogonal projection of $\mathcal{C}$ onto the real axis. Then there are intervals $I_1, \ldots, I_t$ on the real line which together cover $\mathcal{R}$ and satisfy*

$$\ell(I_1) + \ldots + \ell(I_t) \leq 4.$$

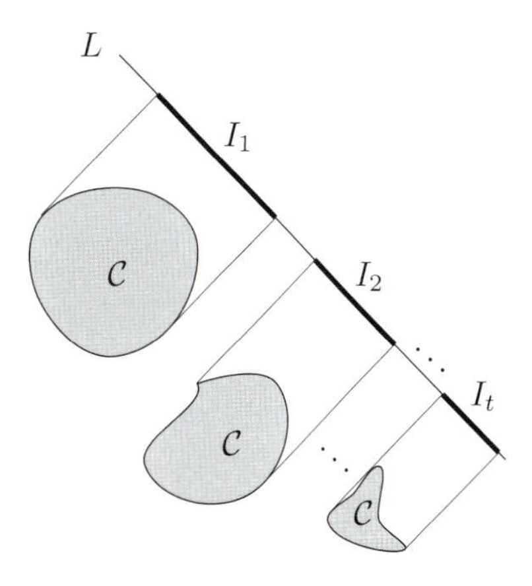

Clearly the bound of 4 in the theorem is attained for $n = 1$. To get more of a feeling for the problem let us look at the polynomial $f(z) = z^2 - 2$, which also attains the bound of 4. If $z = x + iy$ is a complex number, then x is its orthogonal projection onto the real line. Hence

$$\mathcal{R} = \{x \in \mathbb{R} : x + iy \in \mathcal{C} \text{ for some } y\}.$$

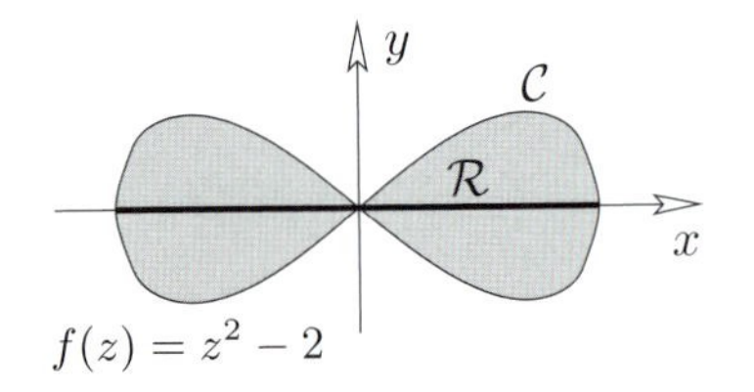

The reader can easily prove that for $f(z) = z^2 - 2$ we have $x + iy \in \mathcal{C}$ if and only if

$$(x^2 + y^2)^2 \le 4(x^2 - y^2).$$

It follows that $x^4 \le (x^2 + y^2)^2 \le 4x^2$, and thus $x^2 \le 4$, that is, $|x| \le 2$. On the other hand, any $z = x \in \mathbb{R}$ with $|x| \le 2$ satisfies $|z^2 - 2| \le 2$, and we find that $\mathcal{R}$ is precisely the interval $[-2, 2]$ of length 4.

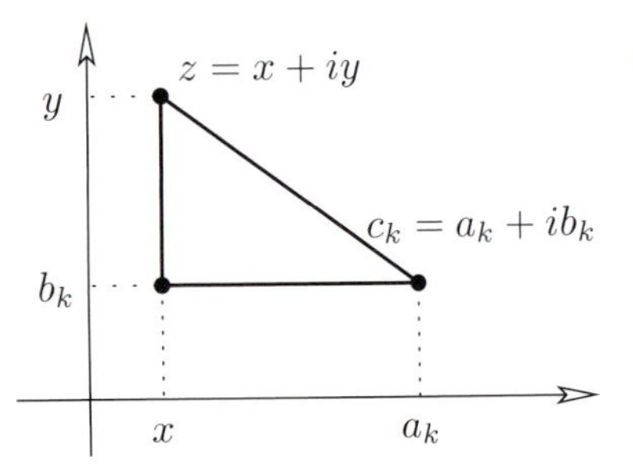

As a first step towards the proof write $f(z) = (z - c_1) \cdots (z - c_n)$ with $c_k = a_k + ib_k$, and consider the *real* polynomial $p(x) = (x - a_1) \cdots (x - a_n)$. Let $z = x + iy \in \mathcal{C}$, then by the theorem of Pythagoras

$$|x - a_k|^2 + |y - b_k|^2 = |z - c_k|^2$$

and hence $|x - a_k| \le |z - c_k|$ for all k, that is,

$$|p(x)| = |x - a_1| \cdots |x - a_n| \le |z - c_1| \cdots |z - c_n| = |f(z)| \le 2.$$

Thus we find that $\mathcal{R}$ is contained in the set $\mathcal{P} = \{x \in \mathbb{R} : |p(x)| \le 2\}$, and if we can show that this latter set is covered by intervals of total length at most 4, then we are done. Accordingly, our main Theorem 1 will be a consequence of the following result.

Theorem 2. *Let $p(x)$ be a real polynomial of degree $n \ge 1$ with leading coefficient 1, and all roots real. Then the set $\mathcal{P} = \{x \in \mathbb{R} : |p(x)| \le 2\}$ can be covered by intervals of total length at most 4.*

As Pólya shows in his paper [2], Theorem 2 is, in turn, a consequence of the following famous result due to Chebyshev. To make this chapter self-contained, we have included a proof in the appendix (following the beautiful exposition by Pólya and Szegő).

Chebyshev's Theorem.
Let $p(x)$ be a real polynomial of degree $n \ge 1$ with leading coefficient 1. Then

$$\max_{-1 \le x \le 1} |p(x)| \ge \frac{1}{2^{n-1}}.$$

Pavnuty Chebyshev on a Soviet stamp from 1946

Let us first note the following immediate consequence.

Corollary. *Let $p(x)$ be a real polynomial of degree $n \ge 1$ with leading coefficient 1, and suppose that $|p(x)| \le 2$ for all x in the interval $[a, b]$. Then $b - a \le 4$.*

■ **Proof.** Consider the substitution $y = \frac{2}{b-a}(x - a) - 1$. This maps the x-interval $[a, b]$ onto the y-interval $[-1, 1]$. The corresponding polynomial

$$q(y) = p(\tfrac{b-a}{2}(y + 1) + a)$$

has leading coefficient $(\frac{b-a}{2})^n$ and satisfies

$$\max_{-1 \le y \le 1} |q(y)| = \max_{a \le x \le b} |p(x)|.$$

By Chebyshev's theorem we deduce

$$2 \geq \max_{a \leq x \leq b} |p(x)| \geq \left(\tfrac{b-a}{2}\right)^n \tfrac{1}{2^{n-1}} = 2\left(\tfrac{b-a}{4}\right)^n,$$

and thus $b - a \leq 4$, as desired. □

This corollary brings us already very close to the statement of Theorem 2. If the set $\mathcal{P} = \{x : |p(x)| \leq 2\}$ is an *interval*, then the length of $\mathcal{P}$ is at most 4. The set $\mathcal{P}$ may, however, not be an interval, as in the example depicted here, where $\mathcal{P}$ consists of two intervals.

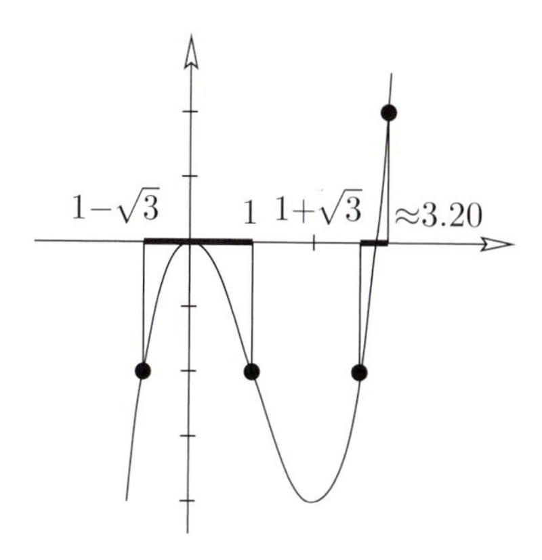

For the polynomial $p(x) = x^2(x-3)$ we get $\mathcal{P} = [1-\sqrt{3}, 1] \cup [1+\sqrt{3}, \approx 3.2]$

What can we say about $\mathcal{P}$? Since $p(x)$ is a continuous function, we know at any rate that $\mathcal{P}$ is the union of disjoint closed intervals $I_1, I_2, \ldots$, and that $p(x)$ assumes the value 2 or -2 at each endpoint of an interval I_j. This implies that there are only finitely many intervals $I_1, \ldots, I_t$, since $p(x)$ can assume any value only finitely often.

Pólya's wonderful idea was to construct another polynomial $\tilde{p}(x)$ of degree n, again with leading coefficient 1, such that $\widetilde{\mathcal{P}} = \{x : |\tilde{p}(x)| \leq 2\}$ is an *interval* of length at least $\ell(I_1) + \ldots + \ell(I_t)$. The corollary then proves $\ell(I_1) + \ldots + \ell(I_t) \leq \ell(\widetilde{\mathcal{P}}) \leq 4$, and we are done.

■ **Proof of Theorem 2.** Consider $p(x) = (x - a_1) \cdots (x - a_n)$ with $\mathcal{P} = \{x \in \mathbb{R} : |p(x)| \leq 2\} = I_1 \cup \ldots \cup I_t$, where we arrange the intervals I_j such that I_1 is the leftmost and I_t the rightmost interval. First we claim that any interval I_j contains a root of $p(x)$. We know that $p(x)$ assumes the values 2 or -2 at the endpoints of I_j. If one value is 2 and the other -2, then there is certainly a root in I_j. So assume $p(x) = 2$ at both endpoints (the case -2 being analogous). Suppose $b \in I_j$ is a point where $p(x)$ assumes its minimum in I_j. Then $p'(b) = 0$ and $p''(b) \geq 0$. If $p''(b) = 0$, then b is a multiple root of $p'(x)$, and hence a root of $p(x)$ by Fact 1 from the box on the next page. If, on the other hand, $p''(b) > 0$, then we deduce $p(b) \leq 0$ from Fact 2 from the same box. Hence either $p(b) = 0$, and we have our root, or $p(b) < 0$, and we obtain a root in the interval from b to either endpoint of I_j.

Here is the final idea of the proof. Let $I_1, \ldots, I_t$ be the intervals as before, and suppose the rightmost interval I_t contains m roots of $p(x)$, counted with their multiplicities. If $m = n$, then I_t is the only interval (by what we just proved), and we are finished. So assume $m < n$, and let d be the distance between I_{t-1} and I_t as in the figure. Let $b_1, \ldots, b_m$ be the roots of $p(x)$ which lie in I_t and $c_1, \ldots c_{n-m}$ the remaining roots. We now write $p(x) = q(x)r(x)$ where $q(x) = (x - b_1) \cdots (x - b_m)$ and $r(x) = (x - c_1) \cdots (x - c_{n-m})$, and set $p_1(x) = q(x + d)r(x)$. The polynomial $p_1(x)$ is again of degree n with leading coefficient 1. For $x \in I_1 \cup \ldots \cup I_{t-1}$ we have $|x + d - b_i| < |x - b_i|$ for all i, and hence $|q(x + d)| < |q(x)|$. It follows that

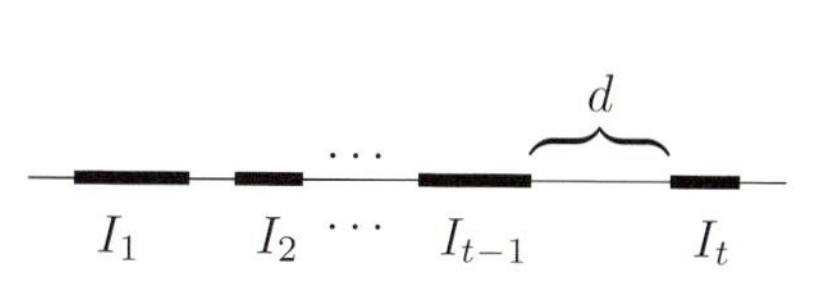

$$|p_1(x)| \leq |p(x)| \leq 2 \qquad \text{for } x \in I_1 \cup \ldots \cup I_{t-1}.$$

If, on the other hand, $x \in I_t$, then we find $|r(x - d)| \leq |r(x)|$ and thus

$$|p_1(x-d)| = |q(x)||r(x-d)| \leq |p(x)| \leq 2,$$

which means that $I_t - d \subseteq \mathcal{P}_1 = \{x : |p_1(x)| \le 2\}$.

In summary, we see that $\mathcal{P}_1$ contains $I_1 \cup \ldots \cup I_{t-1} \cup (I_t - d)$ and hence has total length at least as large as $\mathcal{P}$. Notice now that with the passage from $p(x)$ to $p_1(x)$ the intervals I_{t-1} and $I_t - d$ merge into a single interval. We conclude that the intervals $J_1, \ldots, J_s$ of $p_1(x)$ making up $\mathcal{P}_1$ have total length at least $\ell(I_1) + \ldots + \ell(I_t)$, and that the rightmost interval J_s contains more than m roots of $p_1(x)$. Repeating this procedure at most $t-1$ times, we finally arrive at a polynomial $\tilde{p}(x)$ with $\widetilde{\mathcal{P}} = \{x : |\tilde{p}(x)| \le 2\}$ being an interval of length $\ell(\widetilde{\mathcal{P}}) \ge \ell(I_1) + \ldots \ell(I_t)$, and the proof is complete. □

Two facts about polynomials with real roots

Let $p(x)$ be a non-constant polynomial with only real roots.

Fact 1. *If b is a multiple root of $p'(x)$, then b is also a root of $p(x)$.*

■ **Proof.** Let $b_1, \ldots, b_r$ be the roots of $p(x)$ with multiplicities $s_1, \ldots, s_r$, $\sum_{j=1}^{r} s_j = n$. From $p(x) = (x - b_j)^{s_j} h(x)$ we infer that b_j is a root of $p'(x)$ if $s_j \ge 2$, and the multiplicity of b_j in $p'(x)$ is $s_j - 1$. Furthermore, there is a root of $p'(x)$ between b_1 and b_2, another root between b_2 and b_3, …, and one between b_{r-1} and b_r, and all these roots must be *single* roots, since $\sum_{j=1}^{r}(s_j - 1) + (r-1)$ counts already up to the degree $n-1$ of $p'(x)$. Consequently, the *multiple* roots of $p'(x)$ can only occur among the roots of $p(x)$. □

Fact 2. *We have $p'(x)^2 \ge p(x)p''(x)$ for all $x \in \mathbb{R}$.*

■ **Proof.** If $x = a_i$ is a root of $p(x)$, then there is nothing to show. Assume then x is not a root. The product rule of differentiation yields

$$p'(x) = \sum_{k=1}^{n} \frac{p(x)}{x - a_k}, \quad \text{that is,} \quad \frac{p'(x)}{p(x)} = \sum_{k=1}^{n} \frac{1}{x - a_k}.$$

Differentiating this again we have

$$\frac{p''(x)p(x) - p'(x)^2}{p(x)^2} = -\sum_{k=1}^{n} \frac{1}{(x - a_k)^2} < 0. \quad \square$$

Appendix: Chebyshev's theorem

Theorem. *Let $p(x)$ be a real polynomial of degree $n \geq 1$ with leading coefficient 1. Then*

$$\max_{-1 \leq x \leq 1} |p(x)| \geq \frac{1}{2^{n-1}}.$$

Before we start, let us look at some examples where we have equality. The margin depicts the graphs of polynomials of degrees 1, 2 and 3, where we have equality in each case. Indeed, we will see that for every degree there is precisely one polynomial with equality in Chebyshev's theorem.

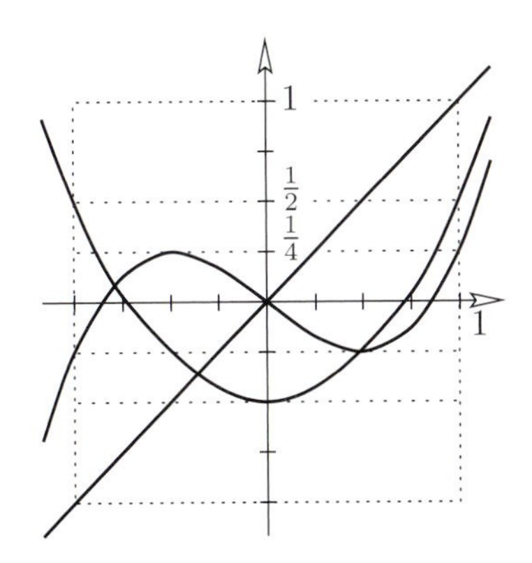

The polynomials $p_1(x) = x$, $p_2(x) = x^2 - \frac{1}{2}$ and $p_3(x) = x^3 - \frac{3}{4}x$ achieve equality in Chebyshev's theorem.

■ **Proof.** Consider a real polynomial $p(x) = x^n + a_{n-1}x^{n-1} + \ldots + a_0$ with leading coefficient 1. Since we are interested in the range $-1 \leq x \leq 1$, we set $x = \cos\vartheta$ and denote by $g(\vartheta) := p(\cos\vartheta)$ the resulting polynomial in $\cos\vartheta$,

$$g(\vartheta) = (\cos\vartheta)^n + a_{n-1}(\cos\vartheta)^{n-1} + \ldots + a_0. \tag{1}$$

The proof proceeds now in the following two steps which are both classical results and interesting in their own right.

(A) We express $g(\vartheta)$ as a so-called *cosine polynomial*, that is, a polynomial of the form

$$g(\vartheta) \;=\; b_n \cos n\vartheta + b_{n-1}\cos(n-1)\vartheta + \ldots + b_1 \cos\vartheta + b_0 \tag{2}$$

with $b_k \in \mathbb{R}$, and show that its leading coefficient is $b_n = \frac{1}{2^{n-1}}$.

(B) Given any cosine polynomial $h(\vartheta)$ of order n (meaning that λ_n is the highest nonvanishing coefficient)

$$h(\vartheta) \;=\; \lambda_n \cos n\vartheta + \lambda_{n-1}\cos(n-1)\vartheta + \ldots + \lambda_0, \tag{3}$$

we show $|\lambda_n| \leq \max |h(\vartheta)|$, which when applied to $g(\vartheta)$ will then prove the theorem.

Proof of (A). To pass from (1) to the representation (2), we have to express all powers $(\cos\vartheta)^k$ as cosine polynomials. For example, the addition theorem for the cosine gives

$$\cos 2\vartheta \;=\; \cos^2\vartheta - \sin^2\vartheta \;=\; 2\cos^2\vartheta - 1,$$

so that $\cos^2\vartheta = \frac{1}{2}\cos 2\vartheta + \frac{1}{2}$. To do this for an arbitrary power $(\cos\vartheta)^k$ we go into the complex numbers, via the relation $e^{ix} = \cos x + i\sin x$. The e^{ix} are the complex numbers of absolute value 1 (see the box on complex unit roots on page 25). In particular, this yields

$$e^{in\vartheta} \;=\; \cos n\vartheta + i \sin n\vartheta. \tag{4}$$

On the other hand,

$$e^{in\vartheta} \;=\; (e^{i\vartheta})^n \;=\; (\cos\vartheta + i\sin\vartheta)^n. \tag{5}$$

Equating the real parts in (4) and (5) we obtain by $i^{4\ell+2} = -1$, $i^{4\ell} = 1$ and $\sin^2\theta = 1 - \cos^2\theta$

$$\begin{aligned}\cos n\vartheta &= \sum_{\ell\geq 0}\binom{n}{4\ell}(\cos\vartheta)^{n-4\ell}(1-\cos^2\vartheta)^{2\ell} \\ &\quad -\sum_{\ell\geq 0}\binom{n}{4\ell+2}(\cos\vartheta)^{n-4\ell-2}(1-\cos^2\vartheta)^{2\ell+1}.\end{aligned} \tag{6}$$

We conclude that $\cos n\vartheta$ is a polynomial in $\cos\vartheta$,

$$\cos n\vartheta = c_n(\cos\vartheta)^n + c_{n-1}(\cos\vartheta)^{n-1} + \ldots + c_0. \tag{7}$$

From (6) we obtain for the highest coefficient

$$c_n = \sum_{\ell\geq 0}\binom{n}{4\ell} + \sum_{\ell\geq 0}\binom{n}{4\ell+2} = 2^{n-1}.$$

$\sum_{k\geq 0}\binom{n}{2k} = 2^{n-1}$ holds for $n > 0$: Every subset of $\{1, 2, \ldots, n-1\}$ yields an *even* sized subset of $\{1, 2, \ldots, n\}$ if we add the element n "if needed."

Now we turn our argument around. Assuming by induction that for $k < n$, $(\cos\vartheta)^k$ can be expressed as a cosine polynomial of order k, we infer from (7) that $(\cos\vartheta)^n$ can be written as a cosine polynomial of order n with leading coefficient $b_n = \frac{1}{2^{n-1}}$.

Proof of (B). Let $h(\vartheta)$ be a cosine polynomial of order n as in (3), and assume without loss of generality $\lambda_n > 0$. Now we set $m(\vartheta) := \lambda_n \cos n\vartheta$ and find

$$m(\tfrac{k}{n}\pi) = (-1)^k\lambda_n \qquad \text{for } k = 0, 1, \ldots, n.$$

Suppose, for a proof by contradiction, that $\max|h(\vartheta)| < \lambda_n$. Then

$$m(\tfrac{k}{n}\pi) - h(\tfrac{k}{n}\pi) = (-1)^k\lambda_n - h(\tfrac{k}{n}\pi)$$

is positive for even k and negative for odd k in the range $0 \leq k \leq n$. We conclude that $m(\vartheta) - h(\vartheta)$ has at least n roots in the interval $[0, \pi]$. But this cannot be since $m(\vartheta) - h(\vartheta)$ is a cosine polynomial of order $n-1$, which can be written in the form (1) and thus has at most $n-1$ roots.

The proof of **(B)** and thus of Chebyshev's theorem is complete. □

The reader can now easily complete the analysis, showing that $g_n(\vartheta) := \frac{1}{2^{n-1}}\cos n\vartheta$ is the *only* cosine polynomial of order n with leading coefficient 1 that achieves the equality $\max|g(\vartheta)| = \frac{1}{2^{n-1}}$.

The polynomials $T_n(x) = \cos n\vartheta$, $x = \cos\vartheta$, are called the *Chebyshev polynomials* (of the first kind); thus $\frac{1}{2^{n-1}}T_n(x)$ is the unique monic polynomial of degree n where equality holds in Chebyshev's theorem.

References

[1] P. L. CEBYCEV: *Œuvres,* Vol. I, Acad. Imperiale des Sciences, St. Petersburg 1899, pp. 387-469.

[2] G. PÓLYA: *Beitrag zur Verallgemeinerung des Verzerrungssatzes auf mehrfach zusammenhängenden Gebieten,* Sitzungsber. Preuss. Akad. Wiss. Berlin (1928), 228-232; Collected Papers Vol. I, MIT Press 1974, 347-351.

[3] G. PÓLYA & G. SZEGŐ: *Problems and Theorems in Analysis, Vol. II,* Springer-Verlag, Berlin Heidelberg New York 1976; Reprint 1998.

On a lemma of Littlewood and Offord

Chapter 19

In their work on the distribution of roots of algebraic equations, Littlewood and Offord proved in 1943 the following result:

> *Let $a_1, a_2, \ldots, a_n$ be complex numbers with $|a_i| \geq 1$ for all i, and consider the 2^n linear combinations $\sum_{i=1}^{n} \varepsilon_i a_i$ with $\varepsilon_i \in \{1, -1\}$. Then the number of sums $\sum_{i=1}^{n} \varepsilon_i a_i$ which lie in the interior of any circle of radius 1 is not greater than*
>
> $$c \frac{2^n}{\sqrt{n}} \log n \quad \textit{for some constant } c > 0.$$

John E. Littlewood

A few years later Paul Erdős improved this bound by removing the $\log n$ term, but what is more interesting, he showed that this is, in fact, a simple consequence of the theorem of Sperner (see page 151).

To get a feeling for his argument, let us look at the case when all a_i are real. We may assume that all a_i are positive (by changing a_i to $-a_i$ and ε_i to $-\varepsilon_i$ whenever $a_i < 0$). Now suppose that a set of combinations $\sum \varepsilon_i a_i$ lies in the interior of an interval of length 2. Let $N = \{1, 2, \ldots, n\}$ be the index set. For every $\sum \varepsilon_i a_i$ we set $I := \{i \in N : \varepsilon_i = 1\}$. Now if $I \subsetneqq I'$ for two such sets, then we conclude that

$$\sum \varepsilon_i' a_i - \sum \varepsilon_i a_i = 2 \sum_{i \in I' \setminus I} a_i \geq 2,$$

which is a contradiction. Hence the sets I form an antichain, and we conclude from the theorem of Sperner that there are at most $\binom{n}{\lfloor n/2 \rfloor}$ such combinations. By Stirling's formula (see page 11) we have

Sperner's theorem. *Any antichain of subsets of an n-set has size at most $\binom{n}{\lfloor n/2 \rfloor}$.*

$$\binom{n}{\lfloor n/2 \rfloor} \leq c \frac{2^n}{\sqrt{n}} \quad \text{for some } c > 0.$$

For n even and all $a_i = 1$ we obtain $\binom{n}{n/2}$ combinations $\sum_{i=1}^{n} \varepsilon_i a_i$ that sum to 0. Looking at the interval $(-1, 1)$ we thus find that the binomial number gives the *exact* bound.

In the same paper Erdős conjectured that $\binom{n}{\lfloor n/2 \rfloor}$ was the right bound for complex numbers as well (he could only prove $c\, 2^n n^{-1/2}$ for some c) and indeed that the same bound is valid for vectors $\boldsymbol{a}_1, \ldots, \boldsymbol{a}_n$ with $|\boldsymbol{a}_i| \geq 1$ in a real Hilbert space, when the circle of radius 1 is replaced by an open ball of radius 1.

Erdős was right, but it took twenty years until Gyula Katona and Daniel Kleitman independently came up with a proof for the complex numbers (or, what is the same, for the plane $\mathbb{R}^2$). Their proofs used explicitly the 2-dimensionality of the plane, and it was not at all clear how they could be extended to cover finite dimensional real vector spaces.

But then in 1970 Kleitman proved the full conjecture on Hilbert spaces with an argument of stunning simplicity. In fact, he proved even more. His argument is a prime example of what you can do when you find the right induction hypothesis.

A word of comfort for all readers who are not familiar with the notion of a Hilbert space: We do not really need general Hilbert spaces. Since we only deal with finitely many vectors $\boldsymbol{a}_i$, it is enough to consider the real space $\mathbb{R}^d$ with the usual scalar product. Here is Kleitman's result.

Theorem. *Let $\boldsymbol{a}_1, \ldots, \boldsymbol{a}_n$ be vectors in $\mathbb{R}^d$, each of length at least 1, and let $R_1, \ldots, R_k$ be k open regions of $\mathbb{R}^d$, where $|\boldsymbol{x} - \boldsymbol{y}| < 2$ for any $\boldsymbol{x}, \boldsymbol{y}$ that lie in the same region R_i.*
Then the number of linear combinations $\sum_{i=1}^n \varepsilon_i \boldsymbol{a}_i$, $\varepsilon_i \in \{1, -1\}$, that can lie in the union $\bigcup_i R_i$ of the regions is at most the sum of the k largest binomial coefficients $\binom{n}{j}$.
In particular, we get the bound $\binom{n}{\lfloor n/2 \rfloor}$ for $k = 1$.

Before turning to the proof note that the bound is exact for

$$\boldsymbol{a}_1 = \ldots = \boldsymbol{a}_n = \boldsymbol{a} = (1, 0, \ldots, 0)^T.$$

Indeed, for even n we obtain $\binom{n}{n/2}$ sums equal to 0, $\binom{n}{n/2-1}$ sums equal to $(-2)\boldsymbol{a}$, $\binom{n}{n/2+1}$ sums equal to $2\boldsymbol{a}$, and so on. Choosing balls of radius 1 around

$$-2\lceil \tfrac{k-1}{2} \rceil \boldsymbol{a}, \quad \ldots \quad (-2)\boldsymbol{a}, \quad \boldsymbol{0}, \quad 2\boldsymbol{a}, \quad \ldots \quad 2\lfloor \tfrac{k-1}{2} \rfloor \boldsymbol{a},$$

we obtain

$$\binom{n}{\lfloor \frac{n-k+1}{2} \rfloor} + \ldots + \binom{n}{\frac{n-2}{2}} + \binom{n}{\frac{n}{2}} + \binom{n}{\frac{n+2}{2}} + \ldots + \binom{n}{\lfloor \frac{n+k-1}{2} \rfloor}$$

sums lying in these k balls, and this is our promised expression, since the largest binomial coefficients are centered around the middle (see page 12). A similar reasoning works when n is odd.

■ **Proof.** We may assume, without loss of generality, that the regions R_i are disjoint, and will do so from now on. The key to the proof is the recursion of the binomial coefficients, which tells us how the largest binomial coefficients of n and $n-1$ are related. Set $r = \lfloor \frac{n-k+1}{2} \rfloor$, $s = \lfloor \frac{n+k-1}{2} \rfloor$, then $\binom{n}{r}, \binom{n}{r+1}, \ldots, \binom{n}{s}$ are the k largest binomial coefficients for n. The recursion $\binom{n}{i} = \binom{n-1}{i} + \binom{n-1}{i-1}$ implies

$$\begin{aligned}
\sum_{i=r}^{s}\binom{n}{i} &= \sum_{i=r}^{s}\binom{n-1}{i}+\sum_{i=r}^{s}\binom{n-1}{i-1}\\
&= \sum_{i=r}^{s}\binom{n-1}{i}+\sum_{i=r-1}^{s-1}\binom{n-1}{i} \qquad (1)\\
&= \sum_{i=r-1}^{s}\binom{n-1}{i}+\sum_{i=r}^{s-1}\binom{n-1}{i},
\end{aligned}$$

and an easy calculation shows that the first sum adds the $k+1$ largest binomial coefficients $\binom{n-1}{i}$, and the second sum the largest $k-1$.

Kleitman's proof proceeds by induction on n, the case $n = 1$ being trivial. In the light of (1) we need only show for the induction step that the linear combinations of $\boldsymbol{a}_1, \ldots, \boldsymbol{a}_n$ that lie in k disjoint regions can be mapped *bijectively* onto combinations of $\boldsymbol{a}_1, \ldots, \boldsymbol{a}_{n-1}$ that lie in $k+1$ or $k-1$ regions.

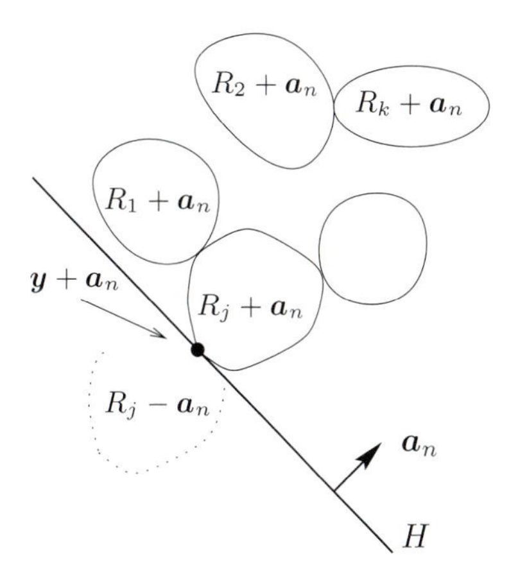

Claim. *At least one of the translated regions $R_j - \boldsymbol{a}_n$ is disjoint from all the translated regions $R_1 + \boldsymbol{a}_n, \ldots, R_k + \boldsymbol{a}_n$.*

To prove this, consider the hyperplane $H = \{\boldsymbol{x} : \langle \boldsymbol{a}_n, \boldsymbol{x}\rangle = c\}$ orthogonal to $\boldsymbol{a}_n$, which contains all translates $R_i + \boldsymbol{a}_n$ on the side that is given by $\langle \boldsymbol{a}_n, \boldsymbol{x}\rangle \geq c$, and which touches the closure of some region, say $R_j + \boldsymbol{a}_n$. Such a hyperplane exists since the regions are bounded. Now $|\boldsymbol{x} - \boldsymbol{y}| < 2$ holds for any $\boldsymbol{x} \in R_j$ and $\boldsymbol{y}$ in the closure of R_j, since R_j is open. We want to show that $R_j - \boldsymbol{a}_n$ lies on the other side of H. Suppose, on the contrary, that $\langle \boldsymbol{a}_n, \boldsymbol{x} - \boldsymbol{a}_n\rangle \geq c$ for some $\boldsymbol{x} \in R_j$, that is, $\langle \boldsymbol{a}_n, \boldsymbol{x}\rangle \geq |\boldsymbol{a}_n|^2 + c$.

Let $\boldsymbol{y} + \boldsymbol{a}_n$ be a point where H touches $R_j + \boldsymbol{a}_n$, then $\boldsymbol{y}$ is in the closure of R_j, and $\langle \boldsymbol{a}_n, \boldsymbol{y} + \boldsymbol{a}_n\rangle = c$, that is, $\langle \boldsymbol{a}_n, -\boldsymbol{y}\rangle = |\boldsymbol{a}_n|^2 - c$. Hence

$$\langle \boldsymbol{a}_n, \boldsymbol{x} - \boldsymbol{y}\rangle \;\geq\; 2|\boldsymbol{a}_n|^2,$$

and we infer from the Cauchy-Schwarz inequality

$$2|\boldsymbol{a}_n|^2 \;\leq\; \langle \boldsymbol{a}_n, \boldsymbol{x} - \boldsymbol{y}\rangle \;\leq\; |\boldsymbol{a}_n||\boldsymbol{x} - \boldsymbol{y}|,$$

and thus (with $|\boldsymbol{a}_n| \geq 1$) we get $2 \leq 2|\boldsymbol{a}_n| \leq |\boldsymbol{x} - \boldsymbol{y}|$, a contradiction.

The rest is easy. We classify the combinations $\sum \varepsilon_i \boldsymbol{a}_i$ which come to lie in $R_1 \cup \ldots \cup R_k$ as follows. Into Class 1 we put all $\sum_{i=1}^{n} \varepsilon_i \boldsymbol{a}_i$ with $\varepsilon_n = -1$ and all $\sum_{i=1}^{n} \varepsilon_i \boldsymbol{a}_i$ with $\varepsilon_n = 1$ lying in R_j, and into Class 2 we throw in the remaining combinations $\sum_{i=1}^{n} \varepsilon_i \boldsymbol{a}_i$ with $\varepsilon_n = 1$, not in R_j. It follows that the combinations $\sum_{i=1}^{n-1} \varepsilon_i \boldsymbol{a}_i$ corresponding to Class 1 lie in the $k+1$ disjoint regions $R_1 + \boldsymbol{a}_n, \ldots, R_k + \boldsymbol{a}_n$ and $R_j - \boldsymbol{a}_n$, and the combinations $\sum_{i=1}^{n-1} \varepsilon_i \boldsymbol{a}_i$ corresponding to Class 2 lie in the $k-1$ disjoint regions $R_1 - \boldsymbol{a}_n, \ldots, R_k - \boldsymbol{a}_n$ without $R_j - \boldsymbol{a}_n$. By induction, Class 1 contains at most $\sum_{i=r-1}^{s} \binom{n-1}{i}$ combinations, while Class 2 contains at most $\sum_{i=r}^{s-1} \binom{n-1}{i}$ combinations — and by (1) this is the whole proof, straight from The Book. □

References

[1] P. ERDŐS: *On a lemma of Littlewood and Offord,* Bulletin Amer. Math. Soc. **51** (1945), 898-902.

[2] G. KATONA: *On a conjecture of Erdős and a stronger form of Sperner's theorem,* Studia Sci. Math. Hungar. **1** (1966), 59-63.

[3] D. KLEITMAN: *On a lemma of Littlewood and Offord on the distribution of certain sums,* Math. Zeitschrift **90** (1965), 251-259.

[4] D. KLEITMAN: *On a lemma of Littlewood and Offord on the distributions of linear combinations of vectors,* Advances Math. **5** (1970), 155-157.

[5] J. E. LITTLEWOOD & A. C. OFFORD: *On the number of real roots of a random algebraic equation III,* Mat. USSR Sb. **12** (1943), 277-285.

Cotangent and the Herglotz trick

Chapter 20

What is the most interesting formula involving elementary functions? In his beautiful article [2], whose exposition we closely follow, Jürgen Elstrodt nominates as a first candidate the partial fraction expansion of the cotangent function:

$$\pi \cot \pi x \quad = \quad \frac{1}{x} + \sum_{n=1}^{\infty} \Big(\frac{1}{x+n} + \frac{1}{x-n} \Big) \qquad (x \in \mathbb{R} \backslash \mathbb{Z}).$$

Gustav Herglotz

This elegant formula was proved by Euler in §178 of his *Introductio in Analysin Infinitorum* from 1748 and it certainly counts among his finest achievements. We can also write it even more elegantly as

$$\pi \cot \pi x \quad = \quad \lim_{N \to \infty} \sum_{n=-N}^{N} \frac{1}{x+n} \tag{1}$$

but one has to note that the evaluation of the sum $\sum_{n \in \mathbb{Z}} \frac{1}{x+n}$ is a bit dangerous, since the sum is only conditionally convergent, so its value depends on the "right" order of summation.

We shall derive (1) by an argument of stunning simplicity which is attributed to Gustav Herglotz — the "Herglotz trick." To get started, set

$$f(x) \quad := \quad \pi \cot \pi x, \qquad g(x) \quad := \quad \lim_{N \to \infty} \sum_{n=-N}^{N} \frac{1}{x+n},$$

and let us try to derive enough common properties of these functions to see in the end that they must coincide...

(A) The functions f and g are defined for all non-integral values and are continuous there.

For the cotangent function $f(x) = \pi \cot \pi x = \pi \frac{\cos \pi x}{\sin \pi x}$, this is clear (see the figure). For $g(x)$, we first use the identity $\frac{1}{x+n} + \frac{1}{x-n} = -\frac{2x}{n^2 - x^2}$ to rewrite Euler's formula as

$$\pi \cot \pi x \quad = \quad \frac{1}{x} - \sum_{n=1}^{\infty} \frac{2x}{n^2 - x^2}\,. \tag{2}$$

Thus for **(A)** we have to prove that for every $x \notin \mathbb{Z}$ the series

$$\sum_{n=1}^{\infty} \frac{1}{n^2 - x^2}$$

converges uniformly in a neighborhood of x.

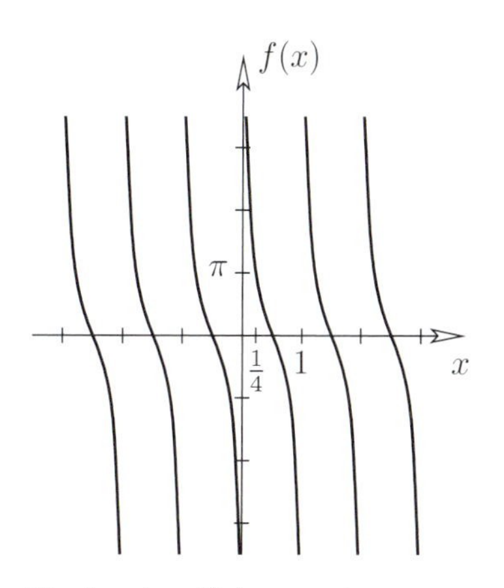

The function $f(x) = \pi \cot \pi x$

For this, we don't get any problem with the first term, for $n = 1$, or with the terms with $2n - 1 \le x^2$, since there is only a finite number of them. On the other hand, for $n \ge 2$ and $2n - 1 > x^2$, that is $n^2 - x^2 > (n-1)^2 > 0$, the summands are bounded by

$$0 \;<\; \frac{1}{n^2 - x^2} \;<\; \frac{1}{(n-1)^2},$$

and this bound is not only true for x itself, but also for values in a neighborhood of x. Finally the fact that $\sum \frac{1}{(n-1)^2}$ converges (to $\frac{\pi^2}{6}$, see page 35) provides the uniform convergence needed for the proof of **(A)**.

(B) Both f and g are *periodic* of period 1, that is, $f(x+1) = f(x)$ and $g(x+1) = g(x)$ hold for all $x \in \mathbb{R}\backslash\mathbb{Z}$.

Since the cotangent has period π, we find that f has period 1 (see again the figure above). For g we argue as follows. Let

$$g_N(x) \;:=\; \sum_{n=-N}^{N} \frac{1}{x+n},$$

then

$$\begin{aligned} g_N(x+1) &= \sum_{n=-N}^{N} \frac{1}{x+1+n} \;=\; \sum_{n=-N+1}^{N+1} \frac{1}{x+n} \\ &= g_{N-1}(x) + \frac{1}{x+N} + \frac{1}{x+N+1}. \end{aligned}$$

Hence $g(x+1) = \lim_{N\to\infty} g_N(x+1) = \lim_{N\to\infty} g_{N-1}(x) = g(x)$.

(C) Both f and g are *odd* functions, that is, we have $f(-x) = -f(x)$ and $g(-x) = -g(x)$ for all $x \in \mathbb{R}\backslash\mathbb{Z}$.

The function f obviously has this property, and for g we just have to observe that $g_N(-x) = -g_N(x)$.

The final two facts constitute the Herglotz trick: First we show that f and g satisfy the same functional equation, and secondly that $h := f - g$ can be continuously extended to all of $\mathbb{R}$.

(D) The two functions f and g satisfy the same functional equation: $f(\frac{x}{2}) + f(\frac{x+1}{2}) = 2f(x)$ and $g(\frac{x}{2}) + g(\frac{x+1}{2}) = 2g(x)$.

For $f(x)$ this results from the addition theorems for the sine and cosine functions:

$$\begin{aligned} f(\tfrac{x}{2}) + f(\tfrac{x+1}{2}) &= \pi \left[\frac{\cos\frac{\pi x}{2}}{\sin\frac{\pi x}{2}} - \frac{\sin\frac{\pi x}{2}}{\cos\frac{\pi x}{2}} \right] \\ &= 2\pi \frac{\cos(\frac{\pi x}{2} + \frac{\pi x}{2})}{\sin(\frac{\pi x}{2} + \frac{\pi x}{2})} \;=\; 2f(x). \end{aligned}$$

Addition theorems:
$\sin(x+y) = \sin x \cos y + \cos x \sin y$
$\cos(x+y) = \cos x \cos y - \sin x \sin y$

$$\begin{aligned} \Longrightarrow \quad \sin(x + \tfrac{\pi}{2}) &= \cos x \\ \cos(x + \tfrac{\pi}{2}) &= -\sin x \\ \sin x &= 2 \sin\tfrac{x}{2} \cos\tfrac{x}{2} \\ \cos x &= \cos^2\tfrac{x}{2} - \sin^2\tfrac{x}{2}. \end{aligned}$$

The functional equation for g follows from

$$g_N(\tfrac{x}{2}) + g_N(\tfrac{x+1}{2}) = 2\,g_{2N}(x) + \frac{2}{x+2N+1}.$$

which in turn follows from

$$\frac{1}{\frac{x}{2}+n} + \frac{1}{\frac{x+1}{2}+n} = 2\Big(\frac{1}{x+2n} + \frac{1}{x+2n+1}\Big).$$

Now let us look at

$$h(x) = f(x) - g(x) = \pi\cot\pi x - \Big(\frac{1}{x} - \sum_{n=1}^{\infty}\frac{2x}{n^2-x^2}\Big). \qquad (3)$$

We know by now that h is a continuous function on $\mathbb{R}\backslash\mathbb{Z}$ that satisfies the properties **(B)**, **(C)**, **(D)**. What happens at the integral values? From the sine and cosine series expansions, or by applying de l'Hospital's rule twice, we find

$$\lim_{x\to 0}\left(\cot x - \frac{1}{x}\right) = \lim_{x\to 0}\frac{x\cos x - \sin x}{x\sin x} = 0,$$

$\cos x = 1 - \frac{x^2}{2!} + \frac{x^4}{4!} - \frac{x^6}{6!} \pm \ldots$

$\sin x = x - \frac{x^3}{3!} + \frac{x^5}{5!} - \frac{x^7}{7!} \pm \ldots$

and hence also

$$\lim_{x\to 0}\left(\pi\cot\pi x - \frac{1}{x}\right) = 0.$$

But since the last sum $\sum_{n=1}^{\infty}\frac{2x}{n^2-x^2}$ in (3) converges to 0 with $x \longrightarrow 0$, we have in fact $\lim\limits_{x\to 0} h(x) = 0$, and thus by periodicity

$$\lim_{x\to n} h(x) = 0 \qquad \text{for all } n\in\mathbb{Z}.$$

In summary, we have shown the following:

(E) By setting $h(x) := 0$ for $x\in\mathbb{Z}$, h becomes a continuous function on all of $\mathbb{R}$ that shares the properties given in **(B)**, **(C)** and **(D)**.

We are ready for the *coup de grâce*. Since h is a periodic continuous function, it possesses a maximum m. Let x_0 be a point in $[0,1]$ with $h(x_0) = m$. It follows from **(D)** that

$$h(\tfrac{x_0}{2}) + h(\tfrac{x_0+1}{2}) = 2m,$$

and hence that $h(\frac{x_0}{2}) = m$. Iteration gives $h(\frac{x_0}{2^n}) = m$ for all n, and hence $h(0) = m$ by continuity. But $h(0) = 0$, and so $m = 0$, that is, $h(x) \le 0$ for all $x\in\mathbb{R}$. As $h(x)$ is an *odd* function, $h(x) < 0$ is impossible, hence $h(x) = 0$ for all $x\in\mathbb{R}$, and Euler's theorem is proved. $\square$

A great many corollaries can be derived from (1), the most famous of which concerns the values of Riemann's zeta function at even positive integers (see the Appendix to Chapter 6),

$$\zeta(2k) = \sum_{n=1}^{\infty}\frac{1}{n^{2k}} \qquad (k\in\mathbb{N}). \qquad (4)$$

So to finish our story let us see how Euler — a few years later, in 1755 — treated the series (4). We start with formula (2). Multiplying (2) by x and setting $y = \pi x$ we find for $|y| < \pi$:

$$\begin{aligned} y \cot y &= 1 - 2\sum_{n=1}^{\infty} \frac{y^2}{\pi^2 n^2 - y^2} \\ &= 1 - 2\sum_{n=1}^{\infty} \frac{y^2}{\pi^2 n^2} \frac{1}{1 - \left(\frac{y}{\pi n}\right)^2}. \end{aligned}$$

The last factor is the sum of a geometric series, hence

$$\begin{aligned} y \cot y &= 1 - 2\sum_{n=1}^{\infty}\sum_{k=1}^{\infty} \left(\frac{y}{\pi n}\right)^{2k} \\ &= 1 - 2\sum_{k=1}^{\infty} \left(\frac{1}{\pi^{2k}} \sum_{n=1}^{\infty} \frac{1}{n^{2k}}\right) y^{2k}, \end{aligned}$$

and we have proved the remarkable result:

For all $k \in \mathbb{N}$, the coefficient of y^{2k} in the power series expansion of $y \cot y$ equals

$$\left[y^{2k}\right] y \cot y = -\frac{2}{\pi^{2k}} \sum_{n=1}^{\infty} \frac{1}{n^{2k}} = -\frac{2}{\pi^{2k}} \zeta(2k). \tag{5}$$

There is another, perhaps much more "canonical," way to obtain a series expansion of $y \cot y$. We know from analysis that $e^{iy} = \cos y + i \sin y$, and thus

$$\cos y = \frac{e^{iy} + e^{-iy}}{2}, \qquad \sin y = \frac{e^{iy} - e^{-iy}}{2i},$$

which yields

$$y \cot y = iy \frac{e^{iy} + e^{-iy}}{e^{iy} - e^{-iy}} = iy \frac{e^{2iy} + 1}{e^{2iy} - 1}.$$

We now substitute $z = 2iy$, and get

$$y \cot y = \frac{z}{2} \frac{e^z + 1}{e^z - 1} = \frac{z}{2} + \frac{z}{e^z - 1}. \tag{6}$$

Thus all we need is a power series expansion of the function $\frac{z}{e^z - 1}$; note that this function is defined and continuous on all of $\mathbb{R}$ (for $z = 0$ use the power series of the exponential function, or alternatively de l'Hospital's rule, which yields the value 1). We write

$$\frac{z}{e^z - 1} =: \sum_{n \geq 0} B_n \frac{z^n}{n!}. \tag{7}$$

The coefficients B_n are known as the *Bernoulli numbers*. The left-hand side of (6) is an *even* function (that is, $f(z) = f(-z)$), and thus we see that $B_n = 0$ for odd $n \geq 3$, while $B_1 = -\frac{1}{2}$ corresponds to the term of $\frac{z}{2}$ in (6).

From

$$\Big(\sum_{n\geq 0} B_n \frac{z^n}{n!}\Big)(e^z - 1) \;=\; \Big(\sum_{n\geq 0} B_n \frac{z^n}{n!}\Big)\Big(\sum_{n\geq 1} \frac{z^n}{n!}\Big) \;=\; z$$

we obtain by comparing coefficients for z^n:

$$\sum_{k=0}^{n-1} \frac{B_k}{k!(n-k)!} \;=\; \begin{cases} 1 & \text{for } n = 1, \\ 0 & \text{for } n \neq 1. \end{cases} \tag{8}$$

We may compute the Bernoulli numbers recursively from (8). The value $n = 1$ gives $B_0 = 1$, $n = 2$ yields $\frac{B_0}{2} + B_1 = 0$, that is $B_1 = -\frac{1}{2}$, and so on.

n	0	1	2	3	4	5	6	7	8
B_n	1	$-\frac{1}{2}$	$\frac{1}{6}$	0	$-\frac{1}{30}$	0	$\frac{1}{42}$	0	$-\frac{1}{30}$

The first few Bernoulli numbers

Now we are almost done: The combination of (6) and (7) yields

$$y \cot y \;=\; \sum_{k=0}^{\infty} B_{2k} \frac{(2iy)^{2k}}{(2k)!} \;=\; \sum_{k=0}^{\infty} \frac{(-1)^k 2^{2k} B_{2k}}{(2k)!} y^{2k},$$

and out comes, with (5), Euler's formula for $\zeta(2k)$:

$$\sum_{n=1}^{\infty} \frac{1}{n^{2k}} \;=\; \frac{(-1)^{k-1} 2^{2k-1} B_{2k}}{(2k)!} \pi^{2k} \qquad (k \in \mathbb{N}). \tag{9}$$

Looking at our table of the Bernoulli numbers, we thus obtain once again the sum $\sum \frac{1}{n^2} = \frac{\pi^2}{6}$ from Chapter 6, and further

$$\sum_{n=1}^{\infty} \frac{1}{n^4} = \frac{\pi^4}{90}, \qquad \sum_{n=1}^{\infty} \frac{1}{n^6} = \frac{\pi^6}{945}, \qquad \sum_{n=1}^{\infty} \frac{1}{n^8} = \frac{\pi^8}{9450},$$

$$\sum_{n=1}^{\infty} \frac{1}{n^{10}} = \frac{\pi^{10}}{93555}, \qquad \sum_{n=1}^{\infty} \frac{1}{n^{12}} = \frac{691\,\pi^{12}}{638512875}, \qquad \ldots$$

The Bernoulli number $B_{10} = \frac{5}{66}$ that gets us $\zeta(10)$ looks innocuous enough, but the next value $B_{12} = -\frac{691}{2730}$, needed for $\zeta(12)$, contains the large prime factor 691 in the numerator. Euler had first computed some values $\zeta(2k)$ without noticing the connection to the Bernoulli numbers. Only the appearance of the strange prime 691 put him on the right track.

Incidentally, since $\zeta(2k)$ converges to 1 for $k \longrightarrow \infty$, equation (9) tells us that the numbers $|B_{2k}|$ grow very fast — something that is not clear from the first few values.

In contrast to all this, one knows very little about the values of the Riemann zeta function at the odd integers $k \geq 3$; see page 41.

IN DEFINIEND. SUMMIS SERIER. INFINIT. 131

lem. Quo autem valor harum ſummarum clarius perſpiciatur, plures hujuſmodi Serierum ſummas commodiori modo expreſſas hic adjiciam.

$1+\frac{1}{2^2}+\frac{1}{3^2}+\frac{1}{4^2}+\frac{1}{5^2}+\&c. = \frac{2^0}{1.2.3}\cdot\frac{1}{1}\pi^2$

$1+\frac{1}{2^4}+\frac{1}{3^4}+\frac{1}{4^4}+\frac{1}{5^4}+\&c. = \frac{2^2}{1.2.3.4.5}\cdot\frac{1}{3}\pi^4$

$1+\frac{1}{2^6}+\frac{1}{3^6}+\frac{1}{4^6}+\frac{1}{5^6}+\&c. = \frac{2^4}{1.2.3....7}\cdot\frac{1}{3}\pi^6$

$1+\frac{1}{2^8}+\frac{1}{3^8}+\frac{1}{4^8}+\frac{1}{5^8}+\&c. = \frac{2^6}{1.2.3....9}\cdot\frac{3}{5}\pi^8$

$1+\frac{1}{2^{10}}+\frac{1}{3^{10}}+\frac{1}{4^{10}}+\frac{1}{5^{10}}+\&c. = \frac{2^8}{1.2.3.....11}\cdot\frac{5}{3}\pi^{10}$

$1+\frac{1}{2^{12}}+\frac{1}{3^{12}}+\frac{1}{4^{12}}+\frac{1}{5^{12}}+\&c. = \frac{2^{10}}{1.2.3.....13}\cdot\frac{691}{105}\pi^{12}$

$1+\frac{1}{2^{14}}+\frac{1}{3^{14}}+\frac{1}{4^{14}}+\frac{1}{5^{14}}+\&c. = \frac{2^{12}}{1.2.3.....15}\cdot\frac{35}{1}\pi^{14}$

$1+\frac{1}{2^{16}}+\frac{1}{3^{16}}+\frac{1}{4^{16}}+\frac{1}{5^{16}}+\&c. = \frac{2^{14}}{1.2.3.....17}\cdot\frac{3617}{15}\pi^{16}$

$1+\frac{1}{2^{18}}+\frac{1}{3^{18}}+\frac{1}{4^{18}}+\frac{1}{5^{18}}+\&c. = \frac{2^{16}}{1.2.3.....19}\cdot\frac{43867}{21}\pi^{18}$

$1+\frac{1}{2^{20}}+\frac{1}{3^{20}}+\frac{1}{4^{20}}+\frac{1}{5^{20}}+\&c. = \frac{2^{18}}{1.2.3....21}\cdot\frac{1222277}{55}\pi^{20}$

$1+\frac{1}{2^{22}}+\frac{1}{3^{22}}+\frac{1}{4^{22}}+\frac{1}{5^{22}}+\&c. = \frac{2^{20}}{1.2.3...23}\cdot\frac{854513}{3}\pi^{22}$

$1+\frac{1}{2^{24}}+\frac{1}{3^{24}}+\frac{1}{4^{24}}+\frac{1}{5^{24}}+\&c. = \frac{2^{22}}{1.\,2.\,3.........25}\cdot\frac{1181820455}{273}\pi^{24}$

$1+\frac{1}{2^{26}}+\frac{1}{3^{26}}+\frac{1}{4^{26}}+\frac{1}{5^{26}}+\&c. = \frac{2^{24}}{1.\,2.\,3.........27}\cdot\frac{76977927}{1}\pi^{26}.$

Hucuſque iſtos Poteſtatum ipſius π Exponentes artificio alibi exponendo continuare licuit, quod ideo hic adjunxi, quod

R 2 Seriei

Page 131 of Euler's 1748 "Introductio in Analysin Infinitorum"

References

[1] S. BOCHNER: *Book review of "Gesammelte Schriften" by Gustav Herglotz,* Bulletin Amer. Math. Soc. **1** (1979), 1020-1022.

[2] J. ELSTRODT: *Partialbruchzerlegung des Kotangens, Herglotz-Trick und die Weierstraßsche stetige, nirgends differenzierbare Funktion,* Math. Semesterberichte **45** (1998), 207-220.

[3] L. EULER: *Introductio in Analysin Infinitorum,* Tomus Primus, Lausanne 1748; Opera Omnia, Ser. 1, Vol. 8. In English: *Introduction to Analysis of the Infinite,* Book I (translated by J. D. Blanton), Springer-Verlag, New York 1988.

[4] L. EULER: *Institutiones calculi differentialis cum ejus usu in analysi finitorum ac doctrina serierum,* Petersburg 1755; Opera Omnia, Ser. 1, Vol. 10.

Buffon's needle problem

Chapter 21

A French nobleman, Georges Louis Leclerc, Comte de Buffon, posed the following problem in 1777:

> *Suppose that you drop a short needle on ruled paper — what is then the probability that the needle comes to lie in a position where it crosses one of the lines?*

Le Comte de Buffon

The probability depends on the distance d between the lines of the ruled paper, and it depends on the length ℓ of the needle that we drop — or rather it depends only on the ratio $\frac{\ell}{d}$. A *short* needle for our purpose is one of length $\ell \leq d$. In other words, a short needle is one that cannot cross two lines at the same time (and will come to touch two lines only with probability zero). The answer to Buffon's problem may come as a surprise: It involves the number π.

Theorem ("Buffon's needle problem")
If a short needle, of length ℓ, is dropped on paper that is ruled with equally spaced lines of distance $d \geq \ell$, then the probability that the needle comes to lie in a position where it crosses one of the lines is exactly

$$p = \frac{2}{\pi}\frac{\ell}{d}.$$

The result means that from an experiment one can get approximate values for π: If you drop a needle N times, and get a positive answer (an intersection) in P cases, then $\frac{P}{N}$ should be approximately $\frac{2}{\pi}\frac{\ell}{d}$, that is, π should be approximated by $\frac{2\ell N}{dP}$. The most extensive (and exhaustive) test was perhaps done by Lazzarini in 1901, who allegedly even built a machine in order to drop a stick 3408 times (with $\frac{\ell}{d} = \frac{5}{6}$). He found that it came to cross a line 1808 times, which yields the approximation $\pi \approx 2 \cdot \frac{5}{6}\frac{3408}{1808} = 3.1415929....$, which is correct to six digits of π, and much too good to be true! (The values that Lazzarini chose lead directly to the well-known approximation $\pi \approx \frac{355}{113}$; see page 31. This explains the more than suspicious choices of 3408 and $\frac{5}{6}$, where $\frac{5}{6}\,3408$ is a multiple of 355. See [5] for a discussion of Lazzarini's hoax.)

The needle problem can be solved by evaluating an integral. We will do that below, and by this method we will also solve the problem for a long needle. But the Book Proof, presented by E. Barbier in 1860, needs no integrals. It just drops a different needle . . .

If you drop *any* needle, short or long, then the expected number of crossings will be

$$E \;=\; p_1 + 2p_2 + 3p_3 + \ldots,$$

where p_1 is the probability that the needle will come to lie with exactly one crossing, p_2 is the probability that we get exactly two crossings, p_3 is the probability for three crossings, etc. The probability that we get at least one crossing, which Buffon's problem asks for, is thus

$$p \;=\; p_1 + p_2 + p_3 + \ldots$$

(Events where the needle comes to lie exactly on a line, or with an endpoint on one of the lines, have probability zero — so they can be ignored throughout our discussion.)

On the other hand, if the needle is *short* then the probability of more than one crossing is zero, $p_2 = p_3 = \ldots = 0$, and thus we get $E = p$: The probability that we are looking for is just the expected number of crossings. This reformulation is extremely useful, because now we can use linearity of expectation (cf. page 84). Indeed, let us write $E(\ell)$ for the expected number of crossings that will be produced by dropping a straight needle of length ℓ. If this length is $\ell = x + y$, and we consider the "front part" of length x and the "back part" of length y of the needle separately, then we get

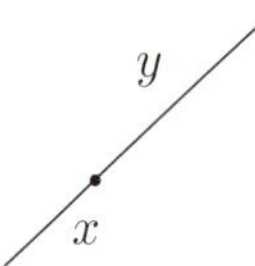

$$E(x+y) \;=\; E(x) + E(y),$$

since the crossings produced are always just those produced by the front part, plus those of the back part.

By induction on n this "functional equation" implies that $E(nx) = nE(x)$ for all $n \in \mathbb{N}$, and then that $mE(\frac{n}{m}x) = E(m\frac{n}{m}x) = E(nx) = nE(x)$, so that $E(rx) = rE(x)$ holds for all *rational* $r \in \mathbb{Q}$. Furthermore, $E(x)$ is clearly monotone in $x \geq 0$, from which we get that $E(x) = cx$ for all $x \geq 0$, where $c = E(1)$ is some constant.

But what is the constant?

For that we use needles of different shape. Indeed, let's drop a "polygonal" needle of total length ℓ, which consists of straight pieces. Then the number of crossings it produces is (with probability 1) the sum of the numbers of crossings produced by its straight pieces. Hence, the expected number of crossings is again

$$E \;=\; c\ell,$$

by linearity of expectation. (For that it is not even important whether the straight pieces are joined together in a rigid or in a flexible way!)

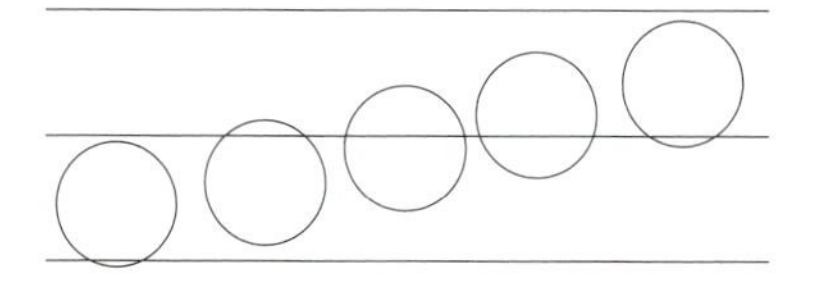

The key to Barbier's solution of Buffon's needle problem is to consider a needle that is a perfect circle C of diameter d, which has length $x = d\pi$. Such a needle, if dropped onto ruled paper, produces exactly two intersections, always!

The circle can be approximated by polygons. Just imagine that together with the circular needle C we are dropping an inscribed polygon P_n, as well as a circumscribed polygon P^n. Every line that intersects P_n will also intersect C, and if a line intersects C then it also hits P^n. Thus the expected numbers of intersections satisfy

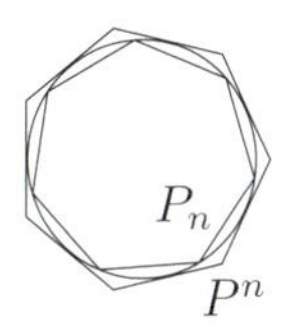

$$E(P_n) \;\leq\; E(C) \;\leq\; E(P^n).$$

Now both P_n and P^n are polygons, so the number of crossings that we may expect is "c times length" for both of them, while for C it is 2, whence

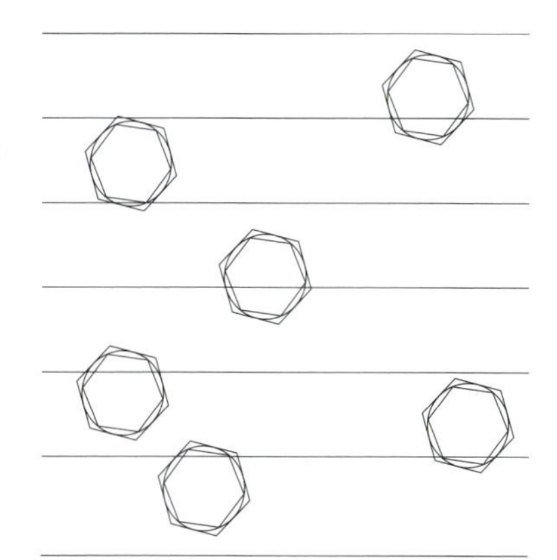

$$c\,\ell(P_n) \;\leq\; 2 \;\leq\; c\,\ell(P^n). \tag{1}$$

Both P_n and P^n approximate C for $n \longrightarrow \infty$. In particular,

$$\lim_{n\to\infty} \ell(P_n) \;=\; d\pi \;=\; \lim_{n\to\infty} \ell(P^n),$$

and thus for $n \longrightarrow \infty$ we infer from (1) that

$$c\,d\pi \;\leq\; 2 \;\leq c\,d\pi,$$

which gives $c = \frac{2}{\pi}\frac{1}{d}$. □

But we *could* also have done it by calculus! The trick to obtain an "easy" integral is to first consider the slope of the needle; let's say it drops to lie with an angle of α away from horizontal, where α will be in the range $0 \leq \alpha \leq \frac{\pi}{2}$. (We will ignore the case where the needle comes to lie with negative slope, since that case is symmetric to the case of positive slope, and produces the same probability.) A needle that lies with angle α has height $\ell \sin\alpha$, and the probability that such a needle crosses one of the horizontal lines of distance d is $\frac{\ell \sin\alpha}{d}$. Thus we get the probability by averaging over the possible angles α, as

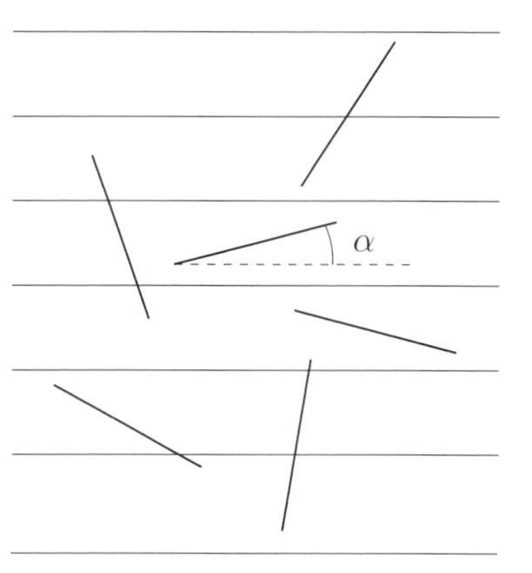

$$p \;=\; \frac{2}{\pi}\int_0^{\pi/2} \frac{\ell \sin\alpha}{d}\,d\alpha \;=\; \frac{2}{\pi}\frac{\ell}{d}\Big[-\cos\alpha\Big]_0^{\pi/2} \;=\; \frac{2}{\pi}\frac{\ell}{d}.$$

For a long needle, we get the same probability $\frac{\ell \sin\alpha}{d}$ as long as $\ell \sin\alpha \leq d$, that is, in the range $0 \leq \alpha \leq \arcsin\frac{d}{\ell}$. However, for larger angles α the needle *must* cross a line, so the probability is 1. Hence we compute

$$\begin{aligned} p \;&=\; \frac{2}{\pi}\Big(\int_0^{\arcsin(d/\ell)} \frac{\ell \sin\alpha}{d}\,d\alpha \;+\; \int_{\arcsin(d/\ell)}^{\pi/2} 1\,d\alpha\Big) \\ &=\; \frac{2}{\pi}\Big(\frac{\ell}{d}\Big[-\cos\alpha\Big]_0^{\arcsin(d/\ell)} + \Big(\frac{\pi}{2} - \arcsin\frac{d}{\ell}\Big)\Big) \\ &=\; 1 + \frac{2}{\pi}\Big(\frac{\ell}{d}\Big(1 - \sqrt{1 - \frac{d^2}{\ell^2}}\Big) - \arcsin\frac{d}{l}\Big) \end{aligned}$$

for $\ell \geq d$.

So the answer isn't that pretty for a longer needle, but it provides us with a nice exercise: Show ("just for safety") that the formula yields $\frac{2}{\pi}$ for $\ell = d$, that it is strictly increasing in ℓ, and that it tends to 1 for $\ell \longrightarrow \infty$.

References

[1] E. BARBIER: *Note sur le problème de l'aiguille et le jeu du joint couvert,* J. Mathématiques Pures et Appliquées (2) **5** (1860), 273-286.

[2] L. BERGGREN, J. BORWEIN & P. BORWEIN, EDS.: *Pi: A Source Book,* Springer-Verlag, New York 1997.

[3] G. L. LECLERC, COMTE DE BUFFON: *Essai d'arithmétique morale,* Appendix to "Histoire naturelle générale et particulière," Vol. 4, 1777.

[4] D. A. KLAIN & G.-C. ROTA: *Introduction to Geometric Probability,* "Lezioni Lincee," Cambridge University Press 1997.

[5] T. H. O'BEIRNE: *Puzzles and Paradoxes,* Oxford University Press, London 1965.

"Got a problem?"

Combinatorics

"A melancholic Latin square"

Pigeon-hole and double counting

Chapter 22

Some mathematical principles, such as the two in the title of this chapter, are so obvious that you might think they would only produce equally obvious results. To convince you that "It ain't necessarily so" we illustrate them with examples that were suggested by Paul Erdős to be included in The Book. We will encounter instances of them also in later chapters.

> **Pigeon-hole principle.**
> *If n objects are placed in r boxes, where $r < n$, then at least one of the boxes contains more than one object.*

"The pigeon-holes from a bird's perspective"

Well, this is indeed obvious, there is nothing to prove. In the language of mappings our principle reads as follows: Let N and R be two finite sets with

$$|N| = n > r = |R|,$$

and let $f : N \longrightarrow R$ be a mapping. Then there exists some $a \in R$ with $|f^{-1}(a)| \geq 2$. We may even state a stronger inequality: There exists some $a \in R$ with

$$|f^{-1}(a)| \;\geq\; \left\lceil \frac{n}{r} \right\rceil. \qquad (1)$$

In fact, otherwise we would have $|f^{-1}(a)| < \frac{n}{r}$ for all a, and hence $n = \sum_{a \in R} |f^{-1}(a)| < r\,\frac{n}{r} = n$, which cannot be.

1. Numbers

> ***Claim.*** *Consider the numbers $1, 2, 3, \ldots, 2n$, and take any $n + 1$ of them. Then there are two among these $n + 1$ numbers which are relatively prime.*

This is again obvious. There must be two numbers which are only 1 apart, and hence relatively prime.

But let us now turn the condition around.

> ***Claim.*** *Suppose again $A \subseteq \{1, 2, \ldots, 2n\}$ with $|A| = n + 1$. Then there are always two numbers in A such that one divides the other.*

This is not so clear. As Erdős told us, he put this question to young Lajos Pośa during dinner, and when the meal was over, Lajos had the answer. It has remained one of Erdős' favorite "initiation" questions to mathematics. The (affirmative) solution is provided by the pigeon-hole principle. Write every number $a \in A$ in the form $a = 2^k m$, where m is an odd number between 1 and $2n - 1$. Since there are $n + 1$ numbers in A, but only n different odd parts, there must be two numbers in A with the *same* odd part. Hence one is a multiple of the other. $\square$

Both results are no longer true if one replaces $n{+}1$ by n: For this consider the sets $\{2, 4, 6, \ldots, 2n\}$, respectively $\{n{+}1, n{+}2, \ldots, 2n\}$.

2. Sequences

Here is another one of Erdős' favorites, contained in a paper of Erdős and Szekeres on Ramsey problems.

Claim. *In any sequence $a_1, a_2, \ldots, a_{mn+1}$ of $mn+1$ distinct real numbers, there exists an increasing subsequence*

$$a_{i_1} < a_{i_2} < \ldots < a_{i_{m+1}} \qquad (i_1 < i_2 < \ldots < i_{m+1})$$

of length $m + 1$, or a decreasing subsequence

$$a_{j_1} > a_{j_2} > \ldots > a_{j_{n+1}} \qquad (j_1 < j_2 < \ldots < j_{n+1})$$

of length $n + 1$, or both.

This time the application of the pigeon-hole principle is not immediate. Associate to each a_i the number t_i which is the length of a *longest increasing* subsequence starting at a_i. If $t_i \geq m + 1$ for some i, then we have an increasing subsequence of length $m + 1$. Suppose then that $t_i \leq m$ for all i. The function $f : a_i \longmapsto t_i$ mapping $\{a_1, \ldots, a_{mn+1}\}$ to $\{1, \ldots, m\}$ tells us by (1) that there is some $s \in \{1, \ldots, m\}$ such that $f(a_i) = s$ for $\frac{mn}{m} + 1 = n + 1$ numbers a_i. Let $a_{j_1}, a_{j_2}, \ldots, a_{j_{n+1}}$ $(j_1 < \ldots < j_{n+1})$ be these numbers. Now look at two consecutive numbers $a_{j_i}, a_{j_{i+1}}$. If $a_{j_i} < a_{j_{i+1}}$, then we would obtain an increasing subsequence of length s starting at $a_{j_{i+1}}$, and consequently an increasing subsequence of length $s + 1$ starting at a_{j_i}, which cannot be since $f(a_{j_i}) = s$. We thus obtain a decreasing subsequence $a_{j_1} > a_{j_2} > \ldots > a_{j_{n+1}}$ of length $n + 1$. $\square$

The reader may have fun in proving that for mn numbers the statement remains no longer true in general.

This simple-sounding result on monotone subsequences has a highly non-obvious consequence on the *dimension of graphs*. We don't need here the notion of dimension for general graphs, but only for complete graphs K_n. It can be phrased in the following way. Let $N = \{1, \ldots, n\}$, $n \geq 3$, and consider m permutations $\pi_1, \ldots, \pi_m$ of N. We say that the permutations π_i *represent* K_n if to every three distinct numbers i, j, k there exists a permutation π in which k comes *after* both i and j. The dimension of K_n is then the smallest m for which a representation $\pi_1, \ldots, \pi_m$ exists.

As an example we have $\dim(K_3) = 3$ since any one of the three numbers must come last, as in $\pi_1 = (1, 2, 3)$, $\pi_2 = (2, 3, 1)$, $\pi_3 = (3, 1, 2)$. What about K_4? Note first $\dim(K_n) \le \dim(K_{n+1})$: just delete $n+1$ in a representation of K_{n+1}. So, $\dim(K_4) \ge 3$, and, in fact, $\dim(K_4) = 3$, by taking

$$\pi_1 = (1,2,3,4), \quad \pi_2 = (2,4,3,1), \quad \pi_3 = (1,4,3,2).$$

It is not quite so easy to prove $\dim(K_5) = 4$, but then, surprisingly, the dimension stays at 4 up to $n = 12$, while $\dim(K_{13}) = 5$. So $\dim(K_n)$ seems to be a pretty wild function. Well, it is not! With n going to infinity, $\dim(K_n)$ is, in fact, a very well-behaved function — and the key for finding a lower bound is the pigeon-hole principle. We claim

$$\dim(K_n) \ \ge\ \log_2 \log_2 n. \tag{2}$$

π_1: 1 2 3 5 6 7 8 9 10 11 12 4
π_2: 2 3 4 8 7 6 5 12 11 10 9 1
π_3: 3 4 1 11 12 9 10 6 5 8 7 2
π_4: 4 1 2 10 9 12 11 7 8 5 6 3

These four permutations represent K_{12}

Since, as we have seen, $\dim(K_n)$ is a monotone function in n, it suffices to verify (2) for $n = 2^{2^p}+1$, that is, we have to show that

$$\dim(K_n) \ \ge\ p+1 \qquad \text{for} \quad n = 2^{2^p}+1.$$

Suppose, on the contrary, $\dim(K_n) \le p$, and let $\pi_1, \dots, \pi_p$ be representing permutations of $N = \{1,2,\dots,2^{2^p}+1\}$. Now we use our result on monotone subsequences p times. In π_1 there exists a monotone subsequence A_1 of length $2^{2^{p-1}}+1$ (it does not matter whether increasing or decreasing). Look at this set A_1 in π_2. Using our result again, we find a monotone subsequence A_2 of A_1 in π_2 of length $2^{2^{p-2}}+1$, and A_2 is, of course, also monotone in π_1. Continuing, we eventually find a subsequence A_p of size $2^{2^0}+1 = 3$ which is monotone in *all* permutations π_i. Let $A_p = (a,b,c)$, then either $a<b<c$ or $a>b>c$ in *all* π_i. But this cannot be, since there must be a permutation where b comes after a and c. $\square$

The right asymptotic growth was provided by Joel Spencer (upper bound) and by Erdős, Szemerédi and Trotter (lower bound):

$$\dim(K_n) \ = \ \log_2 \log_2 n + (\frac{1}{2} + o(1)) \log_2 \log_2 \log_2 n.$$

But this is not the whole story: Very recently, Morris and Hoşten found a method which, in principle, establishes the *precise* value of $\dim(K_n)$. Using their result and a computer one can obtain the values given in the margin. This is truly astounding! Just consider how many permutations of size 1422564 there are. How does one decide whether 7 or 8 of them are required to represent $K_{1422564}$?

$$\begin{aligned}
\dim(K_n) \le 4 &\iff n \le 12\\
\dim(K_n) \le 5 &\iff n \le 81\\
\dim(K_n) \le 6 &\iff n \le 2646\\
\dim(K_n) \le 7 &\iff n \le 1422564
\end{aligned}$$

3. Sums

Paul Erdős attributes the following nice application of the pigeon-hole principle to Andrew Vázsonyi and Marta Sved:

> ***Claim.*** *Suppose we are given n integers $a_1, \dots, a_n$, which need not be distinct. Then there is always a set of consecutive numbers $a_{k+1}, a_{k+2}, \dots, a_\ell$ whose sum $\sum_{i=k+1}^{\ell} a_i$ is a multiple of n.*

For the proof we set $N = \{0, 1, \ldots, n\}$ and $R = \{0, 1, \ldots, n-1\}$. Consider the map $f : N \to R$, where $f(m)$ is the remainder of $a_1 + \ldots + a_m$ upon division by n. Since $|N| = n+1 > n = |R|$, it follows that there are two sums $a_1 + \ldots + a_k$, $a_1 + \ldots + a_\ell$ $(k < \ell)$ with the *same* remainder, where the first sum may be the empty sum denoted by 0. It follows that

$$\sum_{i=k+1}^{\ell} a_i = \sum_{i=1}^{\ell} a_i - \sum_{i=1}^{k} a_i$$

has remainder 0 — end of proof. □

Let us turn to the second principle: counting in two ways. By this we mean the following.

Double counting.
Suppose that we are given two finite sets R and C and a subset $S \subseteq R \times C$. Whenever $(p, q) \in S$, then we say p and q are incident. If r_p denotes the number of elements that are incident to $p \in R$, and c_q denotes the number of elements that are incident to $q \in C$, then

$$\sum_{p \in R} r_p = |S| = \sum_{q \in C} c_q. \tag{3}$$

Again, there is nothing to prove. The first sum classifies the pairs in S according to the first entry, while the second sum classifies the same pairs according to the second entry.

There is a useful way to picture the set S. Consider the matrix $A = (a_{pq})$, the *incidence matrix* of S, where the rows and columns of A are indexed by the elements of R and C, respectively, with

$$a_{pq} = \begin{cases} 1 & \text{if } (p, q) \in S \\ 0 & \text{if } (p, q) \notin S. \end{cases}$$

With this set-up, r_p is the sum of the p-th row of A and c_q is the sum of the q-th column. Hence the first sum in (3) adds the entries of A (that is, counts the elements in S) by rows, and the second sum by columns.

The following example should make this correspondence clear. Let $R = C = \{1, 2, \ldots, 8\}$, and set $S = \{(i, j) : i \text{ divides } j\}$. We then obtain the matrix in the margin, which only displays the 1's.

R \ C	1	2	3	4	5	6	7	8
1	1	1	1	1	1	1	1	1
2		1		1		1		1
3			1			1		
4				1				1
5					1			
6						1		
7							1	
8								1

4. Numbers again

Look at the table on the left. The number of 1's in column j is precisely the number of divisors of j; let us denote this number by $t(j)$. Let us ask how large this number $t(j)$ is on the *average* when j ranges from 1 to n. Thus, we ask for the quantity

$$\bar{t}(n) = \frac{1}{n}\sum_{j=1}^{n} t(j).$$

n	1	2	3	4	5	6	7	8
$\bar{t}(n)$	1	$\frac{3}{2}$	$\frac{5}{3}$	2	2	$\frac{7}{3}$	$\frac{16}{7}$	$\frac{5}{2}$

The first few values of $\bar{t}(n)$

How large is $\bar{t}(n)$ for arbitrary n? At first glance, this seems hopeless. For prime numbers p we have $t(p) = 2$, while for 2^k we obtain a large number $t(2^k) = k + 1$. So, $t(n)$ is a wildly jumping function, and we surmise that the same is true for $\bar{t}(n)$. Wrong guess, the opposite is true! Counting in two ways provides an unexpected and simple answer.

Consider the matrix A (as above) for the integers 1 up to n. Counting by columns we get $\sum_{j=1}^{n} t(j)$. How many 1's are in row i? Easy enough, the 1's correspond to the multiples of i: $1i, 2i, \ldots$, and the last multiple not exceeding n is $\lfloor \frac{n}{i} \rfloor i$. Hence we obtain

$$\bar{t}(n) \;=\; \frac{1}{n}\sum_{j=1}^{n} t(j) \;=\; \frac{1}{n}\sum_{i=1}^{n} \left\lfloor \frac{n}{i} \right\rfloor \;\leq\; \frac{1}{n}\sum_{i=1}^{n} \frac{n}{i} \;=\; \sum_{i=1}^{n} \frac{1}{i},$$

where the error in each summand, when passing from $\lfloor \frac{n}{i} \rfloor$ to $\frac{n}{i}$, is less than 1. Now the last sum is the n-th harmonic number H_n, so we obtain $H_n - 1 < \bar{t}(n) < H_n$, and together with the estimates of H_n on page 11 this gives

$$\log n - 1 \;<\; H_n - 1 - \frac{1}{n} \;<\; \bar{t}(n) \;<\; H_n \;<\; \log n + 1.$$

Thus we have proved the remarkable result that, while $t(n)$ is totally erratic, the average $\bar{t}(n)$ behaves beautifully: It differs from $\log n$ by less than 1.

5. Graphs

Let G be a finite simple graph with vertex set V and edge set E. We have defined in Chapter 11 the *degree* $d(v)$ of a vertex v as the number of edges which have v as an end-vertex. In the example of the figure, the vertices $1, 2, \ldots, 7$ have degrees $3, 2, 4, 3, 3, 2, 3$, respectively.

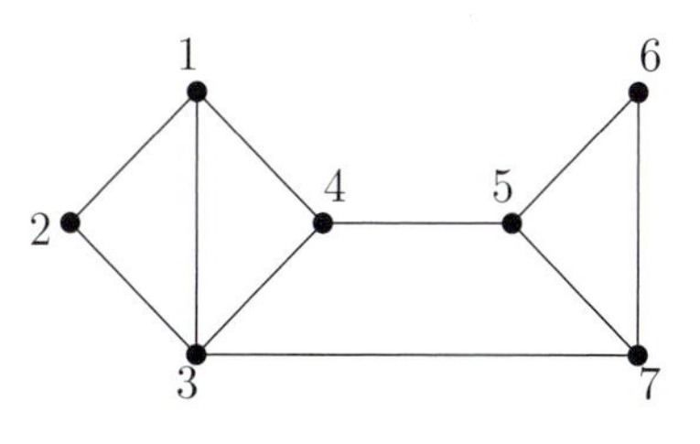

Almost every book in graph theory starts with the following result (that we have already encountered in Chapters 11 and 17):

$$\sum_{v \in V} d(v) \;=\; 2|E|. \tag{4}$$

For the proof consider $S \subseteq V \times E$, where S is the set of pairs (v, e) such that $v \in V$ is an end-vertex of $e \in E$. Counting S in two ways gives on the one hand $\sum_{v \in V} d(v)$, since every vertex contributes $d(v)$ to the count, and on the other hand $2|E|$, since every edge has two ends. □

As simple as the result (4) appears, it has many important consequences, some of which will be discussed as we go along. We want to single out in this section the following beautiful application to an *extremal problem* on graphs. Here is the problem:

> *Suppose $G = (V, E)$ has n vertices and contains no cycle of length 4 (denoted by C_4), that is, no subgraph. How many edges can G have at most?*

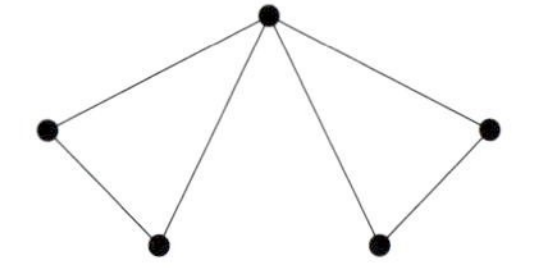

As an example, the graph in the margin on 5 vertices contains no 4-cycle and has 6 edges. The reader may easily show that on 5 vertices the maximal number of edges is 6, and that this graph is indeed the only graph on 5 vertices with 6 edges that has no 4-cycle.

Let us tackle the general problem. Let G be a graph on n vertices without a 4-cycle. As above we denote by $d(u)$ the degree of u. Now we count the following set S in two ways: S is the set of pairs $(u, \{v, w\})$ where u is adjacent to v and to w, with $v \neq w$. In other words, we count all occurrences of

Summing over u, we find $|S| = \sum_{u \in V} \binom{d(u)}{2}$. On the other hand, every pair $\{v, w\}$ has at most one common neighbor (by the C_4-condition). Hence $|S| \leq \binom{n}{2}$, and we conclude

$$\sum_{u \in V} \binom{d(u)}{2} \leq \binom{n}{2}$$

or

$$\sum_{u \in V} d(u)^2 \leq n(n-1) + \sum_{u \in V} d(u). \tag{5}$$

Next (and this is quite typical for this sort of extremal problems) we apply the Cauchy-Schwarz inequality to the vectors $(d(u_1), \ldots, d(u_n))$ and $(1, 1, \ldots, 1)$, obtaining

$$\Big(\sum_{u \in V} d(u)\Big)^2 \leq n \sum_{u \in V} d(u)^2,$$

and hence by (5)

$$\Big(\sum_{u \in V} d(u)\Big)^2 \leq n^2(n-1) + n \sum_{u \in V} d(u).$$

Invoking (4) we find

$$4\,|E|^2 \leq n^2(n-1) + 2n\,|E|$$

or

$$|E|^2 - \frac{n}{2}|E| - \frac{n^2(n-1)}{4} \leq 0.$$

Solving the corresponding quadratic equation we thus obtain the following result of Istvan Reiman.

Theorem. *If the graph G on n vertices contains no 4-cycles, then*

$$|E| \leq \left\lfloor \frac{n}{4}\left(1+\sqrt{4n-3}\right)\right\rfloor. \tag{6}$$

For $n = 5$ this gives $|E| \leq 6$, and the graph above shows that equality can hold.

Counting in two ways has thus produced in an easy way an upper bound on the number of edges. But how good is the bound (6) in general? The following beautiful example [2] [3] [6] shows that it is almost sharp. As is often the case in such problems, finite geometry leads the way.

In presenting the example we assume that the reader is familiar with the finite field $\mathbb{Z}_p$ of integers modulo a prime p (see page 18). Consider the 3-dimensional vector space X over $\mathbb{Z}_p$. We construct from X the following graph G_p. The vertices of G_p are the one-dimensional subspaces $[\boldsymbol{v}] := \text{span}_{\mathbb{Z}_p}\{\boldsymbol{v}\}$, $\boldsymbol{0} \neq \boldsymbol{v} \in X$, and we connect two such subspaces $[\boldsymbol{v}], [\boldsymbol{w}]$ by an edge if

$$\langle \boldsymbol{v}, \boldsymbol{w}\rangle = v_1w_1 + v_2w_2 + v_3w_3 = 0.$$

Note that it does not matter which vector $\neq \boldsymbol{0}$ we take from the subspace. In the language of geometry, the vertices are the *points* of the projective plane over $\mathbb{Z}_p$, and $[\boldsymbol{w}]$ is adjacent to $[\boldsymbol{v}]$ if $\boldsymbol{w}$ lies on the *polar line* of $\boldsymbol{v}$.

As an example, the graph G_2 has no 4-cycle and contains 9 edges, which almost reaches the bound 10 given by (6). We want to show that this is true for any prime p.

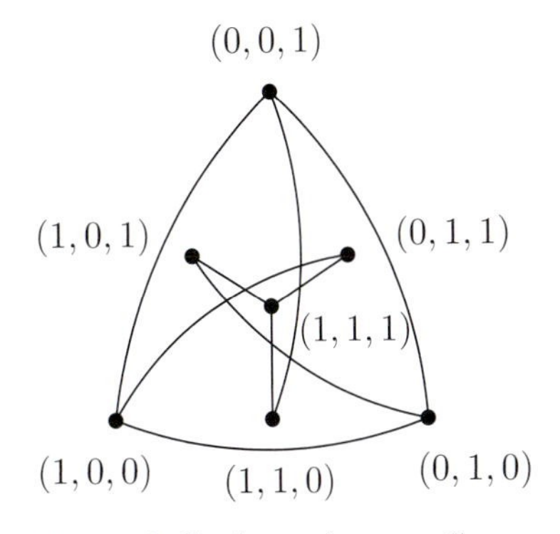

The graph G_2: its vertices are all seven nonzero triples (x, y, z).

Let us first prove that G_p satisfies the C_4-condition. If $[\boldsymbol{u}]$ is a common neighbor of $[\boldsymbol{v}]$ and $[\boldsymbol{w}]$, then $\boldsymbol{u}$ is a solution of the linear equations

$$\begin{aligned} v_1x + v_2y + v_3z &= 0 \\ w_1x + w_2y + w_3z &= 0. \end{aligned}$$

Since $\boldsymbol{v}$ and $\boldsymbol{w}$ are linearly independent, we infer that the solution space has dimension 1, and hence that the common neighbor $[\boldsymbol{u}]$ is unique.

Next, we ask how many vertices G_p has. It's double counting again. The space X contains $p^3 - 1$ vectors $\neq \boldsymbol{0}$. Since every one-dimensional subspace contains $p-1$ vectors $\neq \boldsymbol{0}$, we infer that X has $\frac{p^3-1}{p-1} = p^2+p+1$ one-dimensional subspaces, that is, G_p has $n = p^2+p+1$ vertices. Similarly, any two-dimensional subspace contains p^2-1 vectors $\neq \boldsymbol{0}$, and hence $\frac{p^2-1}{p-1} = p+1$ one-dimensional subspaces.

It remains to determine the number of edges in G_p, or, what is the same by (4), the degrees. By the construction of G_p, the vertices adjacent to $[\boldsymbol{u}]$ are the solutions of the equation

$$u_1x + u_2y + u_3z = 0. \tag{7}$$

The solution space of (7) is a two-dimensional subspace, and hence there are $p+1$ vertices adjacent to $[\boldsymbol{u}]$. But beware, it may happen that $\boldsymbol{u}$ itself is a solution of (7). In this case there are only p vertices adjacent to $[\boldsymbol{u}]$.

In summary, we obtain the following result: If $\boldsymbol{u}$ lies on the *conic* given by $x^2 + y^2 + z^2 = 0$, then $d([\boldsymbol{u}]) = p$, and, if not, then $d([\boldsymbol{u}]) = p + 1$. So it remains to find the number of one-dimensional subspaces on the conic

$$x^2 + y^2 + z^2 = 0.$$

Let us anticipate the result which we shall prove in a moment.

Claim. *There are precisely p^2 solutions (x, y, z) of the equation $x^2+y^2+z^2 = 0$, and hence (excepting the zero solution) precisely $\frac{p^2-1}{p-1} = p+1$ vertices in G_p of degree p.*

With this, we complete our analysis of G_p. There are $p + 1$ vertices of degree p, hence $(p^2 + p + 1) - (p + 1) = p^2$ vertices of degree $p + 1$. Using (4), we obtain

$$\begin{aligned} |E| &= \frac{(p+1)p}{2} + \frac{p^2(p+1)}{2} = \frac{(p+1)^2 p}{2} \\ &= \frac{(p+1)p}{4}(1 + (2p+1)) = \frac{p^2+p}{4}(1 + \sqrt{4p^2+4p+1}). \end{aligned}$$

Setting $n = p^2 + p + 1$, the last equation reads

$$|E| = \frac{n-1}{4}(1 + \sqrt{4n-3}),$$

and we see that this almost agrees with (6).

Now to the proof of the claim. The following argument is a beautiful application of linear algebra involving symmetric matrices and their eigenvalues. We will encounter the same method in Chapter 34, which is no coincidence: both proofs are from the same paper by Erdős, Rényi and Sós.

We represent the one-dimensional subspaces of X as before by vectors $\boldsymbol{v}_1, \boldsymbol{v}_2, \ldots, \boldsymbol{v}_{p^2+p+1}$, any two of which are linearly independent. Similarly, we may represent the two-dimensional subspaces by the *same* set of vectors, where the subspace corresponding to $\boldsymbol{u} = (u_1, u_2, u_3)$ is the set of solutions of the equation $u_1 x + u_2 y + u_3 z = 0$ as in (7). (Of course, this is just the duality principle of linear algebra.) Hence, by (7), a one-dimensional subspace, represented by $\boldsymbol{v}_i$, is contained in the two-dimensional subspace, represented by $\boldsymbol{v}_j$, if and only if $\langle \boldsymbol{v}_i, \boldsymbol{v}_j \rangle = 0$.

$$A = \begin{pmatrix} 0&1&1&1&0&0&0 \\ 1&0&1&0&1&0&0 \\ 1&1&0&0&0&1&0 \\ 1&0&0&1&0&0&1 \\ 0&1&0&0&1&0&1 \\ 0&0&1&0&0&1&1 \\ 0&0&0&1&1&1&0 \end{pmatrix}$$

The matrix for G_2

Consider now the matrix $A = (a_{ij})$ of size $(p^2+p+1)\times(p^2+p+1)$, defined as follows: The rows and columns of A correspond to $\boldsymbol{v}_1, \ldots, \boldsymbol{v}_{p^2+p+1}$ (we use the same numbering for rows and columns) with

$$a_{ij} := \begin{cases} 1 & \text{if } \langle \boldsymbol{v}_i, \boldsymbol{v}_j \rangle = 0, \\ 0 & \text{otherwise.} \end{cases}$$

A is thus a real symmetric matrix, and we have $a_{ii} = 1$ if $\langle \boldsymbol{v}_i, \boldsymbol{v}_i \rangle = 0$, that is, precisely when $\boldsymbol{v}_i$ lies on the conic $x^2 + y^2 + z^2 = 0$. Thus, all that remains to show is that

$$\operatorname{trace} A = p + 1.$$

From linear algebra we know that the trace equals the sum of the eigenvalues. And here comes the trick: While A looks complicated, the matrix A^2 is easy to analyze. We note two facts:

- Any row of A contains precisely $p+1$ 1's. This implies that $p+1$ is an eigenvalue of A, since $A\mathbf{1} = (p+1)\mathbf{1}$, where $\mathbf{1}$ is the vector consisting of 1's.

- For any two distinct rows $\boldsymbol{v}_i, \boldsymbol{v}_j$ there is exactly one column with a 1 in both rows (the column corresponding to the unique subspace spanned by $\boldsymbol{v}_i, \boldsymbol{v}_j$).

Using these facts we find

$$A^2 = \begin{pmatrix} p+1 & 1 & \cdots & 1 \\ 1 & p+1 & & \vdots \\ \vdots & & \ddots & \\ 1 & \cdots & & p+1 \end{pmatrix} = pI + J,$$

where I is the identity matrix and J is the all-ones-matrix. Now, J has the eigenvalue p^2+p+1 (of multiplicity 1) and 0 (of multiplicity p^2+p). Hence A^2 has the eigenvalues $p^2+2p+1 = (p+1)^2$ of multiplicity 1 and p of multiplicity p^2+p. Since A is real and symmetric, hence diagonalizable, we find that A has the eigenvalue $p+1$ or $-(p+1)$ and p^2+p eigenvalues $\pm\sqrt{p}$. From Fact 1 above, the first eigenvalue must be $p+1$. Suppose that $\sqrt{p}$ has multiplicity r, and $-\sqrt{p}$ multiplicity s, then

$$\text{trace } A = (p+1) + r\sqrt{p} - s\sqrt{p}.$$

But now we are home: Since the trace is an integer, we must have $r = s$, so $\text{trace } A = p+1$. □

6. Sperner's Lemma

In 1911, Luitzen Brouwer published his famous fixed point theorem:

> *Every continuous function $f\colon B^n \longrightarrow B^n$ of an n-dimensional ball to itself has a fixed point (a point $\boldsymbol{x} \in B^n$ with $f(\boldsymbol{x}) = \boldsymbol{x}$).*

For dimension 1, that is for an interval, this follows easily from the intermediate value theorem, but for higher dimensions Brouwer's proof needed some sophisticated machinery. It was therefore quite a surprise when in 1928 young Emanuel Sperner (he was 23 at the time) produced a simple combinatorial result from which both Brouwer's fixed point theorem and the invariance of the dimension under continuous bijective maps could be deduced. And what's more, Sperner's ingenious lemma is matched by an equally beautiful proof — it is just double counting.

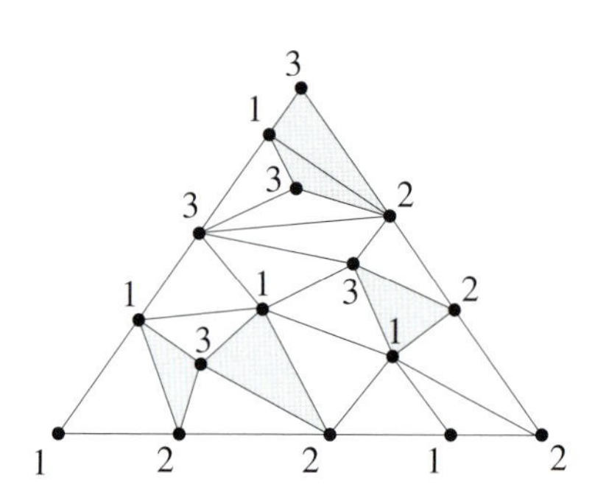

The triangles with three different colors are shaded

We discuss Sperner's lemma, and Brouwer's theorem as a consequence, for the first interesting case, that of dimension $n = 2$. The reader should have no difficulty to extend the proofs to higher dimensions (by induction on the dimension).

Sperner's Lemma.
Suppose that some "big" triangle with vertices V_1, V_2, V_3 is triangulated (that is, decomposed into a finite number of "small" triangles that fit together edge-by-edge).
Assume that the vertices in the triangulation get "colors" from the set $\{1,2,3\}$ such that V_i receives the color i (for each i), and only the colors i and j are used for vertices along the edge from V_i to V_j (for $i \neq j$), while the interior vertices are colored arbitrarily with 1, 2 or 3.
Then in the triangulation there must be a small "tricolored" triangle, which has all three different vertex colors.

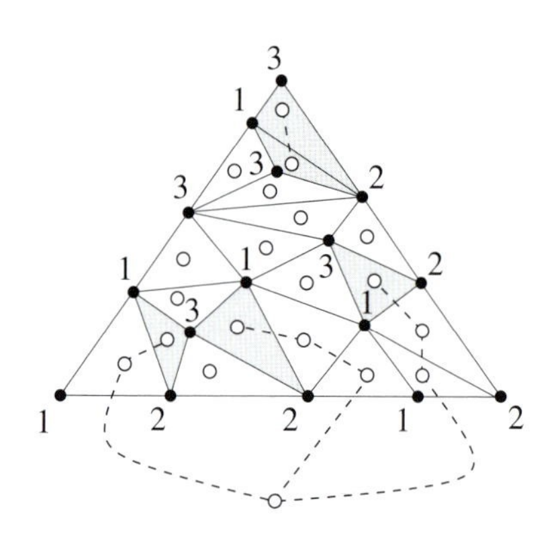

■ **Proof.** We will prove a stronger statement: the number of tricolored triangles is not only nonzero, it is always *odd*.

Consider the dual graph to the triangulation, but don't take all its edges — only those which cross an edge that has endvertices with the (different) colors 1 and 2. Thus we get a "partial dual graph" which has degree 1 at all vertices that correspond to tricolored triangles, degree 2 for all triangles in which the two colors 1 and 2 appear, and degree 0 for triangles that do not have both colors 1 and 2. Thus only the tricolored triangles correspond to vertices of odd degree (of degree 1).

However, the vertex of the dual graph which corresponds to the outside of the triangulation has odd degree: in fact, along the big edge from V_1 to V_2, there is an odd number of changes between 1 and 2. Thus an odd number of edges of the partial dual graph crosses this big edge, while the other big edges cannot have both 1 and 2 occurring as colors.

Now since the number of odd vertices in any finite graph is even (by equation (4)), we find that the number of small triangles with three different colors (corresponding to odd inside vertices of our dual graph) is odd. □

With this lemma, it is easy to derive Brouwer's theorem.

■ **Proof of Brouwer's fixed point theorem (for $n = 2$).** Let Δ be the triangle in $\mathbb{R}^3$ with vertices $\boldsymbol{e}_1 = (1,0,0)$, $\boldsymbol{e}_2 = (0,1,0)$, and $\boldsymbol{e}_3 = (0,0,1)$. It suffices to prove that every continuous map $f\colon \Delta \longrightarrow \Delta$ has a fixed point, since Δ is homeomorphic to the two-dimensional ball B_2.

We use $\delta(\mathcal{T})$ to denote the maximal length of an edge in a triangulation $\mathcal{T}$. One can easily construct an infinite sequence of triangulations $\mathcal{T}_1, \mathcal{T}_2, \ldots$ of Δ such that the sequence of maximal diameters $\delta(\mathcal{T}_k)$ converges to 0. Such a sequence can be obtained by explicit construction, or inductively, for example by taking $\mathcal{T}_{k+1}$ to be the barycentric subdivision of $\mathcal{T}_k$.

For each of these triangulations, we define a 3-coloring of their vertices $\boldsymbol{v}$ by setting $\lambda(\boldsymbol{v}) := \min\{i : f(\boldsymbol{v})_i < v_i\}$, that is, $\lambda(\boldsymbol{v})$ is the smallest index i such that the i-th coordinate of $f(\boldsymbol{v}) - \boldsymbol{v}$ is negative. Assuming that f has no fixed point, this is well-defined. To see this, note that every $\boldsymbol{v} \in \Delta$ lies in the plane $x_1 + x_2 + x_3 = 1$, hence $\sum_i v_i = 1$. So if $f(\boldsymbol{v}) \neq \boldsymbol{v}$, then at least one of the coordinates of $f(\boldsymbol{v}) - \boldsymbol{v}$ must be negative (and at least one must be positive).

Let us check that this coloring satisfies the assumptions of Sperner's lemma. First, the vertex $\boldsymbol{e}_i$ must receive color i, since the only possible negative component of $f(\boldsymbol{e}_i) - \boldsymbol{e}_i$ is the i-th component. Moreover, if $\boldsymbol{v}$ lies on the edge opposite to $\boldsymbol{e}_i$, then $v_i = 0$, so the i-th component of $f(\boldsymbol{v}) - \boldsymbol{v}$ cannot be negative, and hence $\boldsymbol{v}$ does not get the color i.

Sperner's lemma now tells us that in each triangulation $\mathcal{T}_k$ there is a tri-colored triangle $\{\boldsymbol{v}^{k:1}, \boldsymbol{v}^{k:2}, \boldsymbol{v}^{k:3}\}$ with $\lambda(\boldsymbol{v}^{k:i}) = i$. The sequence of points $(\boldsymbol{v}^{k:1})_{k \geq 1}$ need not converge, but since the simplex Δ is compact some subsequence has a limit point. After replacing the sequence of triangulations $\mathcal{T}_k$ by the corresponding subsequence (which for simplicity we also denote by $\mathcal{T}_k$) we can assume that $(\boldsymbol{v}^{k:1})_k$ converges to a point $\boldsymbol{v} \in \Delta$. Now the distance of $\boldsymbol{v}^{k:2}$ and $\boldsymbol{v}^{k:3}$ from $\boldsymbol{v}^{k:1}$ is at most the mesh length $\delta(\mathcal{T}_k)$, which converges to 0. Thus the sequences $(\boldsymbol{v}^{k:2})$ and $(\boldsymbol{v}^{k:3})$ converge to the *same* point $\boldsymbol{v}$.

But where is $f(\boldsymbol{v})$? We know that the first coordinate $f(\boldsymbol{v}^{k:1})$ is smaller than that of $\boldsymbol{v}^{k:1}$ for all k. Now since f is continuous, we derive that the first coordinate of $f(\boldsymbol{v})$ is smaller or equal to that of $\boldsymbol{v}$. The same reasoning works for the second and third coordinates. Thus none of the coordinates of $f(\boldsymbol{v}) - \boldsymbol{v}$ is positive — and we have already seen that this contradicts the assumption $f(\boldsymbol{v}) \neq \boldsymbol{v}$. □

References

[1] L. E. J. BROUWER: *Über Abbildungen von Mannigfaltigkeiten,* Math. Annalen **71** (1912), 97-115.

[2] W. G. BROWN: *On graphs that do not contain a Thomsen graph,* Canadian Math. Bull. **9** (1966), 281-285.

[3] P. ERDŐS, A. RÉNYI & V. SÓS: *On a problem of graph theory,* Studia Sci. Math. Hungar. **1** (1966), 215-235.

[4] P. ERDŐS & G. SZEKERES: *A combinatorial problem in geometry,* Compositio Math. (1935), 463-470.

[5] S. HOŞTEN & W. D. MORRIS: *The order dimension of the complete graph,* Discrete Math. **201** (1999), 133-139.

[6] I. REIMAN: *Über ein Problem von K. Zarankiewicz,* Acta Math. Acad. Sci. Hungar. **9** (1958), 269-273.

[7] J. SPENCER: *Minimal scrambling sets of simple orders,* Acta Math. Acad. Sci. Hungar. **22** (1971), 349-353.

[8] E. SPERNER: *Neuer Beweis für die Invarianz der Dimensionszahl und des Gebietes,* Abh. Math. Sem. Hamburg **6** (1928), 265-272.

[9] W. T. TROTTER: *Combinatorics and Partially Ordered Sets: Dimension Theory,* John Hopkins University Press, Baltimore and London 1992.

Three famous theorems on finite sets

Chapter 23

Emanuel Sperner

In this chapter we are concerned with a basic theme of combinatorics: properties and sizes of special families $\mathcal{F}$ of subsets of a finite set $N = \{1, 2, \ldots, n\}$. We start with two results which are classics in the field: the theorems of Sperner and of Erdős-Ko-Rado. These two results have in common that they were reproved many times and that each of them initiated a new field of combinatorial set theory. For both theorems, induction seems to be the natural method, but the arguments we are going to discuss are quite different and truly inspired.

In 1928 Emanuel Sperner asked and answered the following question: Suppose we are given the set $N = \{1, 2, \ldots, n\}$. Call a family $\mathcal{F}$ of subsets of N an *antichain* if no set of $\mathcal{F}$ contains another set of the family $\mathcal{F}$. What is the size of a largest antichain? Clearly, the family $\mathcal{F}_k$ of all k-sets satisfies the antichain property with $|\mathcal{F}_k| = \binom{n}{k}$. Looking at the maximum of the binomial coefficients (see page 12) we conclude that there is an antichain of size $\binom{n}{\lfloor n/2 \rfloor} = \max_k \binom{n}{k}$. Sperner's theorem now asserts that there are no larger ones.

Theorem 1. *The size of a largest antichain of an n-set is $\binom{n}{\lfloor n/2 \rfloor}$.*

■ **Proof.** Of the many proofs the following one, due to David Lubell, is probably the shortest and most elegant. Let $\mathcal{F}$ be an arbitrary antichain. Then we have to show $|\mathcal{F}| \leq \binom{n}{\lfloor n/2 \rfloor}$. The key to the proof is that we consider *chains* of subsets $\varnothing = C_0 \subset C_1 \subset C_2 \subset \ldots \subset C_n = N$, where $|C_i| = i$ for $i = 0, \ldots, n$. How many chains are there? Clearly, we obtain a chain by adding one by one the elements of N, so there are just as many chains as there are permutations of N, namely $n!$. Next, for a set $A \in \mathcal{F}$ we ask how many of these chains contain A. Again this is easy. To get from $\varnothing$ to A we have to add the elements of A one by one, and then to pass from A to N we have to add the remaining elements. Thus if A contains k elements, then by considering all these pairs of chains linked together we see that there are precisely $k!(n-k)!$ such chains. Note that no chain can pass through two different sets A and B of $\mathcal{F}$, since $\mathcal{F}$ is an antichain.

To complete the proof, let m_k be the number of k-sets in $\mathcal{F}$. Thus $|\mathcal{F}| = \sum_{k=0}^{n} m_k$. Then it follows from our discussion that the number of chains passing through some member of $\mathcal{F}$ is

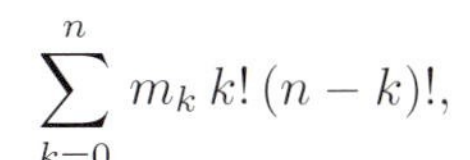

$$\sum_{k=0}^{n} m_k \, k! \, (n-k)!,$$

and this expression cannot exceed the number $n!$ of *all* chains. Hence we conclude

$$\sum_{k=0}^{n} m_k \frac{k!(n-k)!}{n!} \leq 1, \qquad \text{or} \qquad \sum_{k=0}^{n} \frac{m_k}{\binom{n}{k}} \leq 1.$$

Replacing the denominators by the largest binomial coefficient, we therefore obtain

$$\frac{1}{\binom{n}{\lfloor n/2 \rfloor}} \sum_{k=0}^{n} m_k \leq 1, \qquad \text{that is,} \qquad |\mathcal{F}| = \sum_{k=0}^{n} m_k \leq \binom{n}{\lfloor n/2 \rfloor},$$

and the proof is complete. □

Check that the family of all $\frac{n}{2}$-sets for even n respectively the two families of all $\frac{n-1}{2}$-sets and of all $\frac{n+1}{2}$-sets, when n is odd, are indeed the *only* antichains that achieve the maximum size!

Our second result is of an entirely different nature. Again we consider the set $N = \{1, \ldots, n\}$. Call a family $\mathcal{F}$ of subsets an *intersecting family* if any two sets in $\mathcal{F}$ have at least one element in common. It is almost immediate that the size of a largest intersecting family is 2^{n-1}. If $A \in \mathcal{F}$, then the complement $A^c = N \backslash A$ has empty intersection with A and accordingly cannot be in $\mathcal{F}$. Hence we conclude that an intersecting family contains at most half the number 2^n of all subsets, that is, $|\mathcal{F}| \leq 2^{n-1}$. On the other hand, if we consider the family of all sets containing a fixed element, say the family $\mathcal{F}_1$ of all sets containing 1, then clearly $|\mathcal{F}_1| = 2^{n-1}$, and the problem is settled.

But now let us ask the following question: How large can an intersecting family $\mathcal{F}$ be if all sets in $\mathcal{F}$ have the same size, say k ? Let us call such families *intersecting k-families*. To avoid trivialities, we assume $n \geq 2k$ since otherwise any two k-sets intersect, and there is nothing to prove. Taking up the above idea, we certainly obtain such a family $\mathcal{F}_1$ by considering all k-sets containing a fixed element, say 1. Clearly, we obtain all sets in $\mathcal{F}_1$ by adding to 1 all $(k-1)$-subsets of $\{2, 3, \ldots, n\}$, hence $|\mathcal{F}_1| = \binom{n-1}{k-1}$. Can we do better? No — and this is the theorem of Erdős-Ko-Rado.

Theorem 2. *The largest size of an intersecting k-family in an n-set is $\binom{n-1}{k-1}$ when $n \geq 2k$.*

Paul Erdős, Chao Ko and Richard Rado found this result in 1938, but it was not published until 23 years later. Since then multitudes of proofs and variants have been given, but the following argument due to Gyula Katona is particularly elegant.

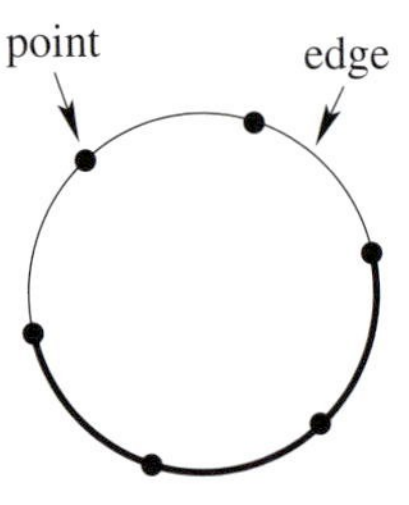

A circle C for $n = 6$. The bold edges depict an arc of length 3.

■ **Proof.** The key to the proof is the following simple lemma, which at first sight seems to be totally unrelated to our problem. Consider a circle C divided by n points into n edges. Let an *arc* of length k consist of $k+1$ consecutive points and the k edges between them.

Lemma. *Let $n \geq 2k$, and suppose we are given t distinct arcs $A_1, \ldots, A_t$ of length k, such that any two arcs have an edge in common. Then $t \leq k$.*

To prove the lemma, note first that any point of C is the endpoint of at most one arc. Indeed, if A_i, A_j had a common endpoint v, then they would have to start in different direction (since they are distinct). But then they cannot have an edge in common as $n \geq 2k$. Let us fix A_1. Since any A_i $(i \geq 2)$ has an edge in common with A_1, one of the endpoints of A_i is an inner point of A_1. Since these endpoints must be distinct as we have just seen, and since A_1 contains $k-1$ inner points, we conclude that there can be at most $k-1$ further arcs, and thus at most k arcs altogether. □

Now we proceed with the proof of the Erdős-Ko-Rado theorem. Let $\mathcal{F}$ be an intersecting k-family. Consider a circle C with n points and n edges as above. We take any cyclic permutation $\pi = (a_1, a_2, \ldots, a_n)$ and write the numbers a_i clockwise next to the edges of C. Let us count the number of sets $A \in \mathcal{F}$ which appear as k *consecutive* numbers on C. Since $\mathcal{F}$ is an intersecting family we see by our lemma that we get at most k such sets. Since this holds for any cyclic permutation, and since there are $(n-1)!$ cyclic permutations, we produce in this way at most

$$k(n-1)!$$

sets of $\mathcal{F}$ which appear as consecutive elements of some cyclic permutation. How often do we count a fixed set $A \in \mathcal{F}$? Easy enough: A appears in π if the k elements of A appear consecutively in some order. Hence we have $k!$ possibilities to write A consecutively, and $(n-k)!$ ways to order the remaining elements. So we conclude that a fixed set A appears in precisely $k!(n-k)!$ cyclic permutations, and hence that

$$|\mathcal{F}| \leq \frac{k(n-1)!}{k!(n-k)!} = \frac{(n-1)!}{(k-1)!(n-1-(k-1))!} = \binom{n-1}{k-1}. \quad \square$$

Again we may ask whether the families containing a fixed element are the only intersecting k-families. This is certainly not true for $n = 2k$. For example, for $n = 4$ and $k = 2$ the family $\{1,2\}, \{1,3\}, \{2,3\}$ also has size $\binom{3}{1} = 3$. More generally, for $n = 2k$ we get the maximal intersecting k-families, of size $\frac{1}{2}\binom{n}{k} = \binom{n-1}{k-1}$, by arbitrarily including one out of every pair of sets formed by a k-set A and its complement $N \backslash A$. But for $n > 2k$ the special families containing a fixed element are indeed the only ones. The reader is invited to try his hand at the proof.

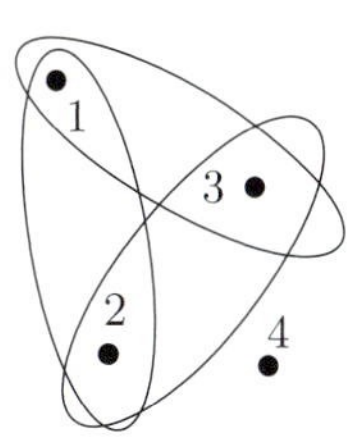

An intersecting family for $n = 4$, $k = 2$

Finally, we turn to the third result which is arguably the most important basic theorem in finite set theory, the "marriage theorem" of Philip Hall proved in 1935. It opened the door to what is today called matching theory, with a wide variety of applications, some of which we shall see as we go along.

Consider a finite set X and a collection $A_1, \ldots, A_n$ of subsets of X (which need not be distinct). Let us call a sequence $x_1, \ldots, x_n$ a *system of distinct representatives* of $\{A_1, \ldots, A_n\}$ if the x_i are distinct elements of X, and if $x_i \in A_i$ for all i. Of course, such a system, abbreviated SDR, need not exist, for example when one of the sets A_i is empty. The content of the theorem of Hall is the precise condition under which an SDR exists.

"A mass wedding"

Before giving the result let us state the human interpretation which gave it the folklore name *marriage theorem*: Consider a set $\{1, \dots, n\}$ of girls and a set X of boys. Whenever $x \in A_i$, then girl i and boy x are inclined to get married, thus A_i is just the set of possible matches of girl i. An SDR represents then a mass-wedding where every girl marries a boy she likes.

Back to sets, here is the statement of the result.

Theorem 3. *Let $A_1, \dots, A_n$ be a collection of subsets of a finite set X. Then there exists a system of distinct representatives if and only if the union of any m sets A_i contains at least m elements, for $1 \le m \le n$.*

The condition is clearly necessary: If m sets A_i contain between them fewer than m elements, then these m sets can certainly not be represented by distinct elements. The surprising fact (resulting in the universal applicability) is that this obvious condition is also sufficient. Hall's original proof was rather complicated, and subsequently many different proofs were given, of which the following one (due to Easterfield and rediscovered by Halmos and Vaughan) may be the most natural.

■ **Proof.** We use induction on n. For $n = 1$ there is nothing to prove. Let $n > 1$, and suppose $\{A_1, \dots, A_n\}$ satisfies the condition of the theorem which we abbreviate by (H). Call a collection of ℓ sets A_i with $1 \le \ell < n$ a *critical family* if its union has cardinality ℓ. Now we distinguish two cases.

Case 1: There is no critical family.

Choose any element $x \in A_n$. Delete x from X and consider the collection $A'_1, \dots, A'_{n-1}$ with $A'_i = A_i \backslash \{x\}$. Since there is no critical family, we find that the union of any m sets A'_i contains at least m elements. Hence by induction on n there exists an SDR $x_1, \dots, x_{n-1}$ of $\{A'_1, \dots, A'_{n-1}\}$, and together with $x_n = x$, this gives an SDR for the original collection.

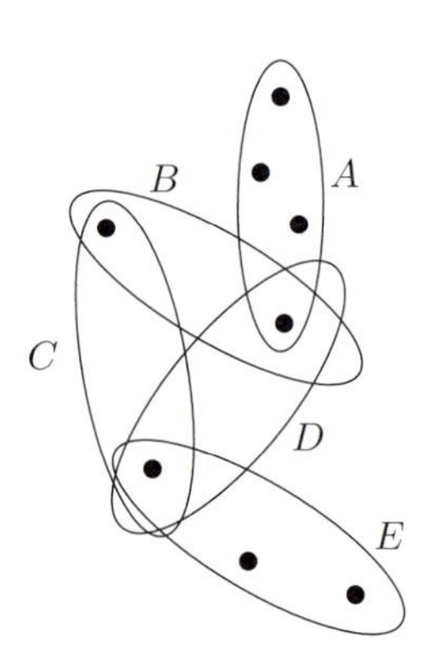

$\{B, C, D\}$ is a critical family

Case 2: There exists a critical family.

After renumbering the sets we may assume that $\{A_1, \dots, A_\ell\}$ is a critical family. Then $\bigcup_{i=1}^{\ell} A_i = \widetilde{X}$ with $|\widetilde{X}| = \ell$. Since $\ell < n$, we infer the existence of an SDR for $A_1, \dots, A_\ell$ by induction, that is, there is a numbering $x_1, \dots, x_\ell$ of $\widetilde{X}$ such that $x_i \in A_i$ for all $i \le \ell$.

Consider now the remaining collection $A_{\ell+1}, \dots, A_n$, and take any m of these sets. Since the union of $A_1, \dots, A_\ell$ and these m sets contains at least $\ell + m$ elements by condition (H), we infer that the m sets contain at least m elements outside $\widetilde{X}$. In other words, condition (H) is satisfied for the family

$$A_{\ell+1} \backslash \widetilde{X}, \; \dots \; , A_n \backslash \widetilde{X}.$$

Induction now gives an SDR for $A_{\ell+1}, \dots, A_n$ that avoids $\widetilde{X}$. Combining it with $x_1, \dots, x_\ell$ we obtain an SDR for all sets A_i. This completes the proof. □

As we mentioned, Hall's theorem was the beginning of the now vast field of matching theory [6]. Of the many variants and ramifications let us state one particularly appealing result which the reader is invited to prove for himself:

> *Suppose the sets* $A_1, \ldots, A_n$ *all have size* $k \geq 1$ *and suppose further that no element is contained in more than* k *sets. Then there exist* k *SDR's such that for any* i *the* k *representatives of* A_i *are distinct and thus together form the set* A_i.

A beautiful result which should open new horizons on marriage possibilities.

References

[1] T. E. EASTERFIELD: *A combinatorial algorithm,* J. London Math. Soc. **21** (1946), 219-226.

[2] P. ERDŐS, C. KO & R. RADO: *Intersection theorems for systems of finite sets,* Quart. J. Math. (Oxford), Ser. (2) **12** (1961), 313-320.

[3] P. HALL: *On representatives of subsets,* J. London Math. Soc. **10** (1935), 26-30.

[4] P. R. HALMOS & H. E. VAUGHAN: *The marriage problem,* Amer. J. Math. **72** (1950), 214-215.

[5] G. KATONA: *A simple proof of the Erdős-Ko-Rado theorem,* J. Combinatorial Theory, Ser. B **13** (1972), 183-184.

[6] L. LOVÁSZ & M. D. PLUMMER: *Matching Theory,* Akadémiai Kiadó, Budapest 1986.

[7] D. LUBELL: *A short proof of Sperner's theorem,* J. Combinatorial Theory **1** (1966), 299.

[8] E. SPERNER: *Ein Satz über Untermengen einer endlichen Menge,* Math. Zeitschrift **27** (1928), 544-548.

Shuffling cards

Chapter 24

How often does one have to shuffle a deck of cards until it is random?

The analysis of random processes is a familiar duty in life ("How long does it take to get to the airport during rush-hour?") as well as in mathematics. Of course, getting meaningful answers to such problems heavily depends on formulating meaningful questions. For the card shuffling problem, this means that we have

- to specify the size of the deck ($n = 52$ cards, say),
- to say how we shuffle (we'll analyze top-in-at-random shuffles first, and then the more realistic and effective riffle shuffles), and finally
- to explain what we mean by "is random" or "is close to random."

So our goal in this chapter is an analysis of the riffle shuffle, due to Edgar N. Gilbert and Claude Shannon (1955, unpublished) and Jim Reeds (1981, unpublished), following the statistician David Aldous and the former magician turned mathematician Persi Diaconis according to [1]. We will not reach the final precise result that 7 riffle shuffles *are* sufficient to get a deck of $n = 52$ cards very close to random, while 6 riffle shuffles do not suffice — but we will obtain an upper bound of 12, and we will see some extremely beautiful ideas on the way: the concepts of stopping rules and of "strong uniform time," the lemma that strong uniform time bounds the variation distance, Reeds' inversion lemma, and thus the interpretation of shuffling as "reversed sorting." In the end, everything will be reduced to two very classical combinatorial problems, namely the coupon collector and the birthday paradox. So let's start with these!

Persi Diaconis' business card as a magician. In a later interview he said: "If you say that you are a professor at Stanford people treat you respectfully. If you say that you invent magic tricks, they don't want to introduce you to their daughter."

The birthday paradox

Take n random people — the participants of a class or seminar, say. What is the probability that they all have different birthdays? With the usual simplifying assumptions (365 days a year, no seasonal effects, no twins present) the probability is

$$p(n) \;=\; \prod_{i=1}^{n-1}\Big(1-\frac{i}{365}\Big),$$

which is smaller than $\frac{1}{2}$ for $n = 23$ (this is the "birthday paradox"!), less than 9 percent for $n = 42$, and exactly 0 for $n > 365$ (the "pigeon-hole principle," see Chapter 22). The formula is easy to see — if we take the persons in some fixed order: If the first i persons have distinct birthdays, then the probability that the $(i+1)$-st person doesn't spoil the series is $1 - \frac{i}{365}$, since there are $365 - i$ birthdays left.

Similarly, if n balls are placed independently and randomly into K boxes, then the probability that no box gets more than one ball is

$$p(n, K) \;=\; \prod_{i=1}^{n-1} \Big(1 - \frac{i}{K}\Big).$$

The coupon collector

Children buy photos of pop stars (or soccer stars) for their albums, but they buy them in little non-transparent envelopes, so they don't know which photo they will get. If there are n different photos, what is the expected number of pictures a kid has to buy until he or she gets every motif at least once?

Equivalently, if you randomly take balls from a bowl that contains n distinguishable balls, and if you put your ball back each time, and then again mix well, how often do you have to draw on average until you have drawn each ball at least once?

If you already have drawn k distinct balls, then the probability not to get a new one in the next drawing is $\frac{k}{n}$. So the probability to need exactly s drawings for the next new ball is $(\frac{k}{n})^{s-1}(1 - \frac{k}{n})$. And thus the expected number of drawings for the next new ball is

$$\sum_{s\geq 1} \Big(\frac{k}{n}\Big)^{s-1} \Big(1 - \frac{k}{n}\Big) s \;=\; \frac{1}{1 - \frac{k}{n}},$$

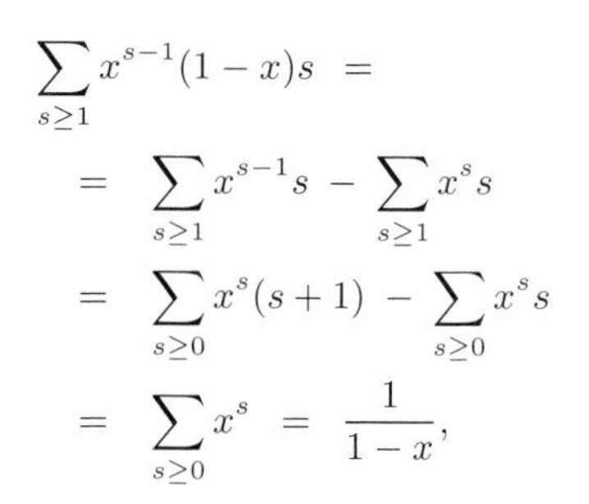

where at the end we sum a geometric series (see page 28).

as we get from the series in the margin. So the expected number of drawings until we have drawn *each* of the n different balls at least once is

$$\sum_{k=0}^{n-1} \frac{1}{1 - \frac{k}{n}} \;=\; \frac{n}{n} + \frac{n}{n-1} + \cdots + \frac{n}{2} + \frac{n}{1} \;=\; nH_n \;\approx\; n \log n,$$

with the bounds on the size of harmonic numbers that we had obtained on page 11. So the answer to the coupon collector's problem is that we have to expect that roughly $n \log n$ drawings are necessary.

The estimate that we need in the following is for the probability that you need significantly more than $n \log n$ trials. If V_n denotes the number of drawings needed (this is the random variable whose expected value is $E[V_n] \approx n \log n$), then for $n \geq 1$ and $c \geq 0$, the probability that we need more than $m := \lceil n \log n + cn \rceil$ drawings is

$$\text{Prob}\big[V_n > m\big] \;\leq\; e^{-c}.$$

Indeed, if A_i denotes the event that the ball i is not drawn in the first m drawings, then

$$\begin{aligned}\text{Prob}\big[V_n > m\big] &= \text{Prob}\Big[\bigcup_i A_i\Big] \le \sum_i \text{Prob}\big[A_i\big] \\ &= n\Big(1-\frac{1}{n}\Big)^m < ne^{-m/n} \le e^{-c}.\end{aligned}$$

A little calculus shows that $\left(1-\frac{1}{n}\right)^n$ is an increasing function in n, which converges to $1/e$. So $\left(1-\frac{1}{n}\right)^n < \frac{1}{e}$ holds for all $n \ge 1$.

Now let's grab a deck of n cards. We number them 1 up to n in the order in which they come — so the card numbered "1" is at the top of the deck, while "n" is at the bottom. Let us denote from now on by $\mathfrak{S}_n$ the set of all permutations of $1, \ldots, n$. *Shuffling* the deck amounts to the application of certain *random permutations* to the order of the cards. Ideally, this might mean that we apply an arbitrary permutation $\pi \in \mathfrak{S}_n$ to our starting order $(1, 2, \ldots, n)$, each of them with the same probability $\frac{1}{n!}$. Thus, after doing this just once, we would have our deck of cards in order $\pi = (\pi(1), \pi(2), \ldots, \pi(n))$, and this would be a perfect random order. But that's not what happens in real life. Rather, when shuffling only "certain" permutations occur, perhaps not all of them with the same probability, and this is repeated a "certain" number of times. After that, we expect or hope the deck to be at least "close to random."

Top-in-at-random shuffles

These are performed as follows: you take the top card from the deck, and insert it into the deck at one of the n distinct possible places, each of them with probability $\frac{1}{n}$. Thus one of the permutations

$$\tau_i = (2, 3, \ldots, \overset{\overset{\displaystyle i}{\downarrow}}{i, 1}, i{+}1, \ldots, n)$$

is applied, $1 \le i \le n$. After one such shuffle the deck doesn't look random, and indeed we expect to need lots of such shuffles until we reach that goal. A typical run of top-in-at-random shuffles may look as follows (for $n = 5$):

Top-in-at-random

1	2	3	2	2	4	
2	3	2	4	4	1	
3	4	4	1	1	5	...
4	1	1	5	5	3	
5	5	5	3	3	2	

How should we measure "being close to random"? Probabilists have cooked up the "variation distance" as a rather unforgiving measure of randomness: We look at the probability distribution on the $n!$ different orderings of our deck, or equivalently, on the $n!$ different permutations $\sigma \in \mathfrak{S}_n$ that yield the orderings.

Two examples are our starting distribution E, which is given by

$$\begin{aligned}\mathsf{E}(\mathrm{id}) &= 1,\\ \mathsf{E}(\pi) &= 0 \quad \text{otherwise,}\end{aligned}$$

and the uniform distribution U given by

$$\mathsf{U}(\pi) = \tfrac{1}{n!} \quad \text{for all } \pi \in \mathfrak{S}_n.$$

The *variation distance* between two probability distributions Q_1 and Q_2 is now defined as

$$\|\mathsf{Q}_1 - \mathsf{Q}_2\| := \tfrac{1}{2} \sum_{\pi \in \mathfrak{S}_n} |\mathsf{Q}_1(\pi) - \mathsf{Q}_2(\pi)|.$$

By setting $S := \{\pi \in \mathfrak{S}_n : \mathsf{Q}_1(\pi) > \mathsf{Q}_2(\pi)\}$ and using $\sum_\pi \mathsf{Q}_1(\pi) = \sum_\pi \mathsf{Q}_2(\pi) = 1$ we can rewrite this as

$$\|\mathsf{Q}_1 - \mathsf{Q}_2\| = \max_{S \subseteq \mathfrak{S}_n} |\mathsf{Q}_1(S) - \mathsf{Q}_2(S)|,$$

with $\mathsf{Q}_i(S) := \sum_{\pi \in S} \mathsf{Q}_i(\pi)$. Clearly we have $0 \leq \|\mathsf{Q}_1 - \mathsf{Q}_2\| \leq 1$. In the following, "being close to random" will be interpreted as "having small variation distance from the uniform distribution." Here the distance between the starting distribution and the uniform distribution is very close to 1:

$$\|\mathsf{E} - \mathsf{U}\| = 1 - \tfrac{1}{n!}.$$

After one top-in-at-random shuffle, this will not be much better:

$$\|\mathsf{Top} - \mathsf{U}\| = 1 - \tfrac{1}{(n-1)!}.$$

For card players, the question is not "exactly how close to uniform is the deck after a million riffle shuffles?", but "is 7 shuffles enough?"

(Aldous & Diaconis [1])

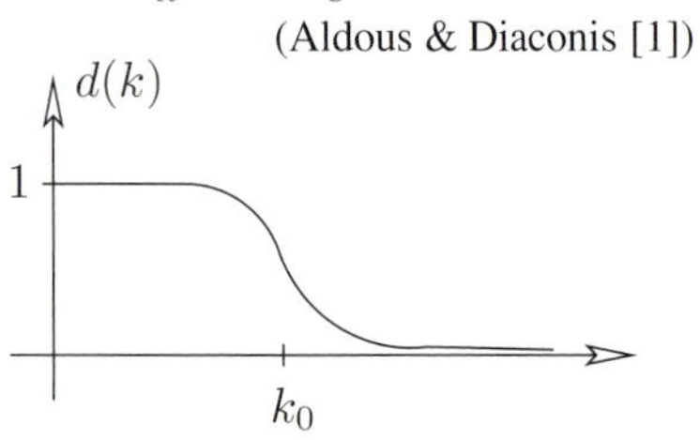

The probability distribution on $\mathfrak{S}_n$ that we obtain by applying the top-in-at-random shuffle k times will be denoted by Top^{*k}. So how does $\|\mathsf{Top}^{*k} - \mathsf{U}\|$ behave if k gets larger, that is, if we repeat the shuffling? And similarly for other types of shuffling? General theory (in particular, Markov chains on finite groups; see e. g. Behrends [3]) implies that for large k the variation distance $d(k) := \|\mathsf{Top}^{*k} - \mathsf{U}\|$ goes to zero exponentially fast, but it does not yield the "cut-off" phenomenon that one observes in practice: After a certain number k_0 of shuffles "suddenly" $d(k)$ goes to zero rather fast. Our margin displays a schematic sketch of the situation.

Strong uniform stopping rules

The amazing idea of strong uniform stopping rules by Aldous and Diaconis captures the essential features. Imagine that the casino manager closely watches the shuffling process, analyzes the specific permutations that are applied to the deck in each step, and after a number of steps that depends on the permutations that he has seen calls "STOP!". So he has a *stopping rule* that ends the shuffling process. It depends only on the (random) shuffles that have already been applied. The stopping rule is *strong uniform* if the following condition holds for all $k \geq 0$:

> ***If** the process is stopped after exactly k steps, **then** the resulting permutations of the deck have uniform distribution (exactly!).*

Let T be the number of steps that are performed until the stopping rule tells the manager to cry "STOP!"; so this is a random variable. Similarly, the ordering of the deck after k shuffles is given by a random variable X_k (with values in $\mathfrak{S}_n$). With this, the stopping rule is strong uniform if for all feasible values of k,

$$\text{Prob}\big[X_k = \pi \mid T = k\big] \;=\; \frac{1}{n!} \qquad \text{for all } \pi \in \mathfrak{S}_n.$$

Three aspects make this interesting, useful, and remarkable:

1. Strong uniform stopping rules exist: For many examples they are quite simple.
2. Moreover, these can be analyzed: Trying to determine $\text{Prob}[T > k]$ leads often to simple combinatorial problems.
3. This yields effective upper bounds on the variation distances such as $d(k) = \|\mathsf{Top}^{*k} - \mathsf{U}\|$.

For example, for the top-in-at-random shuffles a strong uniform stopping rule is

> "STOP after the original bottom card (labelled n) is first inserted back into the deck."

Indeed, if we trace the card n during these shuffles,

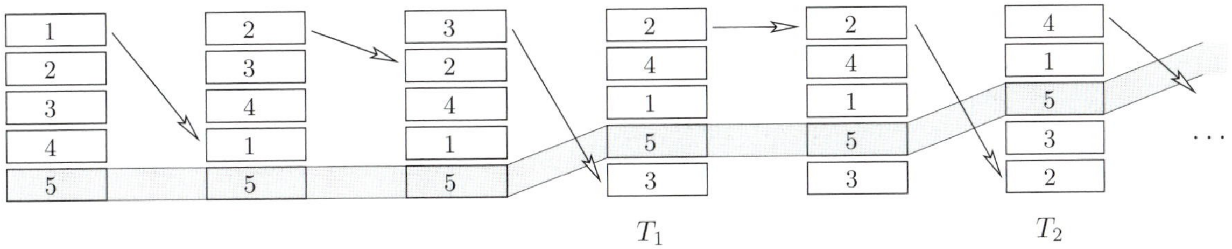

we see that during the whole process the ordering of the cards below this card is completely uniform. So, after the card n rises to the top and then is inserted at random, the deck is uniformly distributed; we just don't know when precisely this happens (but the manager does).

Now let T_i be the random variable which counts the number of shuffles that are performed until for the first time i cards lie below card n. So we have to determine the distribution of

$$T \;=\; T_1 + (T_2 - T_1) + \ldots + (T_{n-1} - T_{n-2}) + (T - T_{n-1}).$$

But each summand in this corresponds to a coupon collector's problem: $T_i - T_{i-1}$ is the time until the top card is inserted at one of the i possible places below the card n. So it is also the time that the coupon collector takes from the $(n-i)$-th coupon to the $(n-i+1)$-st coupon. Let V_i be the number of pictures bought until he has i different pictures. Then

$$V_n \;=\; V_1 + (V_2 - V_1) + \ldots + (V_{n-1} - V_{n-2}) + (V_n - V_{n-1}),$$

Relative probabilities

The *relative probability*

$$\text{Prob}[A \mid B]$$

denotes the probability of the event A under the condition that B happens. This is just the probability that both events happen, divided by the probability that B is true, that is,

$$\text{Prob}[A \mid B] \;=\; \frac{\text{Prob}[A \wedge B]}{\text{Prob}[B]}.$$

and we have seen that $\text{Prob}[T_i - T_{i-1} = j] = \text{Prob}[V_{n-i+1} - V_{n-i} = j]$ for all i and j. Hence the coupon collector and the top-in-at-random shuffler perform equivalent sequences of independent random processes, just in the opposite order (for the coupon collector, it's hard at the end). Thus we know that the strong uniform stopping rule for the top-in-at-random shuffles takes more than $k = \lceil n \log n + cn \rceil$ steps with low probability:

$$\text{Prob}\big[T > k\big] \;\le\; e^{-c}.$$

And this in turn means that after $k = \lceil n \log n + cn \rceil$ top-in-at-random shuffles, our deck is "close to random," with

$$d(k) \;=\; \|\mathsf{Top}^{*k} - \mathsf{U}\| \;\le\; e^{-c},$$

due to the following simple but crucial lemma.

Lemma. *Let* $\mathsf{Q} : \mathfrak{S}_n \longrightarrow \mathbb{R}$ *be any probability distribution that defines a shuffling process* Q^{*k} *with a strong uniform stopping rule whose stopping time is* T. *Then for all* $k \ge 0$,

$$\|\mathsf{Q}^{*k} - \mathsf{U}\| \;\le\; \text{Prob}\big[T > k\big].$$

■ **Proof.** If X is a random variable with values in $\mathfrak{S}_n$, with probability distribution Q, then we write $\mathsf{Q}(S)$ for the probability that X takes a value in $S \subseteq \mathfrak{S}_n$. Thus $\mathsf{Q}(S) = \text{Prob}[X \in S]$, and in the case of the uniform distribution $\mathsf{Q} = \mathsf{U}$ we get

$$\mathsf{U}(S) \;=\; \text{Prob}\big[X \in S\big] \;=\; \frac{|S|}{n!}.$$

For every subset $S \subseteq \mathfrak{S}_n$, we get the probability that after k steps our deck is ordered according to a permutation in S as

$$\begin{aligned}
\mathsf{Q}^{*k}(S) \;&=\; \text{Prob}[X_k \in S] \\
&=\; \sum_{j \le k} \text{Prob}[X_k \in S \,\wedge\, T = j] \;+\; \text{Prob}[X_k \in S \,\wedge\, T > k] \\
&=\; \sum_{j \le k} \mathsf{U}(S)\, \text{Prob}[T = j] \;+\; \text{Prob}[X_k \in S \,|\, T > k] \cdot \text{Prob}[T > k] \\
&=\; \mathsf{U}(S)\,(1 - \text{Prob}[T > k]) \;+\; \text{Prob}[X_k \in S \,|\, T > k] \cdot \text{Prob}[T > k] \\
&=\; \mathsf{U}(S) \;+\; \big(\text{Prob}[X_k \in S \,|\, T > k] - \mathsf{U}(S)\big) \cdot \text{Prob}[T > k].
\end{aligned}$$

This yields

$$|\mathsf{Q}^{*k}(S) - \mathsf{U}(S)| \;\le\; \text{Prob}[T > k]$$

since

$$\text{Prob}[X_k \in S \,|\, T > k] \;-\; \mathsf{U}(S)$$

is a difference of two probabilities, so it has absolute value at most 1. □

This is the point where we have completed our analysis of the top-in-at-random shuffle: We have proved the following upper bound for the number of shuffles needed to get "close to random."

Theorem 1. *Let $c \geq 0$ and $k := \lceil n \log n + cn \rceil$. Then after performing k top-in-at-random shuffles on a deck of n cards, the variation distance from the uniform distribution satisfies*

$$d(k) \;:=\; \|\mathsf{Top}^{*k} - \mathsf{U}\| \;\leq\; e^{-c}.$$

One can also verify that the variation distance $d(k)$ stays large if we do significantly fewer than $n \log n$ top-in-at-random shuffles. The reason is that a smaller number of shuffles will not suffice to destroy the relative ordering on the lowest few cards in the deck.

Of course, top-in-at-random shuffles are extremely ineffective — with the bounds of our theorem, we need roughly $n \log n \approx 205$ top-in-at random shuffles until a deck of $n = 52$ cards is mixed up well. Thus we now switch our attention to a much more interesting and realistic model of shuffling.

Riffle shuffles

This is what dealers do at the casino: They take the deck, split it into two parts, and these are then interleaved, for example by dropping cards from the bottoms of the two half-decks in some irregular pattern.

A riffle shuffle

Again a riffle shuffle performs a certain permutation on the cards in the deck, which we initially assume to be labelled from 1 to n, where 1 is the top card. The riffle shuffles correspond exactly to the permutations $\pi \in \mathfrak{S}_n$ such that the sequence

$$(\pi(1), \pi(2), \; \ldots \; , \pi(n))$$

consists of two interlaced increasing sequences (only for the identity permutation it is one increasing sequence), and that there are exactly $2^n - n$ distinct riffle shuffles on a deck of n cards.

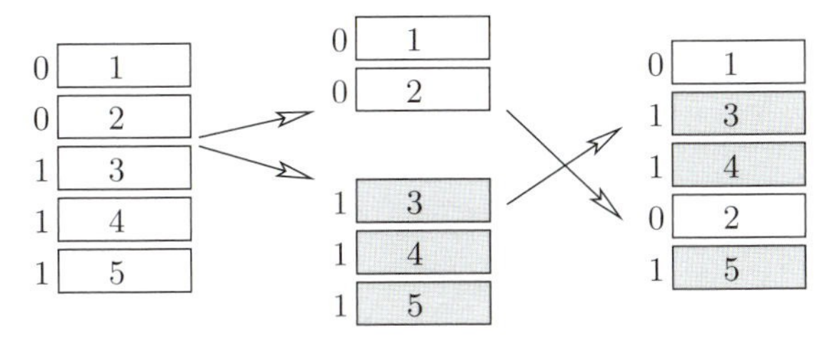

In fact, if the pack is split such that the top t cards are taken into the right hand ($0 \leq t \leq n$) and the other $n - t$ cards into the left hand, then there are $\binom{n}{t}$ ways to interleave the two hands, all of which generate distinct permutations — except that for each t there is one possibility to obtain the identity permutation.

Now it's not clear which probability distribution one should put on the riffle shuffles — there is no unique answer since amateurs and professional dealers would shuffle differently. However, the following model, developed first by Edgar N. Gilbert and Claude Shannon in 1955 (at the legendary

Bell Labs "Mathematics of Communication" department at the time), has several virtues:

- it is elegant, simple, and seems natural,
- it models quite well the way an amateur would perform riffle shuffles,
- and we have a chance to analyze it.

Here are three descriptions — all of them describe the same probability distribution Rif on $\mathfrak{S}_n$:

1. $\mathsf{Rif} : \mathfrak{S}_n \longrightarrow \mathbb{R}$ is defined by

$$\mathsf{Rif}(\pi) \;:=\; \begin{cases} \frac{n+1}{2^n} & \text{if } \pi = \text{id}, \\ \frac{1}{2^n} & \text{if } \pi \text{ consists of two increasing sequences}, \\ 0 & \text{otherwise.} \end{cases}$$

2. Cut off t cards from the deck with probability $\frac{1}{2^n}\binom{n}{t}$, take them into your right hand, and take the rest of the deck into your left hand. Now when you have r cards in the right hand and ℓ in the left, "drop" the bottom card from your right hand with probability $\frac{r}{r+\ell}$, and from your left hand with probability $\frac{\ell}{r+\ell}$. Repeat!

The inverse riffle shuffles correspond to the permutations $\pi = (\pi(1), \ldots, \pi(n))$ that are increasing except for at most one "descent." (Only the identity permutation has no descent.)

3. An *inverse shuffle* would take a subset of the cards in the deck, remove them from the deck, and place them on top of the remaining cards of the deck — while maintaining the relative order in both parts of the deck. Such a move is determined by the subset of the cards: Take all subsets with equal probability.

   Equivalently, assign a label "0" or "1" to each card, randomly and independently with probabilities $\frac{1}{2}$, and move the cards labelled "0" to the top.

It is easy so see that these descriptions yield the same probability distributions. For (1) $\Longleftrightarrow$ (3) just observe that we get the identity permutation whenever all the 0-cards are on top of all the cards that are assigned a 1.

This defines the model. So how can we analyze it? How many riffle shuffles are needed to get close to random? We won't get the precise best-possible answer, but quite a good one, by combining three components:

(1) We analyze inverse riffle shuffles instead,

(2) we describe a strong uniform stopping rule for these,

(3) and show that the key to its analysis is given by the birthday paradox!

Theorem 2. *After performing k riffle shuffles on a deck of n cards, the variation distance from a uniform distribution satisfies*

$$\|\mathsf{Rif}^{*k} - \mathsf{U}\| \;\leq\; 1 - \prod_{i=1}^{n-1}\left(1 - \frac{i}{2^k}\right).$$

■ **Proof.** (1) We may indeed analyze inverse riffle shuffles and try to see how fast they get us from the starting distribution to (close to) uniform. These inverse shuffles correspond to the probability distribution that is given by $\overline{\mathsf{Rif}}(\pi) := \mathsf{Rif}(\pi^{-1})$.

Now the fact that every permutation has its unique inverse, and the fact that $\mathsf{U}(\pi) = \mathsf{U}(\pi^{-1})$, yield

$$\|\mathsf{Rif}^{*k} - \mathsf{U}\| \;=\; \|\overline{\mathsf{Rif}}^{*k} - \mathsf{U}\|.$$

(This is Reeds' inversion lemma!)

(2) In every inverse riffle shuffle, each card gets associated a digit 0 or 1:

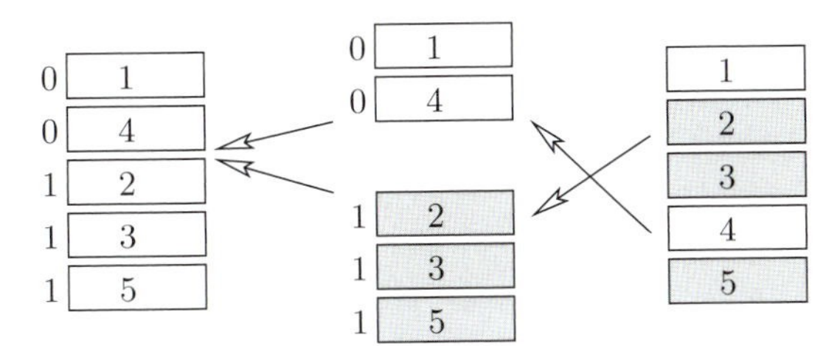

If we remember these digits — say we just write them onto the cards — then after k inverse riffle shuffles, each card has gotten an ordered string of k digits. Our stopping rule is:

"STOP as soon as all cards have distinct strings."

When this happens, the cards in the deck are *sorted* according to the binary numbers $b_k b_{k-1} \dots b_2 b_1$, where b_i is the bit that the card has picked up in the i-th inverse riffle shuffle. Since these bits are perfectly random and independent, this stopping rule is strong uniform!

In the following example, for $n = 5$ cards, we need $T = 3$ inverse shuffles until we stop:

000	4	00	4	0	1	1
001	2	01	2	0	4	2
010	1	01	5	1	2	3
101	5	10	1	1	3	4
111	3	11	3	1	5	5

(3) The expected time T taken by this stopping rule is distributed according to the birthday paradox, for $K = 2^k$: We put two cards into the same box if they have the same label $b_k b_{k-1} \dots b_2 b_1 \in \{0,1\}^k$. So there are $K = 2^k$ boxes, and the probability that some box gets more than one card ist

$$\mathrm{Prob}[T > k] \;=\; 1 - \prod_{i=1}^{n-1}\left(1 - \frac{i}{2^k}\right),$$

and as we have seen this bounds the variation distance $\|\mathsf{Rif}^{*k} - \mathsf{U}\| = \|\overline{\mathsf{Rif}}^{*k} - \mathsf{U}\|$. □

k	$d(k)$
1	1.000
2	1.000
3	1.000
4	1.000
5	0.952
6	0.614
7	0.334
8	0.167
9	0.085
10	0.043

The variation distance after k riffle shuffles, according to [2]

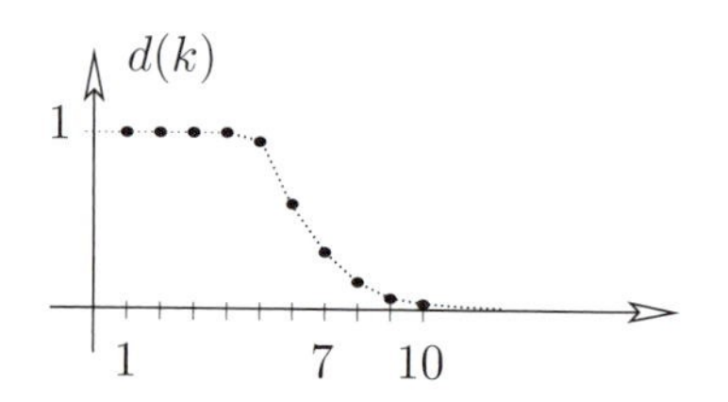

So how often do we have to shuffle? For large n we will need roughly $k = 2\log_2(n)$ shuffles. Indeed, setting $k := 2\log_2(cn)$ for some $c \geq 1$ we find (with a bit of calculus) that $P[T > k] \approx 1 - e^{\frac{1}{2c^2}} \approx \frac{1}{2c^2}$.
Explicitly, for $n = 52$ cards the upper bound of Theorem 2 reads $d(10) \leq 0.73$, $d(12) \leq 0.28$, $d(14) \leq 0.08$ — so $k = 12$ should be "random enough" for all practical purposes. But we don't do 12 shuffles "in practice" — and they are not really necessary, as a more detailed analysis shows (with the results given in the margin). The analysis of riffle shuffles is part of a lively ongoing discussion about the right measure of what is "random enough." Diaconis [4] is a guide to recent developments.

Indeed, does it matter? Yes, it does: Even after three good riffle shuffles a sorted deck of 52 cards looks quite random ... but it isn't. Martin Gardner [5, Chapter 7] describes a number of striking card tricks that are based on the hidden order in such a deck!

References

[1] D. ALDOUS & P. DIACONIS: *Shuffling cards and stopping times,* Amer. Math. Monthly **93** (1986), 333-348.

[2] D. BAYER & P. DIACONIS: *Trailing the dovetail shuffle to its lair,* Annals Applied Probability **2** (1992), 294-313.

[3] E. BEHRENDS: *Introduction to Markov Chains,* Vieweg, Braunschweig/Wiesbaden 2000.

[4] P. DIACONIS: *Mathematical developments from the analysis of riffle shuffling,* in: "Groups, Combinatorics and Geometry. Durham 2001" (A. A. Ivanov, M. W. Liebeck and J. Saxl, eds.), World Scientific, Singapore 2003, pp. 73-97.

[5] M. GARDNER: *Mathematical Magic Show,* Knopf, New York/Allen & Unwin, London 1977.

[6] E. N. GILBERT: *Theory of Shuffling,* Technical Memorandum, Bell Laboratories, Murray Hill NJ, 1955.

Random enough?

Lattice paths and determinants

Chapter 25

The essence of mathematics is proving theorems — and so, that is what mathematicians do: They prove theorems. But to tell the truth, what they really want to prove, once in their lifetime, is a *Lemma*, like the one by Fatou in analysis, the Lemma of Gauss in number theory, or the Burnside-Frobenius Lemma in combinatorics.

Now what makes a mathematical statement a true Lemma? First, it should be applicable to a wide variety of instances, even seemingly unrelated problems. Secondly, the statement should, once you have seen it, be completely obvious. The reaction of the reader might well be one of faint envy: Why haven't I noticed this before? And thirdly, on an esthetic level, the Lemma — including its proof — should be beautiful!

In this chapter we look at one such marvelous piece of mathematical reasoning, a counting lemma that first appeared in a paper by Bernt Lindström in 1972. Largely overlooked at the time, the result became an instant classic in 1985, when Ira Gessel and Gerard Viennot rediscovered it and demonstrated in a wonderful paper how the lemma could be successfully applied to a diversity of difficult combinatorial enumeration problems.

The starting point is the usual permutation representation of the determinant of a matrix. Let $M = (m_{ij})$ be a real $n \times n$-matrix. Then

$$\det M \;=\; \sum_{\sigma} \operatorname{sign}\sigma\, m_{1\sigma(1)}\, m_{2\sigma(2)} \cdots m_{n\sigma(n)}, \qquad (1)$$

where σ runs through all permutations of $\{1, 2, \ldots, n\}$, and the sign of σ is 1 or -1, depending on whether σ is the product of an even or an odd number of transpositions.

Now we pass to graphs, more precisely to *weighted directed bipartite* graphs. Let the vertices $A_1, \ldots, A_n$ stand for the rows of M, and $B_1, \ldots, B_n$ for the columns. For each pair of i and j draw an arrow from A_i to B_j and give it the weight m_{ij}, as in the figure.

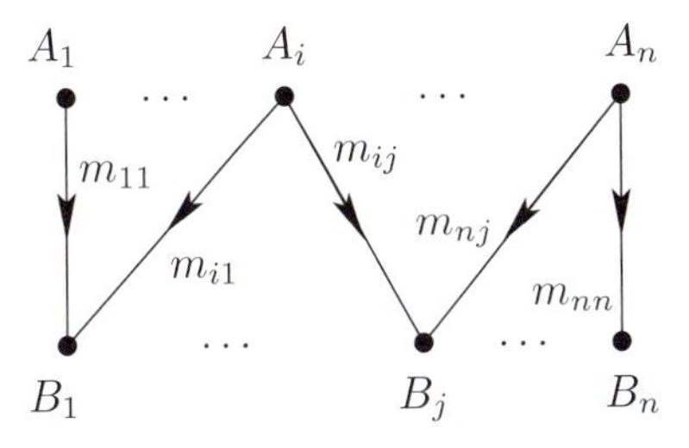

In terms of this graph, the formula (1) has the following interpretation:

- The left-hand side is the determinant of the *path-matrix* M, whose (i, j)-entry is the *weight* of the (unique) directed path from A_i to B_j.
- The right-hand side is the weighted (signed) sum over all *vertex-disjoint path systems* from $\mathcal{A} = \{A_1, \ldots, A_n\}$ to $\mathcal{B} = \{B_1, \ldots, B_n\}$. Such a system $\mathcal{P}_\sigma$ is given by paths

$$A_1 \to B_{\sigma(1)}, \;\ldots\;, A_n \to B_{\sigma(n)},$$

and the *weight* of the path system $\mathcal{P}_\sigma$ is the product of the weights of the individual paths:

$$w(\mathcal{P}_\sigma) = w(A_1 \to B_{\sigma(1)}) \cdots w(A_n \to B_{\sigma(n)}).$$

In this interpretation formula (1) reads

$$\det M = \sum_\sigma \operatorname{sign} \sigma \, w(\mathcal{P}_\sigma).$$

And what is the result of Gessel and Viennot? It is the natural generalization of (1) from bipartite to arbitrary graphs. It is precisely this step which makes the Lemma so widely applicable — and what's more, the proof is stupendously simple and elegant.

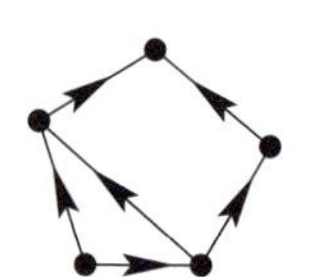

An acyclic directed graph

Let us first collect the necessary concepts. We are given a finite acyclic directed graph $G = (V, E)$, where *acyclic* means that there are no directed cycles in G. In particular, there are only finitely many directed paths between any two vertices A and B, where we include all trivial paths $A \to A$ of length 0. Every edge e carries a weight $w(e)$. If P is a directed path from A to B, written shortly $P : A \to B$, then we define the *weight* of P as

$$w(P) := \prod_{e \in P} w(e),$$

which is defined to be $w(P) = 1$ if P is a path of length 0.

Now let $\mathcal{A} = \{A_1, \ldots, A_n\}$ and $\mathcal{B} = \{B_1, \ldots, B_n\}$ be two sets of n vertices, where $\mathcal{A}$ and $\mathcal{B}$ need not be disjoint. To $\mathcal{A}$ and $\mathcal{B}$ we associate the *path matrix* $M = (m_{ij})$ with

$$m_{ij} := \sum_{P : A_i \to B_j} w(P).$$

A *path system* $\mathcal{P}$ *from* $\mathcal{A}$ *to* $\mathcal{B}$ consists of a permutation σ together with n paths $P_i : A_i \to B_{\sigma(i)}$, for $i = 1, \ldots, n$; we write $\operatorname{sign} \mathcal{P} = \operatorname{sign} \sigma$. The *weight* of $\mathcal{P}$ is the product of the path weights

$$w(\mathcal{P}) = \prod_{i=1}^{n} w(P_i), \qquad (2)$$

which is the product of the weights of all the edges of the path system.

Finally, we say that the path system $\mathcal{P} = (P_1, \ldots, P_n)$ is *vertex-disjoint* if the paths of $\mathcal{P}$ are pairwise vertex-disjoint.

Lemma. *Let $G = (V, E)$ be a finite weighted acyclic directed graph, $\mathcal{A} = \{A_1, \ldots, A_n\}$ and $\mathcal{B} = \{B_1, \ldots, B_n\}$ two n-sets of vertices, and M the path matrix from $\mathcal{A}$ to $\mathcal{B}$. Then*

$$\det M = \sum_{\substack{\mathcal{P} \text{ vertex-disjoint} \\ \text{path system}}} \operatorname{sign} \mathcal{P} \, w(\mathcal{P}). \qquad (3)$$

■ **Proof.** A typical summand of $\det(M)$ is $\operatorname{sign}\sigma\, m_{1\sigma(1)}\cdots m_{n\sigma(n)}$, which can be written as

$$\operatorname{sign}\sigma\ \Big(\sum_{P_1:A_1\to B_{\sigma(1)}} w(P_1)\Big)\ \cdots\ \Big(\sum_{P_n:A_n\to B_{\sigma(n)}} w(P_n)\Big).$$

Summing over σ we immediately find from (2) that

$$\det M \;=\; \sum_{\mathcal{P}} \operatorname{sign}\mathcal{P}\, w(\mathcal{P}),$$

where $\mathcal{P}$ runs through *all* path systems from $\mathcal{A}$ to $\mathcal{B}$ (vertex-disjoint or not). Hence to arrive at (3), all we have to show is

$$\sum_{\mathcal{P}\in N} \operatorname{sign}\mathcal{P}\, w(\mathcal{P}) \;=\; 0\,, \tag{4}$$

where N is the set of all path systems that are *not* vertex-disjoint. And this is accomplished by an argument of singular beauty. Namely, we exhibit an involution $\pi : N \to N$ (without fixed points) such that for $\mathcal{P}$ and $\pi\mathcal{P}$

$$w(\pi\mathcal{P}) \;=\; w(\mathcal{P}) \qquad \text{and} \qquad \operatorname{sign}\pi\mathcal{P} \;=\; -\operatorname{sign}\mathcal{P}\,.$$

Clearly, this will imply (4) and thus the formula (3) of the Lemma.

The involution π is defined in the most natural way. Let $\mathcal{P}\in N$ with paths $P_i : A_i \to B_{\sigma(i)}$. By definition, some pair of paths will intersect:

- Let i_0 be the minimal index such that P_{i_0} shares some vertex with another path.
- Let X be the first such common vertex on the path P_{i_0}.
- Let j_0 be the minimal index ($j_0 > i_0$) such that P_{j_0} has the vertex X in common with P_{i_0}.

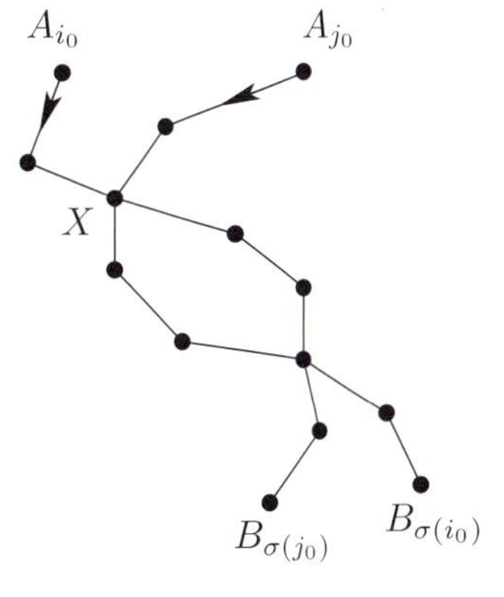

Now we construct the new system $\pi\mathcal{P} = (P_1', \dots, P_n')$ as follows:

- Set $P_k' = P_k$ for all $k \neq i_0, j_0$.
- The new path P_{i_0}' goes from A_{i_0} to X along P_{i_0}, and then continues to $B_{\sigma(j_0)}$ along P_{j_0}. Similarly, P_{j_0}' goes from A_{j_0} to X along P_{j_0} and continues to $B_{\sigma(i_0)}$ along P_{i_0}.

Clearly $\pi(\pi\mathcal{P}) = \mathcal{P}$, since the index i_0, the vertex X, and the index j_0 are the same as before. In other words, applying π twice we switch back to the old paths P_i. Next, since $\pi\mathcal{P}$ and $\mathcal{P}$ use precisely the same edges, we certainly have $w(\pi\mathcal{P}) = w(\mathcal{P})$. And finally, since the new permutation σ' is obtained by multiplying σ with the transposition (i_0, j_0), we find that $\operatorname{sign}\pi\mathcal{P} \;=\; -\operatorname{sign}\mathcal{P}$, and that's it. □

The Gessel-Viennot Lemma can be used to derive all basic properties of determinants, just by looking at appropriate graphs. Let us consider one particularly striking example, the formula of Binet-Cauchy, which gives a very useful generalization of the product rule for determinants.

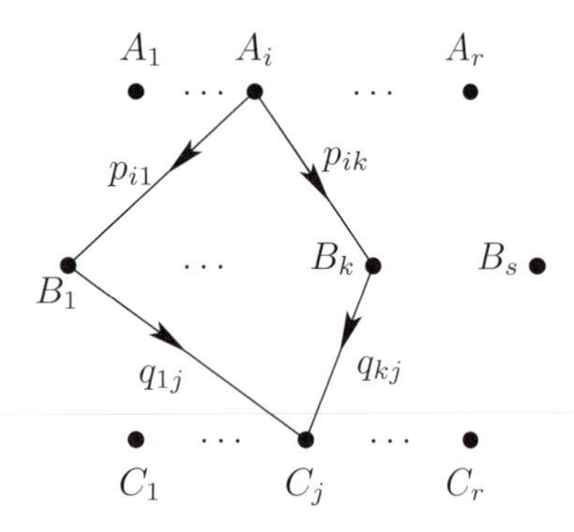

Theorem. *If P is an $(r \times s)$-matrix and Q an $(s \times r)$-matrix, $r \leq s$, then*

$$\det(PQ) \quad = \quad \sum_{\mathcal{Z}} (\det P_{\mathcal{Z}})(\det Q_{\mathcal{Z}}),$$

where $P_{\mathcal{Z}}$ is the $(r \times r)$-submatrix of P with column-set $\mathcal{Z}$, and $Q_{\mathcal{Z}}$ the $(r \times r)$-submatrix of Q with the corresponding rows $\mathcal{Z}$.

■ **Proof.** Let the bipartite graph on $\mathcal{A}$ and $\mathcal{B}$ correspond to P as before, and similarly the bipartite graph on $\mathcal{B}$ and $\mathcal{C}$ to Q. Consider now the concatenated graph as indicated in the figure on the left, and observe that the (i,j)-entry m_{ij} of the path matrix M from $\mathcal{A}$ to $\mathcal{C}$ is precisely $m_{ij} = \sum_k p_{ik} q_{kj}$, thus $M = PQ$.
Since the vertex-disjoint path systems from $\mathcal{A}$ to $\mathcal{C}$ in the concatenated graph correspond to pairs of systems from $\mathcal{A}$ to $\mathcal{Z}$ resp. from $\mathcal{Z}$ to $\mathcal{C}$, the result follows immediately from the Lemma, by noting that sign $(\sigma\tau)$ = (sign σ) (sign τ). □

The Lemma of Gessel-Viennot is also the source of a great number of results that relate determinants to enumerative properties. The recipe is always the same: Interpret the matrix M as a path matrix, and try to compute the right-hand side of (3). As an illustration we will consider the original problem studied by Gessel and Viennot, which led them to their Lemma:

> *Suppose that $a_1 < a_2 < \ldots < a_n$ and $b_1 < b_2 < \ldots < b_n$ are two sets of natural numbers. We wish to compute the determinant of the matrix $M = (m_{ij})$, where m_{ij} is the binomial coefficient $\binom{a_i}{b_j}$.*

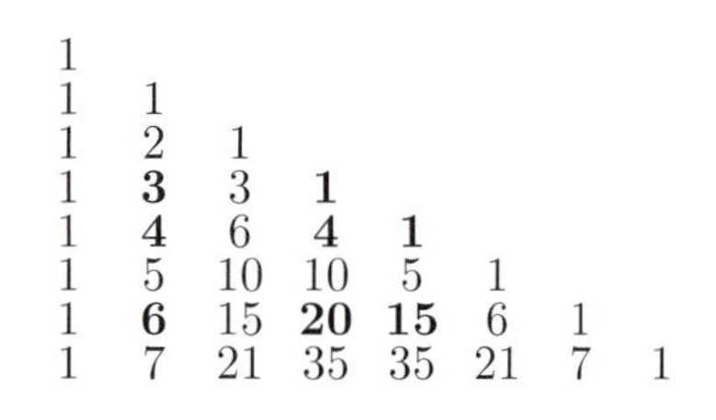

In other words, Gessel and Viennot were looking at the determinants of arbitrary square matrices of Pascal's triangle, such as the matrix

$$\det\begin{pmatrix} \binom{3}{1} & \binom{3}{3} & \binom{3}{4} \\ \binom{4}{1} & \binom{4}{3} & \binom{4}{4} \\ \binom{6}{1} & \binom{6}{3} & \binom{6}{4} \end{pmatrix} = \det\begin{pmatrix} 3 & 1 & 0 \\ 4 & 4 & 1 \\ 6 & 20 & 15 \end{pmatrix}$$

given by the bold entries of Pascal's triangle, as displayed in the margin.

As a preliminary step to the solution of the problem we recall a well-known result which connects binomial coefficients to lattice paths. Consider an $a \times b$-lattice as in the margin. Then the number of paths from the lower left-hand corner to the upper right-hand corner, where the only steps that are allowed for the paths are up (North) and to the right (East), is $\binom{a+b}{a}$.
The proof of this is easy: each path consists of an arbitrary sequence of b "east" and a "north" steps, and thus it can be encoded by a sequence of the form NENEEEN, consisting of $a+b$ letters, a N's and b E's. The number of such strings is the number of ways to choose a positions of letters N from a total of $a+b$ positions, which is $\binom{a+b}{a} = \binom{a+b}{b}$.

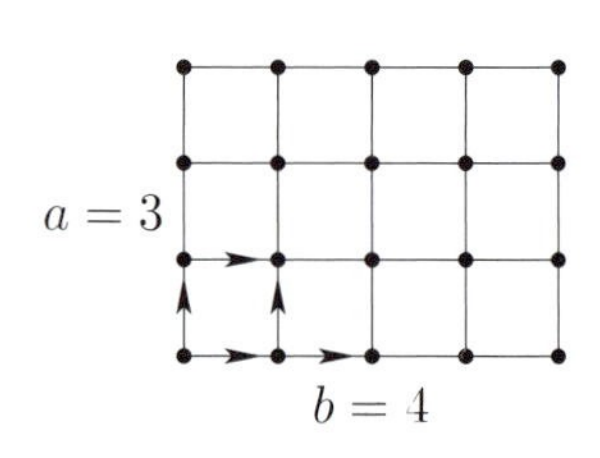

Now look at the figure to the right, where A_i is placed at the point $(0, -a_i)$ and B_j at $(b_j, -b_j)$.

The number of paths from A_i to B_j in this grid that use only steps to the north and east is, by what we just proved, $\binom{b_j+(a_i-b_j)}{b_j} = \binom{a_i}{b_j}$. In other words, the matrix of binomials M is precisely the path matrix from $\mathcal{A}$ to $\mathcal{B}$ in the directed lattice graph for which all edges have weight 1, and all edges are directed to go north or east. Hence to compute $\det M$ we may apply the Gessel-Viennot Lemma. A moment's thought shows that every vertex-disjoint path system $\mathcal{P}$ from $\mathcal{A}$ to $\mathcal{B}$ must consist of paths $P_i : A_i \to B_i$ for all i. Thus the only possible permutation is the identity, which has sign $= 1$, and we obtain the beautiful result

$$\det\left(\binom{a_i}{b_j}\right) \quad = \quad \#\text{ vertex-disjoint path systems from } \mathcal{A} \text{ to } \mathcal{B}.$$

In particular, this implies the far from obvious fact that $\det M$ is always nonnegative, since the right-hand side of the equality *counts* something. More precisely, one gets from the Gessel-Viennot Lemma that $\det M = 0$ if and only if $a_i < b_i$ for some i.

In our previous small example,

$$\det\begin{pmatrix} \binom{3}{1} & \binom{3}{3} & \binom{3}{4} \\ \binom{4}{1} & \binom{4}{3} & \binom{4}{4} \\ \binom{6}{1} & \binom{6}{3} & \binom{6}{4} \end{pmatrix} \quad = \quad \# \begin{array}{l}\text{vertex-disjoint} \\ \text{path systems in}\end{array}$$

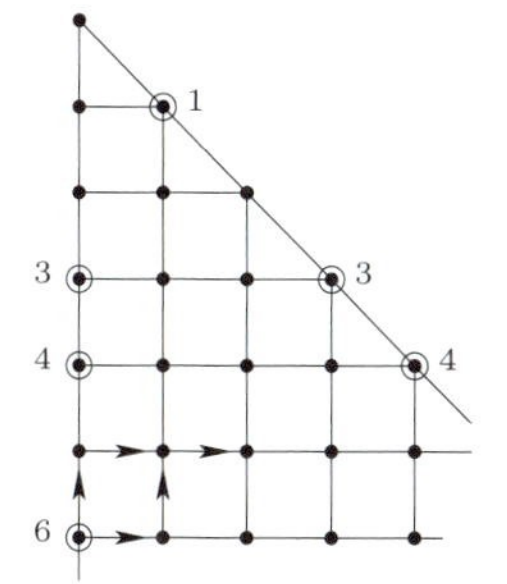

"Lattice paths"

References

[1] I. M. GESSEL & G. VIENNOT: *Binomial determinants, paths, and hook length formulae,* Advances in Math. **58** (1985), 300-321.

[2] B. LINDSTRÖM: *On the vector representation of induced matroids,* Bulletin London Math. Soc. **5** (1973), 85-90.

Cayley's formula for the number of trees

Chapter 26

One of the most beautiful formulas in enumerative combinatorics concerns the number of labeled trees. Consider the set $N = \{1, 2, \ldots, n\}$. How many different trees can we form on this vertex set? Let us denote this number by T_n. Enumeration "by hand" yields $T_1 = 1$, $T_2 = 1$, $T_3 = 3$, $T_4 = 16$, with the trees shown in the following table:

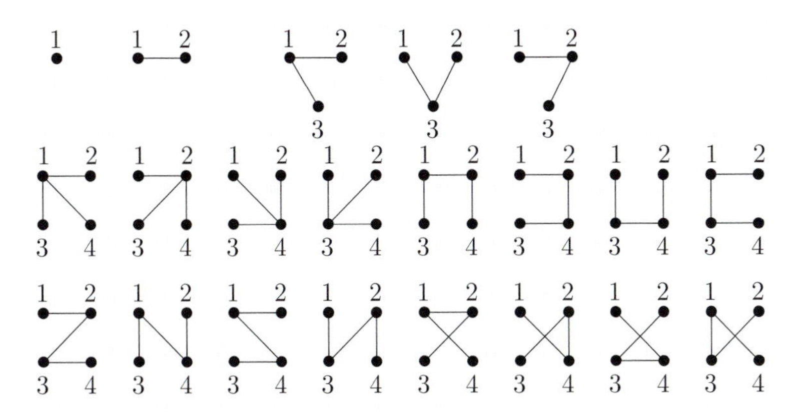

Arthur Cayley

Note that we consider *labeled* trees, that is, although there is only one tree of order 3 in the sense of graph isomorphism, there are 3 different labeled trees obtained by marking the inner vertex 1, 2 or 3. For $n = 5$ there are three non-isomorphic trees:

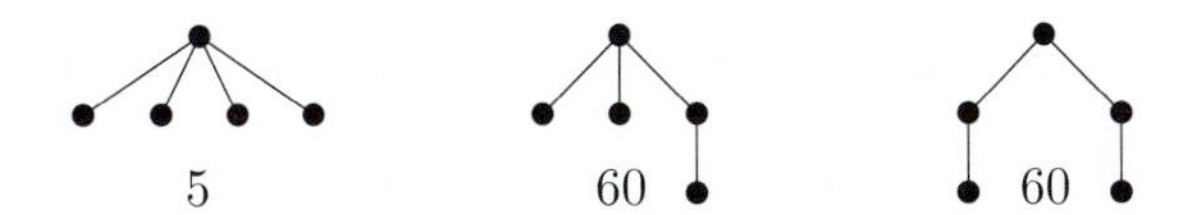

For the first tree there are clearly 5 different labelings, and for the second and third there are $\frac{5!}{2} = 60$ labelings, so we obtain $T_5 = 125$. This should be enough to conjecture $T_n = n^{n-2}$, and that is precisely Cayley's result.

Theorem. *There are n^{n-2} different labeled trees on n vertices.*

This beautiful formula yields to equally beautiful proofs, drawing on a variety of combinatorial and algebraic techniques. We will outline three of them before presenting the proof which is to date the most beautiful of them all.

■ **First proof (Bijection).** The classical and most direct method is to find a bijection from the set of all trees on n vertices onto another set whose cardinality is known to be n^{n-2}. Naturally, the set of all ordered sequences $(a_1, \ldots, a_{n-2})$ with $1 \leq a_i \leq n$ comes into mind. Thus we want to uniquely encode every tree T by a sequence $(a_1 \ldots, a_{n-2})$. Such a code was found by Prüfer and is contained in most books on graph theory.

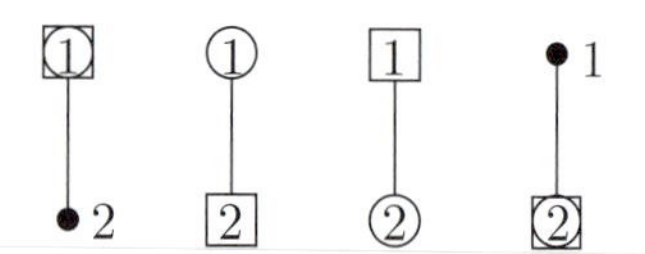

The four trees of $\mathcal{T}_2$

Here we want to discuss another bijection proof, due to Joyal, which is less known but of equal elegance and simplicity. For this, we consider not just trees t on $N = \{1, \ldots, n\}$ but trees together with two distinguished vertices, the *left end* $\bigcirc$ and the *right end* $\square$, which may coincide. Let $\mathcal{T}_n = \{(t; \bigcirc, \square)\}$ be this new set; then, clearly, $|\mathcal{T}_n| = n^2 T_n$.

Our goal is thus to prove $|\mathcal{T}_n| = n^n$. Now there is a set whose size is known to be n^n, namely the set N^N of all mappings from N into N. Thus our formula is proved if we can find a bijection from N^N onto $\mathcal{T}_n$.

Let $f : N \longrightarrow N$ be any map. We represent f as a directed graph $\vec{G}_f$ by drawing arrows from i to $f(i)$.

For example, the map

$$f = \begin{pmatrix} 1 & 2 & 3 & 4 & 5 & 6 & 7 & 8 & 9 & 10 \\ 7 & 5 & 5 & 9 & 1 & 2 & 5 & 8 & 4 & 7 \end{pmatrix}$$

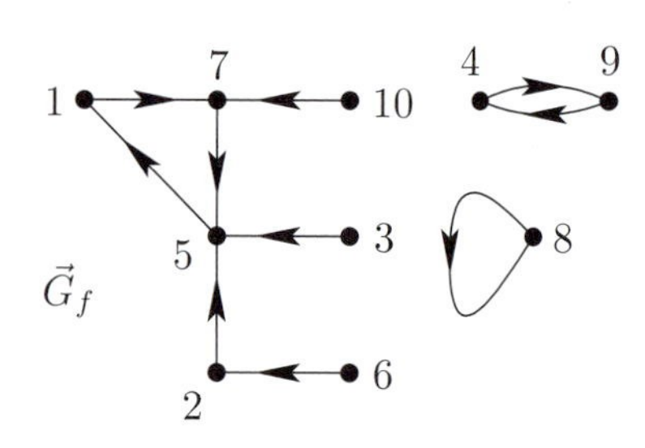

is represented by the directed graph in the margin.

Look at a component of $\vec{G}_f$. Since there is precisely one edge emanating from each vertex, the component contains equally many vertices and edges, and hence precisely one directed cycle. Let $M \subseteq N$ be the union of the vertex sets of these cycles. A moment's thought shows that M is the *unique* maximal subset of N such that the restriction of f onto M acts as a bijection on M. Write $f|_M = \begin{pmatrix} a & b & \ldots & z \\ f(a) & f(b) & \ldots & f(z) \end{pmatrix}$ such that the numbers $a, b, \ldots, z$ in the first row appear in natural order. This gives us an ordering $f(a), f(b), \ldots, f(z)$ of M according to the second row. Now $f(a)$ is our left end and $f(z)$ is our right end.

The tree t corresponding to the map f is now constructed as follows: Draw $f(a), \ldots, f(z)$ in this order as a *path* from $f(a)$ to $f(z)$, and fill in the remaining vertices as in $\vec{G}_f$ (deleting the arrows).

In our example above we obtain $M = \{1, 4, 5, 7, 8, 9\}$

$$f|_M = \begin{pmatrix} 1 & 4 & 5 & 7 & 8 & 9 \\ 7 & 9 & 1 & 5 & 8 & 4 \end{pmatrix}$$

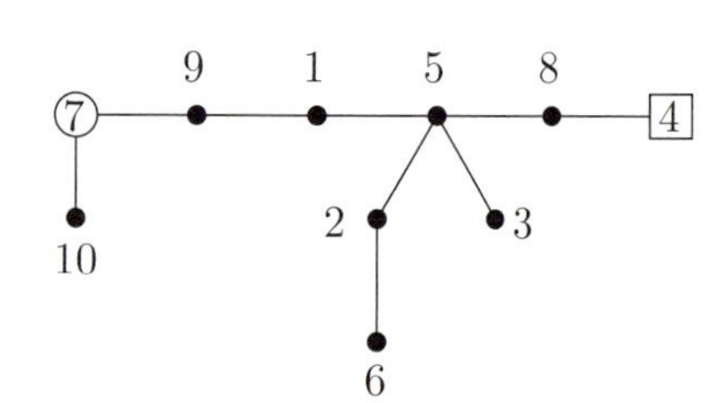

and thus the tree t depicted in the margin.

It is immediate how to reverse this correspondence: Given a tree t, we look at the unique path P from the left end to the right end. This gives us the set M and the mapping $f|_M$. The remaining correspondences $i \to f(i)$ are then filled in according to the unique paths from i to P. □

■ **Second proof (Linear Algebra).** We can think of T_n as the number of spanning trees in the complete graph K_n. Now let us look at an arbitrary connected simple graph G on $V = \{1, 2, \ldots, n\}$, denoting by $t(G)$ the number of spanning trees; thus $T_n = t(K_n)$. The following celebrated result is Kirchhoff's *matrix-tree theorem* (see [1]). Consider the incidence matrix $B = (b_{ie})$ of G, whose rows are labeled by V, the columns by E, where we write $b_{ie} = 1$ or 0 depending on whether $i \in e$ or $i \notin e$. Note that $|E| \geq n - 1$ since G is connected. In every column we replace one of the two 1's by -1 in an arbitrary manner (this amounts to an orientation of G), and call the new matrix C. $M = CC^T$ is then a symmetric $(n \times n)$-matrix with the degrees $d_1, \ldots, d_n$ in the main diagonal.

Proposition. *We have* $t(G) = \det M_{ii}$ *for all* $i = 1, \ldots, n$, *where* M_{ii} *results from* M *by deleting the* i*-th row and the* i*-th column.*

■ **Proof.** The key to the proof is the Binet-Cauchy theorem proved in the previous chapter: When P is an $(r \times s)$-matrix and Q an $(s \times r)$-matrix, $r \leq s$, then $\det(PQ)$ equals the sum of the products of determinants of corresponding $(r \times r)$-submatrices, where "corresponding" means that we take the same indices for the r columns of P and the r rows of Q.

For M_{ii} this means that

$$\det M_{ii} \;=\; \sum\nolimits_N \det N \cdot \det N^T \;=\; \sum\nolimits_N (\det N)^2,$$

where N runs through all $(n-1) \times (n-1)$ submatrices of $C \backslash \{\text{row } i\}$. The $n - 1$ columns of N correspond to a subgraph of G with $n - 1$ edges on n vertices, and it remains to show that

$$\det N = \begin{cases} \pm 1 & \text{if these edges span a tree} \\ 0 & \text{otherwise.} \end{cases}$$

"A nonstandard method of counting trees: Put a cat into each tree, walk your dog, and count how often he barks."

Suppose the $n - 1$ edges do not span a tree. Then there exists a component which does not contain i. Since the corresponding rows of this component add to 0, we infer that they are linearly dependent, and hence $\det N = 0$.

Assume now that the columns of N span a tree. Then there is a vertex $j_1 \neq i$ of degree 1; let e_1 be the incident edge. Deleting j_1, e_1 we obtain a tree with $n - 2$ edges. Again there is a vertex $j_2 \neq i$ of degree 1 with incident edge e_2. Continue in this way until $j_1, j_2, \ldots, j_{n-1}$ and $e_1, e_2, \ldots, e_{n-1}$ with $j_i \in e_i$ are determined. Now permute the rows and columns to bring j_k into the k-th row and e_k into the k-th column. Since by construction $j_k \notin e_\ell$ for $k < \ell$, we see that the new matrix N' is lower triangular with all elements on the main diagonal equal to ± 1. Thus $\det N = \pm \det N' = \pm 1$, and we are done.

For the special case $G = K_n$ we clearly obtain

$$M_{ii} = \begin{pmatrix} n-1 & -1 & \ldots & -1 \\ -1 & n-1 & & -1 \\ \vdots & & \ddots & \vdots \\ -1 & -1 & \ldots & n-1 \end{pmatrix}$$

and an easy computation shows $\det M_{ii} = n^{n-2}$. □

■ **Third proof (Recursion).** Another classical method in enumerative combinatorics is to establish a recurrence relation and to solve it by induction. The following idea is essentially due to Riordan and Rényi. To find the proper recursion, we consider a more general problem (which already appears in Cayley's paper). Let A be an arbitrary k-set of the vertices. By $T_{n,k}$ we denote the number of (labeled) forests on $\{1,\ldots,n\}$ consisting of k trees where the vertices of A appear in different trees. Clearly, the set A does not matter, only the size k. Note that $T_{n,1} = T_n$.

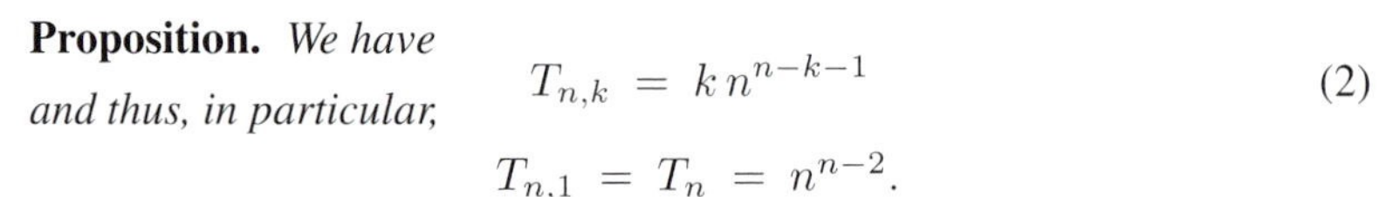

For example, $T_{4,2} = 8$ for $A = \{1,2\}$

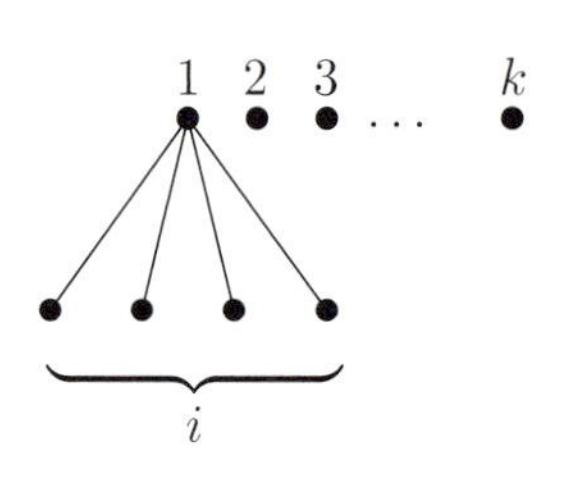

Consider such a forest F with $A = \{1,2,\ldots,k\}$, and suppose 1 is adjacent to i vertices, as indicated in the margin. Deleting 1, the i neighbors together with $2,\ldots,k$ yield one vertex each in the components of a forest that consists of $k-1+i$ trees. As we can (re)construct F by first fixing i, then choosing the i neighbors of 1 and then the forest $F\backslash 1$, this yields

$$T_{n,k} = \sum_{i=0}^{n-k} \binom{n-k}{i} T_{n-1,k-1+i} \tag{1}$$

for all $n \geq k \geq 1$, where we set $T_{0,0} = 1$, $T_{n,0} = 0$ for $n > 0$. Note that $T_{0,0} = 1$ is necessary to ensure $T_{n,n} = 1$.

Proposition. *We have*

$$T_{n,k} = k\,n^{n-k-1} \tag{2}$$

and thus, in particular,

$$T_{n,1} = T_n = n^{n-2}.$$

■ **Proof.** By (1), and using induction, we find

$$\begin{aligned}
T_{n,k} &= \sum_{i=0}^{n-k} \binom{n-k}{i} (k-1+i)(n-1)^{n-1-k-i} \qquad (i \to n-k-i)\\
&= \sum_{i=0}^{n-k} \binom{n-k}{i} (n-1-i)(n-1)^{i-1}\\
&= \sum_{i=0}^{n-k} \binom{n-k}{i} (n-1)^{i} - \sum_{i=1}^{n-k} \binom{n-k}{i} i(n-1)^{i-1}\\
&= n^{n-k} - (n-k)\sum_{i=1}^{n-k} \binom{n-1-k}{i-1} (n-1)^{i-1}\\
&= n^{n-k} - (n-k)\sum_{i=0}^{n-1-k} \binom{n-1-k}{i} (n-1)^{i}\\
&= n^{n-k} - (n-k)n^{n-1-k} = k\,n^{n-1-k}.
\end{aligned}$$

□

■ **Fourth proof (Double Counting).** The following marvelous idea due to Jim Pitman gives Cayley's formula and its generalization (2) without induction or bijection — it is just clever counting in two ways.

A *rooted forest* on $\{1,\dots,n\}$ is a forest together with a choice of a root in each component tree. Let $\mathcal{F}_{n,k}$ be the set of all rooted forests that consist of k rooted trees. Thus $\mathcal{F}_{n,1}$ is the set of all rooted trees.

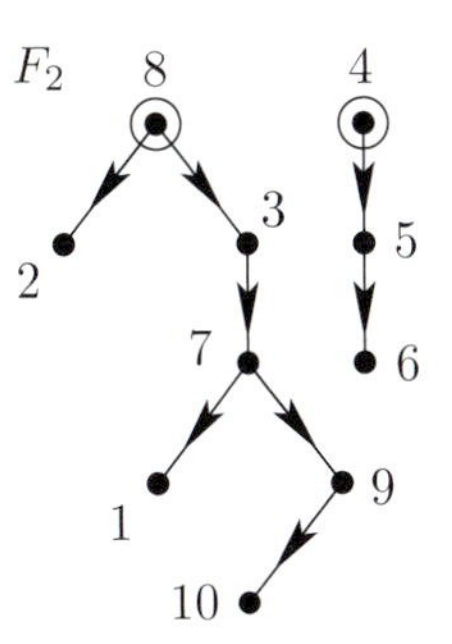

F_2 contains F_3

Note that $|\mathcal{F}_{n,1}| = nT_n$, since in every tree there are n choices for the root. We now regard $F_{n,k} \in \mathcal{F}_{n,k}$ as a *directed* graph with all edges directed away from the roots. Say that a forest F *contains* another forest F' if F contains F' as directed graph. Clearly, if F properly contains F', then F has fewer components than F'. The figure shows two such forests with the roots on top.

Here is the crucial idea. Call a sequence $F_1,\dots,F_k$ of forests a *refining sequence* if $F_i \in \mathcal{F}_{n,i}$ and F_i contains F_{i+1}, for all i.

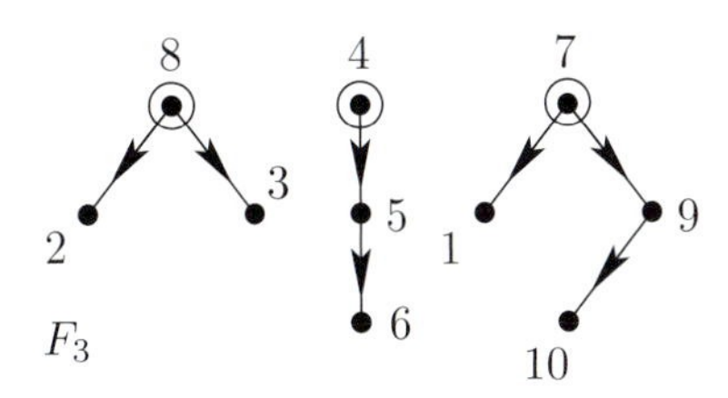

Now let F_k be a fixed forest in $\mathcal{F}_{n,k}$ and denote

- by $N(F_k)$ the number of rooted trees containing F_k, and
- by $N^*(F_k)$ the number of refining sequences ending in F_k.

We count $N^*(F_k)$ in two ways, first by starting at a tree and secondly by starting at F_k. Suppose $F_1 \in \mathcal{F}_{n,1}$ contains F_k. Since we may delete the $k-1$ edges of $F_1 \backslash F_k$ in any possible order to get a refining sequence from F_1 to F_k, we find

$$N^*(F_k) \;=\; N(F_k)\,(k-1)!. \tag{3}$$

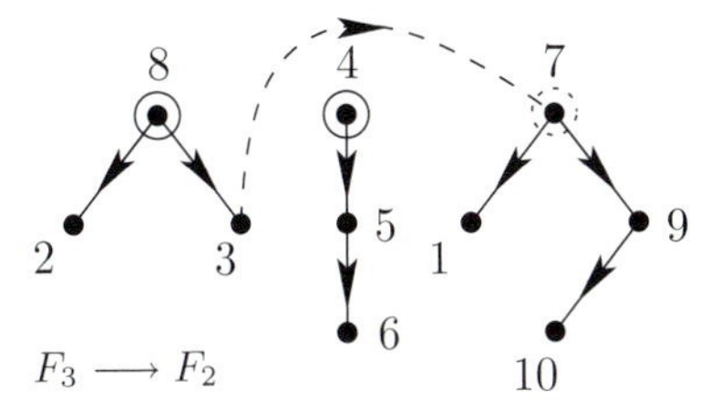

$F_3 \longrightarrow F_2$

Let us now start at the other end. To produce from F_k an F_{k-1} we have to add a directed edge, from any vertex a, to any of the $k-1$ roots of the trees that do not contain a (see the figure on the right, where we pass from F_3 to F_2 by adding the edge $3 \!\bullet\!\!\longrightarrow\!\!\bullet 7$). Thus we have $n(k-1)$ choices. Similarly, for F_{k-1} we may produce a directed edge from any vertex b to any of the $k-2$ roots of the trees not containing b. For this we have $n(k-2)$ choices. Continuing this way, we arrive at

$$N^*(F_k) \;=\; n^{k-1}(k-1)!, \tag{4}$$

and out comes, with (3), the unexpectedly simple relation

$$N(F_k) = n^{k-1} \qquad \text{for } \textit{any } F_k \in \mathcal{F}_{n,k}.$$

For $k=n$, F_n consists just of n isolated vertices. Hence $N(F_n)$ counts the number of *all* rooted trees, and we obtain $|\mathcal{F}_{n,1}| = n^{n-1}$, and thus Cayley's formula. □

But we get even more out of this proof. Formula (4) yields for $k=n$:

$$\#\{\text{refining sequences } (F_1,F_2,\dots,F_n)\} \;=\; n^{n-1}(n-1)!. \tag{5}$$

For $F_k \in \mathcal{F}_{n,k}$, let $N^{**}(F_k)$ denote the number of those refining sequences $F_1,\dots,F_n$ whose k-th term is F_k. Clearly this is $N^*(F_k)$ times the number of ways to choose $(F_{k+1}, \ldots, F_n)$. But this latter number is $(n-k)!$ since we may delete the $n-k$ edges of F_k in any possible way, and so

$$N^{**}(F_k) \;=\; N^*(F_k)(n-k)! \;=\; n^{k-1}(k-1)!(n-k)!. \qquad (6)$$

Since this number does not depend on the choice of F_k, dividing (5) by (6) yields the number of rooted forests with k trees:

$$|\mathcal{F}_{n,k}| \;=\; \frac{n^{n-1}(n-1)!}{n^{k-1}(k-1)!(n-k)!} \;=\; \binom{n}{k} k\, n^{n-1-k}.$$

As we may choose the k roots in $\binom{n}{k}$ possible ways, we have reproved the formula $T_{n,k} = kn^{n-k-1}$ without recourse to induction.

Let us end with a historical note. Cayley's paper from 1889 was anticipated by Carl W. Borchardt (1860), and this fact was acknowledged by Cayley himself. An equivalent result appeared even earlier in a paper of James J. Sylvester (1857), see [2, Chapter 3]. The novelty in Cayley's paper was the use of graph theory terms, and the theorem has been associated with his name ever since.

References

[1] M. AIGNER: *Combinatorial Theory,* Springer-Verlag, Berlin Heidelberg New York 1979; Reprint 1997.

[2] N. L. BIGGS, E. K. LLOYD & R. J. WILSON: *Graph Theory 1736-1936,* Clarendon Press, Oxford 1976.

[3] A. CAYLEY: *A theorem on trees,* Quart. J. Pure Appl. Math. **23** (1889), 376-378; Collected Mathematical Papers Vol. 13, Cambridge University Press 1897, 26-28.

[4] A. JOYAL: *Une théorie combinatoire des séries formelles,* Advances in Math. **42** (1981), 1-82.

[5] J. PITMAN: *Coalescent random forests,* J. Combinatorial Theory, Ser. A **85** (1999), 165-193.

[6] H. PRÜFER: *Neuer Beweis eines Satzes über Permutationen,* Archiv der Math. u. Physik (3) **27** (1918), 142-144.

[7] A. RÉNYI: *Some remarks on the theory of trees.* MTA Mat. Kut. Inst. Kozl. (Publ. math. Inst. Hungar. Acad. Sci.) **4** (1959), 73-85; Selected Papers Vol. 2, Akadémiai Kiadó, Budapest 1976, 363-374.

[8] J. RIORDAN: *Forests of labeled trees,* J. Combinatorial Theory **5** (1968), 90-103.

Completing Latin squares

Chapter 27

Some of the oldest combinatorial objects, whose study apparently goes back to ancient times, are the *Latin squares*. To obtain a Latin square, one has to fill the n^2 cells of an $(n \times n)$-square array with the numbers $1, 2, \ldots, n$ so that that every number appears exactly once in every row and in every column. In other words, the rows and columns each represent permutations of the set $\{1, \ldots, n\}$. Let us call n the *order* of the Latin square.

1	2	3	4
2	1	4	3
4	3	1	2
3	4	2	1

A Latin square of order 4

Here is the problem we want to discuss. Suppose someone started filling the cells with the numbers $\{1, 2, \ldots, n\}$. At some point he stops and asks us to fill in the remaining cells so that we get a Latin square. When is this possible? In order to have a chance at all we must, of course, assume that at the start of our task any element appears at most once in every row and in every column. Let us give this situation a name. We speak of a *partial Latin square* of order n if some cells of an $(n \times n)$-array are filled with numbers from the set $\{1, \ldots, n\}$ such that every number appears at most once in every row and column. So the problem is:

> *When can a partial Latin square be completed to a Latin square of the same order?*

Let us look at a few examples. Suppose the first $n-1$ rows are filled and the last row is empty. Then we can easily fill in the last row. Just note that every element appears $n-1$ times in the partial Latin square and hence is missing from exactly one column. Hence by writing each element below the column where it is missing we have completed the square correctly.

1	4	2	5	3
4	2	5	3	1
2	5	3	1	4
5	3	1	4	2
3	1	4	2	5

A cyclic Latin square

Going to the other end, suppose only the first row is filled. Then it is again easy to complete the square by cyclically rotating the elements one step in each of the following rows.

So, while in our first example the completion is forced, we have lots of possibilities in the second example. In general, the fewer cells are prefilled, the more freedom we should have in completing the square.

However, the margin displays an example of a partial square with only n cells filled which clearly cannot be completed, since there is no way to fill the upper right-hand corner without violating the row or column condition.

1	2	...	n-1	
				n

A partial Latin square that cannot be completed

> *If fewer than n cells are filled in an $(n \times n)$-array, can one then always complete it to obtain a Latin square?*

This question was raised by Trevor Evans in 1960, and the assertion that a completion is always possible quickly became known as the Evans conjecture. Of course, one would try induction, and this is what finally led to success. But Bohdan Smetaniuk's proof from 1981, which answered the question, is a beautiful example of just how subtle an induction proof may be needed in order to do such a job. And, what's more, the proof is constructive, it allows us to complete the Latin square explicitly from any initial partial configuration.

1	3	2
2	1	3
3	2	1

R: 1 1 1 2 2 2 3 3 3
C: 1 2 3 1 2 3 1 2 3
E: 1 3 2 2 1 3 3 2 1

Before proceeding to the proof let us take a closer look at Latin squares in general. We can alternatively view a Latin square as a $(3 \times n^2)$-array, called the *line array* of the Latin square. The figure to the left shows a Latin square of order 3 and its associated line array, where R, C and E stand for rows, columns and elements.

The condition on the Latin square is equivalent to saying that in any two lines of the line array all n^2 ordered pairs appear (and therefore each pair appears exactly once). Clearly, we may permute the symbols in each line arbitrarily (corresponding to permutations of rows, columns or elements) and still obtain a Latin square. But the condition on the $(3 \times n^2)$-array tells us more: There is no special role for the elements. We may also permute the lines of the array (as a whole) and still preserve the conditions on the line array and hence obtain a Latin square.

If we permute the lines of the above example cyclically, $R \longrightarrow C \longrightarrow E \longrightarrow R$, then we obtain the following line array and Latin square:

1	2	3
3	1	2
2	3	1

R: 1 3 2 2 1 3 3 2 1
C: 1 1 1 2 2 2 3 3 3
E: 1 2 3 1 2 3 1 2 3

Latin squares that are connected by any such permutation are called *conjugates*. Here is the observation which will make the proof transparent: A partial Latin square obviously corresponds to a partial line array (every pair appears at most once in any two lines), and any conjugate of a partial Latin square is again a partial Latin square. In particular, a partial Latin square can be completed if and only if any conjugate can be completed (just complete the conjugate and then reverse the permutation of the three lines).

We will need two results, due to Herbert J. Ryser and to Charles C. Lindner, that were known prior to Smetaniuk's theorem. If a partial Latin square is of the form that the first r rows are completely filled and the remaining cells are empty, then we speak of an $(r \times n)$-*Latin rectangle*.

Lemma 1. *Any $(r \times n)$-Latin rectangle, $r < n$, can be extended to an $((r+1) \times n)$-Latin rectangle and hence can be completed to a Latin square.*

■ **Proof.** We apply Hall's theorem (see Chapter 23). Let A_j be the set of numbers that do *not* appear in column j. An admissible $(r+1)$-st row corresponds then precisely to a system of distinct representatives for the collection $A_1, \dots, A_n$. To prove the lemma we therefore have to verify Hall's condition (H). Every set A_j has size $n - r$, and every element is in precisely $n - r$ sets A_j (since it appears r times in the rectangle). Any m of the sets A_j contain together $m(n-r)$ elements and therefore at least m different ones, which is just condition (H). □

Lemma 2. *Let P be a partial Latin square of order n with at most $n-1$ cells filled and at most $\frac{n}{2}$ distinct elements, then P can be completed to a Latin square of order n.*

■ **Proof.** We first transform the problem into a more convenient form. By the conjugacy principle discussed above, we may replace the condition "at most $\frac{n}{2}$ distinct elements" by the condition that the entries appear in at most $\frac{n}{2}$ rows, and we may further assume that these rows are the top rows. So let the rows with filled cells be the rows $1, 2, \ldots, r$, with f_i filled cells in row i, where $r \leq \frac{n}{2}$ and $\sum_{i=1}^{r} f_i \leq n-1$. By permuting the rows, we may assume that $f_1 \geq f_2 \geq \ldots \geq f_r$. Now we complete the rows $1, \ldots, r$ step by step until we reach an $(r \times n)$-rectangle which can then be extended to a Latin square by Lemma 1.

Suppose we have already filled rows $1, 2, \ldots, \ell - 1$. In row ℓ there are f_ℓ filled cells which we may assume to be at the end. The current situation is depicted in the figure, where the shaded part indicates the filled cells.

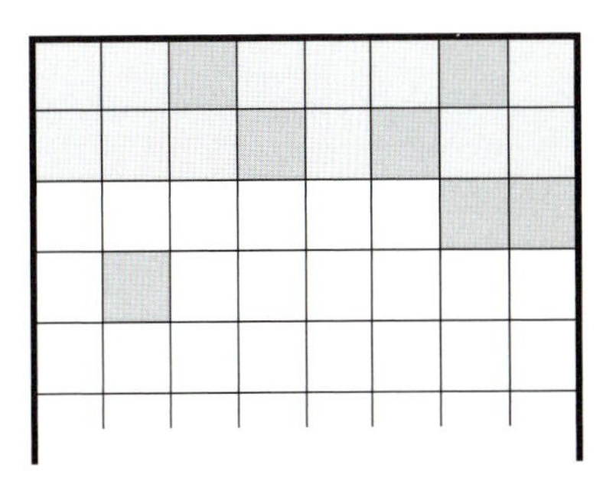

A situation for $n = 8$, with $\ell = 3$, $f_1 = f_2 = f_3 = 2$, $f_4 = 1$. The dark squares represent the pre-filled cells, the lighter ones show the cells that have been filled in the completion process.

The completion of row ℓ is performed by another application of Hall's theorem, but this time it is quite subtle. Let X be the set of elements that do *not* appear in row ℓ, thus $|X| = n - f_\ell$, and for $j = 1, \ldots, n - f_\ell$ let A_j denote the set of those elements in X which do *not* appear in column j (neither above nor below row ℓ). Hence in order to complete row ℓ we must verify condition (H) for the collection $A_1, \ldots, A_{n-f_\ell}$.

First we claim

$$n - f_\ell - \ell + 1 \;>\; \ell - 1 + f_{\ell+1} + \ldots + f_r. \qquad (1)$$

The case $\ell = 1$ is clear. Otherwise $\sum_{i=1}^{r} f_i < n$, $f_1 \geq \ldots \geq f_r$ and $1 < \ell \leq r$ together imply

$$n \;>\; \sum_{i=1}^{r} f_i \;\geq\; (\ell - 1) f_{\ell-1} + f_\ell + \ldots + f_r.$$

Now either $f_{\ell-1} \geq 2$ (in which case (1) holds) or $f_{\ell-1} = 1$. In the latter case, (1) reduces to $n > 2(\ell - 1) + r - \ell + 1 = r + \ell - 1$, which is true because of $\ell \leq r \leq \frac{n}{2}$.

Let us now take m sets A_j, $1 \leq m \leq n - f_\ell$, and let B be their union. We must show $|B| \geq m$. Consider the number c of cells in the m columns corresponding to the A_j's which contain elements of X. There are at most $(\ell - 1)m$ such cells above row ℓ and at most $f_{\ell+1} + \ldots + f_r$ below row ℓ, and thus

$$c \;\leq\; (\ell - 1)m + f_{\ell+1} + \ldots + f_r.$$

On the other hand, each element $x \in X \backslash B$ appears in each of the m columns, hence $c \geq m(|X| - |B|)$, and therefore (with $|X| = n - f_\ell$)

$$|B| \;\geq\; |X| - \tfrac{1}{m} c \;\geq\; n - f_\ell - (\ell - 1) - \tfrac{1}{m}(f_{\ell+1} + \ldots + f_r).$$

It follows that $|B| \geq m$ if

$$n - f_\ell - (\ell - 1) - \tfrac{1}{m}(f_{\ell+1} + \ldots + f_r) \;>\; m - 1,$$

that is, if

$$m(n - f_\ell - \ell + 2 - m) \;>\; f_{\ell+1} + \ldots + f_r. \qquad (2)$$

Inequality (2) is true for $m = 1$ and for $m = n - f_\ell - \ell + 1$ by (1), and hence for all values m between 1 and $n - f_\ell - \ell + 1$, since the left-hand side is a quadratic function in m with leading coefficient -1. The remaining case is $m > n - f_\ell - \ell + 1$. Since any element x of X is contained in at most $\ell - 1 + f_{\ell+1} + \ldots + f_r$ rows, it can also appear in at most that many columns. Invoking (1) once more, we find that x is in one of the sets A_j, so in this case $B = X$, $|B| = n - f_\ell \geq m$, and the proof is complete. □

Let us finally prove Smetaniuk's theorem.

Theorem. *Any partial Latin square of order n with at most $n - 1$ filled cells can be completed to a Latin square of the same order.*

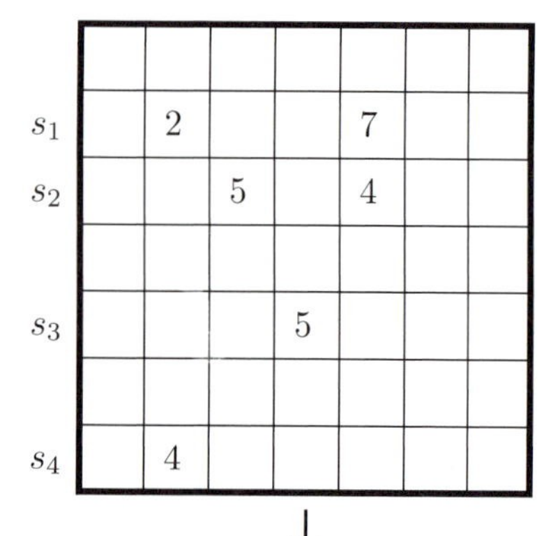

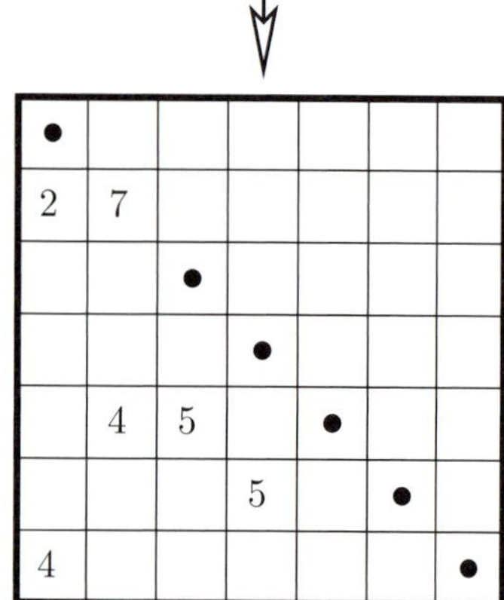

2	3	4	1	6	5	
5	6	1	4	2	3	
1	2	3	6	5	4	
6	**4**	**5**	2	3	1	
3	1	6	**5**	4	2	
4	5	2	3	1	6	

■ **Proof.** We use induction on n, the cases $n \leq 2$ being trivial. Thus we now study a partial Latin square of order $n \geq 3$ with at most $n - 1$ filled cells. With the notation used above these cells lie in $r \leq n - 1$ different rows numbered $s_1, \ldots, s_r$, which contain $f_1, \ldots, f_r > 0$ filled cells, with $\sum_{i=1}^{r} f_i \leq n - 1$. By Lemma 2 we may assume that there are more than $\frac{n}{2}$ different elements; thus there is an element that appears only once: after renumbering and permutation of rows (if necessary) we may assume that the element n occurs only once, and this is in row s_1.

In the next step we want to permute the rows and columns of the partial Latin square such that after the permutations all the filled cells lie below the diagonal — except for the cell filled with n, which will end up on the diagonal. (The diagonal consists of the cells (k, k) with $1 \leq k \leq n$.) We achieve this as follows: First we permute row s_1 into the position f_1. By permutation of columns we move all the filled cells to the left, so that n occurs as the last element in its row, on the diagonal. Next we move row s_2 into position $1 + f_1 + f_2$, and again the filled cells as far to the left as possible. In general, for $1 < i \leq r$ we move the row s_i into position $1 + f_1 + f_2 + \ldots + f_i$ and the filled cells as far left as possible. This clearly gives the desired set-up. The drawing to the left shows an example, with $n = 7$: the rows $s_1 = 2$, $s_2 = 3$, $s_3 = 5$ and $s_4 = 7$ with $f_1 = f_2 = 2$ and $f_3 = f_4 = 1$ are moved into the rows numbered 2, 5, 6 and 7, and the columns are permuted "to the left" so that in the end all entries except for the single 7 come to lie below the diagonal, which is marked by •s.

In order to be able to apply induction we now remove the entry n from the diagonal and ignore the first row and the last column (which do not not contain any filled cells): thus we are looking at a partial Latin square of order $n - 1$ with at most $n - 2$ filled cells, which by induction can be completed to a Latin square of order $n - 1$. The margin shows one (of many) completions of the partial Latin square that arises in our example. In the figure, the original entries are printed bold. They are already final, as are all the elements in shaded cells; some of the other entries will be changed in the following, in order to complete the Latin square of order n.

In the next step we want to move the diagonal elements of the square to the last column and put entries n onto the diagonal in their place. However, in general we cannot do this, since the diagonal elements need not be distinct. Thus we proceed more carefully and perform successively, for $k = 2, 3, \dots, n-1$ (in this order), the following operation:

Put the value n into the cell (k, n). This yields a correct partial Latin square. Now exchange the value x_k in the diagonal cell (k, k) with the value n in the cell (k, n) in the last column.

If the value x_k did not already occur in the last column, then our job for the current k is completed. After this, the current elements in the k-th column will not be changed any more.

In our example this works without problems for $k = 2$, 3 and 4, and the corresponding diagonal elements 3, 1 and 6 move to the last column. The following three figures show the corresponding operations.

2	3	4	1	6	5	7
5	6	1	4	2	3	
1	2	3	6	5	4	
6	**4**	**5**	2	3	1	
3	1	6	**5**	4	2	
4	5	2	3	1	6	

2	7	4	1	6	5	3
5	6	1	4	2	3	7
1	2	3	6	5	4	
6	**4**	**5**	2	3	1	
3	1	6	**5**	4	2	
4	5	2	3	1	6	

2	7	4	1	6	5	3
5	6	7	4	2	3	1
1	2	3	6	5	4	7
6	**4**	**5**	2	3	1	
3	1	6	**5**	4	2	
4	5	2	3	1	6	

Now we have to treat the case in which there is already an element x_k in the last column. In this case we proceed as follows:

If there is already an element x_k in a cell (j, n) with $2 \leq j < k$, then we exchange in row j the element x_k in the n-th column with the element x'_k in the k-th column. If the element x'_k also occurs in a cell (j', n), then we also exchange the elements in the j'-th row that occur in the n-th and in the k-th columns, and so on.

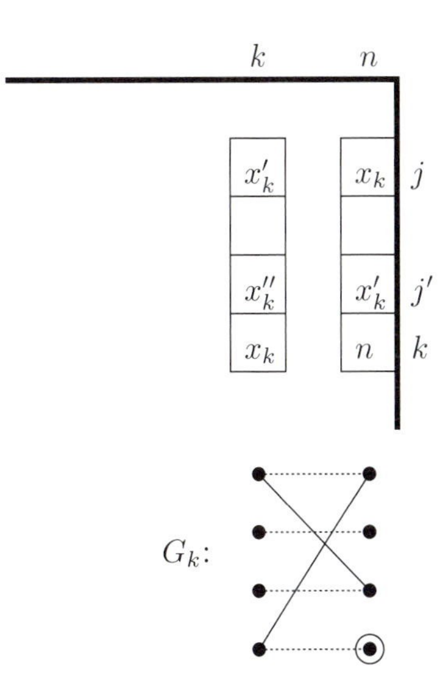

If we proceed like this there will never be two equal entries in a row. Our exchange process ensures that there also will never be two equal elements in a column. So we only have to verify that the exchange process between the k-th and the n-th column does not lead to an infinite loop. This can be seen from the following bipartite graph G_k: Its vertices correspond to the cells (i, k) and (j, n) with $2 \leq i, j \leq k$ whose elements might be exchanged. There is an edge between (i, k) and (j, n) if these two cells lie in the same row (that is, for $i = j$), or if the cells before the exchange process contain the same element (which implies $i \neq j$). In our sketch the edges for $i = j$ are dotted, the others are not. All vertices in G_k have degree 1 or 2. The cell (k, n) corresponds to a vertex of degree 1; this vertex is the beginning of a path which leads to column k on a horizontal edge, then possibly on a sloped edge back to column n, then horizontally back to column k and so on. It ends in column k at a value that does not occur in column n. Thus the exchange operations will end at some point with a step where we move a *new* element into the last column. Then the work on column k is completed, and the elements in the cells (i, k) for $i \geq 2$ are fixed for good.

In our example the "exchange case" happens for $k = 5$: the element $x_5 = 3$ does already occur in the last column, so that entry has to be moved back to column $k = 5$. But the exchange element $x'_5 = 6$ is not new either, it is exchanged by $x''_5 = 5$, and this one is new.

2	7	4	1	6	5	3
5	6	7	4	2	3	1
1	2	3	7	5	4	6
6	**4**	**5**	2	3	1	7
3	1	6	**5**	4	2	
4	5	2	3	1	6	

2	7	4	1	3	5	6
5	6	7	4	2	3	1
1	2	3	7	6	4	5
6	**4**	**5**	2	7	1	3
3	1	6	**5**	4	2	
4	5	2	3	1	6	

Finally, the exchange for $k = 6 = n - 1$ poses no problem, and after that the completion of the Latin square is unique:

2	7	4	1	3	5	6
5	6	7	4	2	3	1
1	2	3	7	6	4	5
6	**4**	**5**	2	7	1	3
3	1	6	**5**	4	2	7
4	5	2	3	1	6	

2	7	4	1	3	5	6
5	6	7	4	2	3	1
1	2	3	7	6	4	5
6	**4**	**5**	2	7	1	3
3	1	6	**5**	4	7	2
4	5	2	3	1	6	

7	3	1	6	4	2	4
2	7	4	1	3	5	6
5	6	7	4	2	3	1
1	2	3	7	6	4	5
6	**4**	**5**	2	7	1	3
3	1	6	**5**	4	7	2
4	5	2	3	1	6	7

... and the same occurs in general: We put an element n into the cell (n, n), and after that the first row can be completed by the missing elements of the respective columns (see Lemma 1), and this completes the proof. In order to get explicitly a completion of the original partial Latin square of order n, we only have to reverse the element, row and column permutations of the first two steps of the proof. □

References

[1] T. EVANS: *Embedding incomplete Latin squares,* Amer. Math. Monthly **67** (1960), 958-961.

[2] C. C. LINDNER: *On completing Latin rectangles,* Canadian Math. Bulletin **13** (1970), 65-68.

[3] H. J. RYSER: *A combinatorial theorem with an application to Latin rectangles,* Proc. Amer. Math. Soc. **2** (1951), 550-552.

[4] B. SMETANIUK: *A new construction on Latin squares I: A proof of the Evans conjecture,* Ars Combinatoria **11** (1981), 155-172.

The Dinitz problem

Chapter 28

The four-color problem was a main driving force for the development of graph theory as we know it today, and coloring is still a topic that many graph theorists like best. Here is a simple-sounding coloring problem, raised by Jeff Dinitz in 1978, which defied all attacks until its astonishingly simple solution by Fred Galvin fifteen years later.

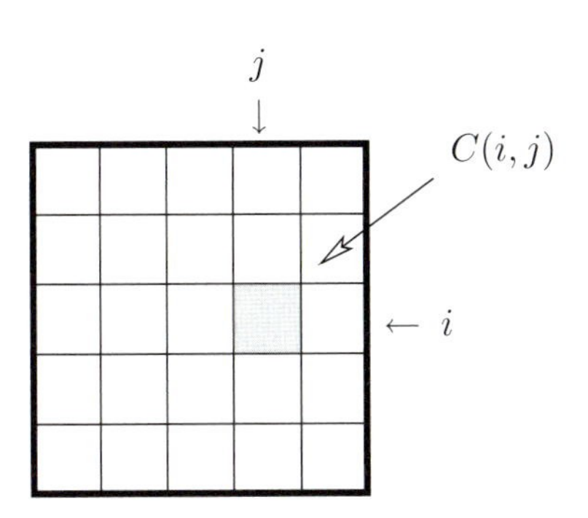

> *Consider n^2 cells arranged in an $(n \times n)$-square, and let (i,j) denote the cell in row i and column j. Suppose that for every cell (i,j) we are given a set $C(i,j)$ of n colors.*
> *Is it then always possible to color the whole array by picking for each cell (i,j) a color from its set $C(i,j)$ such that the colors in each row and each column are distinct?*

As a start consider the case when all color sets $C(i,j)$ are the same, say $\{1,2,\dots,n\}$. Then the Dinitz problem reduces to the following task: Fill the $(n \times n)$-square with the numbers $1,2,\dots,n$ in such a way that the numbers in any row and column are distinct. In other words, any such coloring corresponds to a Latin square, as discussed in the previous chapter. So, in this case, the answer to our question is "yes."

Since this is so easy, why should it be so much harder in the general case when the set $C := \bigcup_{i,j} C(i,j)$ contains even more than n colors? The difficulty derives from the fact that not every color of C is available at each cell. For example, whereas in the Latin square case we can clearly choose an arbitrary permutation of the colors for the first row, this is not so anymore in the general problem. Already the case $n = 2$ illustrates this difficulty.

$\{1,2\}$	$\{2,3\}$
$\{1,3\}$	$\{2,3\}$

Suppose we are given the color sets that are indicated in the figure. If we choose the colors 1 and 2 for the first row, then we are in trouble since we would then have to pick color 3 for both cells in the second row.

Before we tackle the Dinitz problem, let us rephrase the situation in the language of graph theory. As usual we only consider graphs $G = (V,E)$ without loops and multiple edges. Let $\chi(G)$ denote the *chromatic number* of the graph, that is, the smallest number of colors that one can assign to the vertices such that adjacent vertices receive different colors.

In other words, a coloring calls for a partition of V into classes (colored with the same color) such that there are no edges within a class. Calling a set $A \subseteq V$ *independent* if there are no edges within A, we infer that the chromatic number is the smallest number of independent sets which partition the vertex set V.

In 1976 Vizing, and three years later Erdős, Rubin, and Taylor, studied the following coloring variant which leads us straight to the Dinitz problem. Suppose in the graph $G = (V, E)$ we are given a set $C(v)$ of colors for each vertex v. A *list coloring* is a coloring $c : V \longrightarrow \bigcup_{v \in V} C(v)$ where $c(v) \in C(v)$ for each $v \in V$. The definition of the *list chromatic number* $\chi_\ell(G)$ should now be clear: It is the smallest number k such for *any* list of color sets $C(v)$ with $|C(v)| = k$ for all $v \in V$ there always exists a list coloring. Of course, we have $\chi_\ell(G) \leq |V|$ (we never run out of colors). Since ordinary coloring is just the special case of list coloring when all sets $C(v)$ are equal, we obtain for any graph G

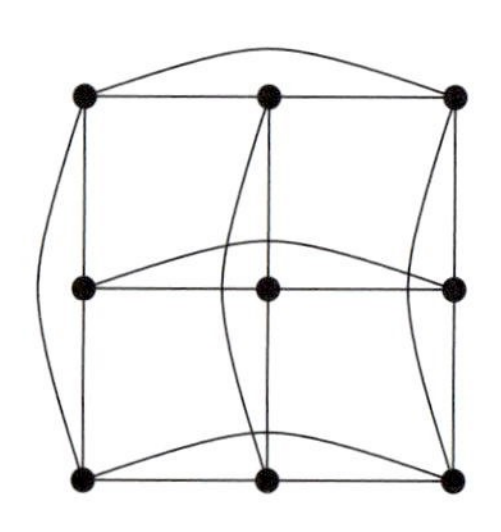

The graph S_3

$$\chi(G) \;\leq\; \chi_\ell(G).$$

To get back to the Dinitz problem, consider the graph S_n which has as vertex set the n^2 cells of our $(n \times n)$-array with two cells adjacent if and only if they are in the same row or column.

Since any n cells in a row are pairwise adjacent we need at least n colors. Furthermore, any coloring with n colors corresponds to a Latin square, with the cells occupied by the same number forming a color class. Since Latin squares, as we have seen, exist, we infer $\chi(S_n) = n$, and the Dinitz problem can now be succinctly stated as

$$\chi_\ell(S_n) \;=\; n?$$

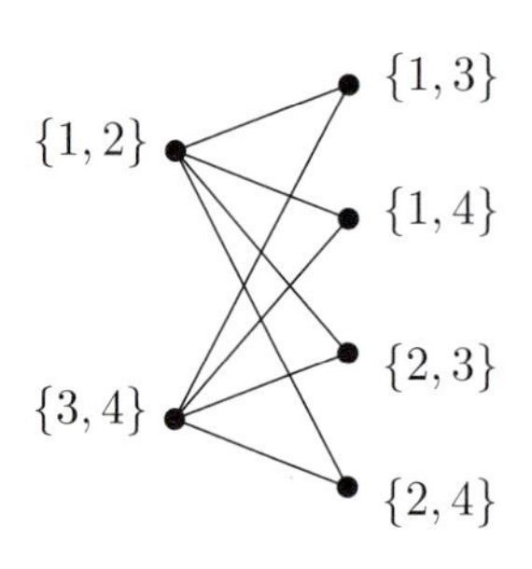

One might think that perhaps $\chi(G) = \chi_\ell(G)$ holds for any graph G, but this is a long shot from the truth. Consider the graph $G = K_{2,4}$. The chromatic number is 2 since we may use one color for the two left vertices and the second color for the vertices on the right. But now suppose that we are given the color sets indicated in the figure.

To color the left vertices we have the four possibilities $1|3$, $1|4$, $2|3$ and $2|4$, but any one of these pairs appears as a color set on the right-hand side, so a list coloring is not possible. Hence $\chi_\ell(G) \geq 3$, and the reader may find it fun to prove $\chi_\ell(G) = 3$ (there is no need to try out all possibilities!). Generalizing this example, it is not hard to find graphs G where $\chi(G) = 2$, but $\chi_\ell(G)$ is arbitrarily large! So the list coloring problem is not as easy as it looks at first glance.

Back to the Dinitz problem. A significant step towards the solution was made by Jeanette Janssen in 1992 when she proved $\chi_\ell(S_n) \leq n + 1$, and the *coup de grâce* was delivered by Fred Galvin by ingeniously combining two results, both of which had long been known. We are going to discuss these two results and show then how they imply $\chi_\ell(S_n) = n$.

First we fix some notation. Suppose v is a vertex of the graph G, then we denote as before by $d(v)$ the *degree* of v. In our square graph S_n every vertex has degree $2n - 2$, accounting for the $n - 1$ other vertices in the same row and in the same column. For a subset $A \subseteq V$ we denote by G_A the subgraph which has A as vertex set and which contains all edges of G between vertices of A. We call G_A the subgraph induced by A, and say that H is an *induced subgraph* of G if $H = G_A$ for some A.

To state our first result we need *directed graphs* $\vec{G} = (V, E)$, that is, graphs where every edge e has an orientation. The notation $e = (u, v)$ means that there is an arc e, also denoted by $u \longrightarrow v$, whose initial vertex is u and whose terminal vertex is v. It then makes sense to speak of the *outdegree* $d^+(v)$ resp. the *indegree* $d^-(v)$, where $d^+(v)$ counts the number of edges with v as initial vertex, and similarly for $d^-(v)$; furthermore, $d^+(v) + d^-(v) = d(v)$. When we write G, we mean the graph $\vec{G}$ without the orientations.

The following concept originated in the analysis of games and will play a crucial role in our discussion.

Definition 1. Let $\vec{G} = (V, E)$ be a directed graph. A *kernel* $K \subseteq V$ is a subset of the vertices such that

(i) K is independent in G, and

(ii) for every $u \notin K$ there exists a vertex $v \in K$ with an edge $u \longrightarrow v$.

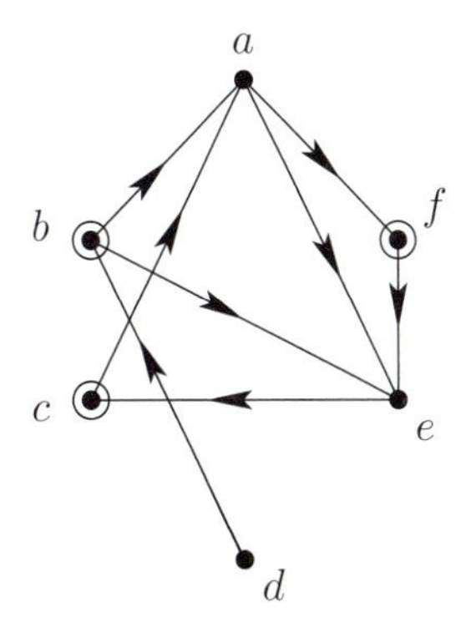

Let us look at the example in the figure. The set $\{b, c, f\}$ constitutes a kernel, but the subgraph induced by $\{a, c, e\}$ does not have a kernel since the three edges cycle through the vertices.

With all these preparations we are ready to state the first result.

Lemma 1. *Let $\vec{G} = (V, E)$ be a directed graph, and suppose that for each vertex $v \in V$ we have a color set $C(v)$ that is larger than the outdegree, $|C(v)| \geq d^+(v) + 1$. If every induced subgraph of $\vec{G}$ possesses a kernel, then there exists a list coloring of G with a color from $C(v)$ for each v.*

■ **Proof.** We proceed by induction on $|V|$. For $|V| = 1$ there is nothing to prove. Choose a color $c \in C = \bigcup_{v \in V} C(v)$ and set

$$A(c) := \{v \in V : c \in C(v)\}.$$

By hypothesis, the induced subgraph $G_{A(c)}$ possesses a kernel $K(c)$. Now we color all $v \in K(c)$ with the color c (this is possible since $K(c)$ is independent), and delete $K(c)$ from G and c from C. Let G' be the induced subgraph of G on $V \backslash K(c)$ with $C'(v) = C(v) \backslash c$ as the new list of color sets. Notice that for each $v \in A(c) \backslash K(c)$, the outdegree $d^+(v)$ is decreased by at least 1 (due to condition (ii) of a kernel). So $d^+(v) + 1 \leq |C'(v)|$ still holds in $\vec{G}'$. The same condition also holds for the vertices outside $A(c)$, since in this case the color sets $C(v)$ remain unchanged. The new graph G' contains fewer vertices than G, and we are done by induction. □

The method of attack for the Dinitz problem is now obvious: We have to find an orientation of the graph S_n with outdegrees $d^+(v) \leq n - 1$ for all v and which ensures the existence of a kernel for all induced subgraphs. This is accomplished by our second result.

Again we need a few preparations. Recall (from Chapter 9) that a *bipartite* graph $G = (X \cup Y, E)$ is a graph with the following property: The vertex set V is split into two parts X and Y such that every edge has one endvertex in X and the other in Y. In other words, the bipartite graphs are precisely those which can be colored with two colors (one for X and one for Y).

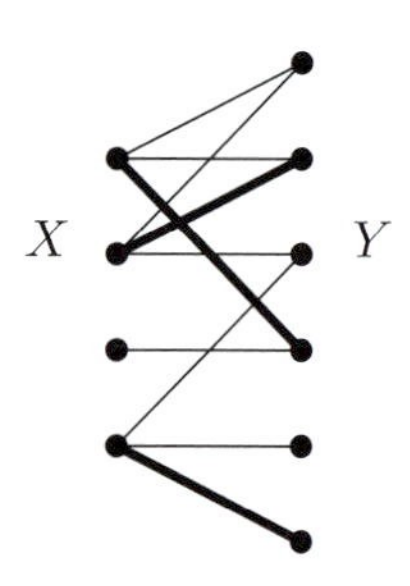

A bipartite graph with a matching

Now we come to an important concept, "stable matchings," with a down-to-earth interpretation. A *matching* M in a bipartite graph $G = (X \cup Y, E)$ is a set of edges such that no two edges in M have a common endvertex. In the displayed graph the edges drawn in bold lines constitute a matching.

Consider X to be a set of men and Y a set of women and interpret $uv \in E$ to mean that u and v might marry. A matching is then a mass-wedding with no person committing bigamy. For our purposes we need a more refined (and more realistic?) version of a matching, suggested by David Gale and Lloyd S. Shapley. Clearly, in real life every person has preferences, and this is what we add to the set-up. In $G = (X \cup Y, E)$ we assume that for every $v \in X \cup Y$ there is a ranking of the set $N(v)$ of vertices adjacent to v, $N(v) = \{z_1 > z_2 > \ldots > z_{d(v)}\}$. Thus z_1 is the top choice for v, followed by z_2, and so on.

Definition 2. A matching M of $G = (X \cup Y, E)$ is called *stable* if the following condition holds: Whenever $uv \in E \backslash M$, $u \in X$, $v \in Y$, then either $uy \in M$ with $y > v$ in $N(u)$ or $xv \in M$ with $x > u$ in $N(v)$, or both.

In our real life interpretation a set of marriages is stable if it never happens that u and v are not married but u prefers v to his partner (if he has one at all) and v prefers u to her mate (if she has one at all), which would clearly be an unstable situation.

Before proving our second result let us take a look at the following example:

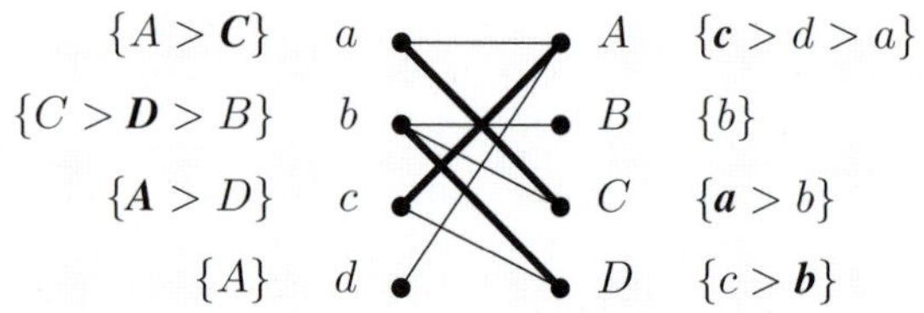

$\{A > \mathbf{C}\}$	a	A	$\{\mathbf{c} > d > a\}$
$\{C > \mathbf{D} > B\}$	b	B	$\{b\}$
$\{\mathbf{A} > D\}$	c	C	$\{\mathbf{a} > b\}$
$\{A\}$	d	D	$\{c > \mathbf{b}\}$

The bold edges constitute a stable matching. In each priority list, the choice leading to a stable matching is printed bold.

Notice that in this example there is a unique largest matching M with four edges, $M = \{aC, bB, cD, dA\}$, but M is not stable (consider cA).

Lemma 2. *A stable matching always exists.*

■ **Proof.** Consider the following algorithm. In the first stage all men $u \in X$ propose to their top choice. If a girl receives more than one proposal she picks the one she likes best and keeps him on a string, and if she receives just one proposal she keeps that one on a string. The remaining men are rejected and form the reservoir R. In the second stage all men in R propose to their next choice. The women compare the proposals (together with the one on the string, if there is one), pick their favorite and put him on the string. The rest is rejected and forms the new set R. Now the men in R propose to their next choice, and so on. A man who has proposed to his last choice and is again rejected drops out from further consideration (as well as from the reservoir). Clearly, after some time the reservoir R is empty, and at this point the algorithm stops.

Claim. *When the algorithm stops, then the men on the strings together with the corresponding girls form a stable matching.*

Notice first that the men on the string of a particular girl move there in increasing preference (of the girl) since at each stage the girl compares the new proposals with the present mate and then picks the new favorite. Hence if $uv \in E$ but $uv \notin M$, then either u never proposed to v in which case he found a better mate before he even got around to v, implying $uy \in M$ with $y > v$ in $N(u)$, or u proposed to v but was rejected, implying $xv \in M$ with $x > u$ in $N(v)$. But this is exactly the condition of a stable matching. □

Putting Lemmas 1 and 2 together, we now get Galvin's solution of the Dinitz problem.

Theorem. *We have* $\chi_\ell(S_n) = n$ *for all* n.

■ **Proof.** As before we denote the vertices of S_n by (i,j), $1 \leq i,j \leq n$. Thus (i,j) and (r,s) are adjacent if and only if $i = r$ or $j = s$. Take any Latin square L with letters from $\{1,2,\ldots,n\}$ and denote by $L(i,j)$ the entry in cell (i,j). Next make S_n into a directed graph $\vec{S}_n$ by orienting the horizontal edges $(i,j) \longrightarrow (i,j')$ if $L(i,j) < L(i,j')$ and the vertical edges $(i,j) \longrightarrow (i',j)$ if $L(i,j) > L(i',j)$. Thus, horizontally we orient from the smaller to the larger element, and vertically the other way around. (In the margin we have an example for $n = 3$.)

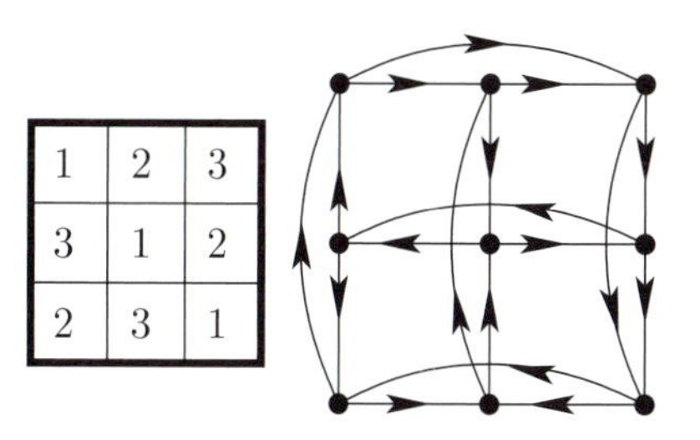

Notice that we obtain $d^+(i,j) = n-1$ for all (i,j). In fact, if $L(i,j) = k$, then $n-k$ cells in row i contain an entry larger than k, and $k-1$ cells in column j have an entry smaller than k.

By Lemma 1 it remains to show that every induced subgraph of $\vec{S}_n$ possesses a kernel. Consider a subset $A \subseteq V$, and let X be the set of rows of L, and Y the set of its columns. Associate to A the bipartite graph $G = (X \cup Y, A)$, where every $(i,j) \in A$ is represented by the edge ij with $i \in X, j \in Y$. In the example in the margin the cells of A are shaded.

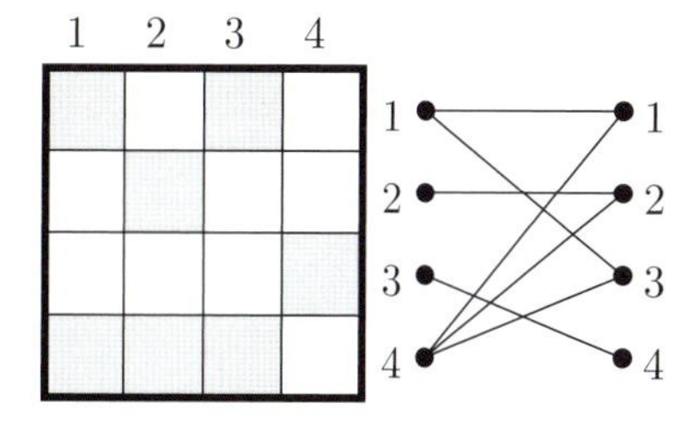

The orientation on S_n naturally induces a ranking on the neighborhoods in $G = (X \cup Y, A)$ by setting $j' > j$ in $N(i)$ if $(i,j) \longrightarrow (i,j')$ in $\vec{S}_n$ respectively $i' > i$ in $N(j)$ if $(i,j) \longrightarrow (i',j)$. By Lemma 2, $G = (X \cup Y, A)$ possesses a stable matching M. This M, viewed as a subset of A, is our desired kernel! To see why, note first that M is independent in A since as edges in $G = (X \cup Y, A)$ they do not share an endvertex i or j. Secondly, if $(i,j) \in A \backslash M$, then by the definition of a stable matching there either exists $(i,j') \in M$ with $j' > j$ or $(i',j) \in M$ with $i' > i$, which for $\vec{S}_n$ means $(i,j) \longrightarrow (i,j') \in M$ or $(i,j) \longrightarrow (i',j) \in M$, and the proof is complete. □

To end the story let us go a little beyond. The reader may have noticed that the graph S_n arises from a bipartite graph by a simple construction. Take the complete bipartite graph, denoted by $K_{n,n}$, with $|X| = |Y| = n$, and *all* edges between X and Y. If we consider the edges of $K_{n,n}$ as vertices of a new graph, joining two such vertices if and only if as edges in $K_{n,n}$ they have a common endvertex, then we clearly obtain the square graph S_n. Let us say that S_n is the *line graph* of $K_{n,n}$. Now this same construction can be performed on any graph G with the resulting graph called the *line graph* $L(G)$ of G.

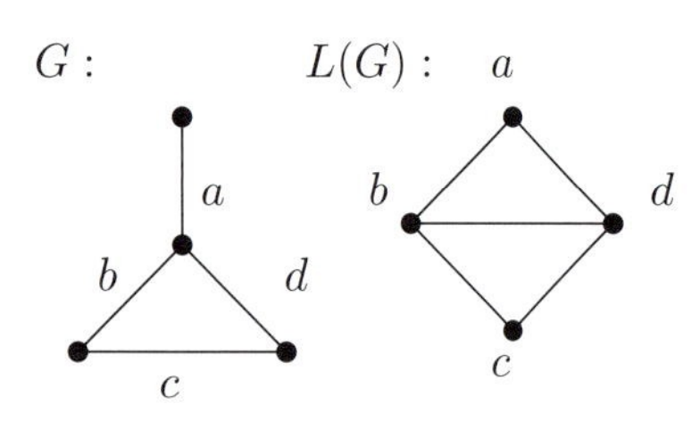

Construction of a line graph

In general, call H a *line graph* if $H = L(G)$ for some graph G. Of course, not every graph is a line graph, an example being the graph $K_{2,4}$ that we considered earlier, and for this graph we have seen $\chi(K_{2,4}) < \chi_\ell(K_{2,4})$. But what if H *is* a line graph? By adapting the proof of our theorem it can easily be shown that $\chi(H) = \chi_\ell(H)$ holds whenever H is the line graph of a *bipartite* graph, and the method may well go some way in verifying the supreme conjecture in this field:

Does $\chi(H) = \chi_\ell(H)$ *hold for every line graph* H*?*

Very little is known about this conjecture, and things look hard — but after all, so did the Dinitz problem twenty years ago.

References

[1] P. Erdős, A. L. Rubin & H. Taylor: *Choosability in graphs,* Proc. West Coast Conference on Combinatorics, Graph Theory and Computing, Congressus Numerantium **26** (1979), 125-157.

[2] D. Gale & L. S. Shapley: *College admissions and the stability of marriage,* Amer. Math. Monthly **69** (1962), 9-15.

[3] F. Galvin: *The list chromatic index of a bipartite multigraph,* J. Combinatorial Theory, Ser. B **63** (1995), 153-158.

[4] J. C. M. Janssen: *The Dinitz problem solved for rectangles,* Bulletin Amer. Math. Soc. **29** (1993), 243-249.

[5] V. G. Vizing: *Coloring the vertices of a graph in prescribed colours (in Russian),* Metody Diskret. Analiz. **101** (1976), 3-10.

Identities versus bijections

Chapter 29

Consider the infinite product $(1+x)(1+x^2)(1+x^3)(1+x^4)\cdots$ and expand it in the usual way into a series $\sum_{n\geq 0} a_n x^n$ by grouping together those products that yield the same power x^n. By inspection we find for the first terms

$$\prod_{k\geq 1}(1+x^k) = 1+x+x^2+2x^3+2x^4+3x^5+4x^6+5x^7+\dots. \quad (1)$$

So we have e. g. $a_6 = 4$, $a_7 = 5$, and we (rightfully) suspect that a_n goes to infinity with $n \longrightarrow \infty$.

Looking at the equally simple product $(1-x)(1-x^2)(1-x^3)(1-x^4)\cdots$ something unexpected happens. Expanding this product we obtain

$$\prod_{k\geq 1}(1-x^k) \;=\; 1-x-x^2+x^5+x^7-x^{12}-x^{15}+x^{22}+x^{26}-\;\dots. \quad (2)$$

It seems that all coefficients are equal to 1, -1 or 0. But is this true? And if so, what is the pattern?

Infinite sums and products and their convergence have played a central role in analysis since the invention of the calculus, and contributions to the subject have been made by some of the greatest names in the field, from Leonhard Euler to Srinivasa Ramanujan.

In explaining identities such as (1) and (2), however, we disregard convergence questions — we simply manipulate the coefficients. In the language of the trade we deal with "formal" power series and products. In this framework we are going to show how combinatorial arguments lead to elegant proofs of seemingly difficult identities.

Our basic notion is that of a *partition* of a natural number. We call any sum

$$\lambda:\; n \;=\; \lambda_1+\lambda_2+\ldots+\lambda_t \quad \text{with} \quad \lambda_1\geq\lambda_2\geq\ldots\geq\lambda_t\geq 1$$

a *partition* of n. $P(n)$ shall be the set of all partitions of n, with $p(n) := |P(n)|$, where we set $p(0)=1$.

$5=5$
$5=4+1$
$5=3+2$
$5=3+1+1$
$5=2+2+1$
$5=2+1+1+1$
$5=1+1+1+1+1.$

The partitions counted by $p(5)=7$

What have partitions got to do with our problem? Well, consider the following product of infinitely many series:

$$(1+x+x^2+x^3+\ldots)(1+x^2+x^4+x^6+\ldots)(1+x^3+x^6+x^9+\ldots)\;\cdots \quad (3)$$

where the k-th factor is $(1+x^k+x^{2k}+x^{3k}+\ldots)$. What is the coefficient of x^n when we expand this product into a series $\sum_{n\geq 0} a_n x^n$? A moment's thought should convince you that this is just the number of ways to write n as a sum

$$\begin{aligned} n &= n_1 \cdot 1 + n_2 \cdot 2 + n_3 \cdot 3 + \ldots \\ &= \underbrace{1 + \ldots + 1}_{n_1} + \underbrace{2 + \ldots + 2}_{n_2} + \underbrace{3 + \ldots + 3}_{n_3} + \ldots. \end{aligned}$$

So the coefficient is nothing else but the number $p(n)$ of partitions of n. Since the geometric series $1 + x^k + x^{2k} + \ldots$ equals $\frac{1}{1-x^k}$, we have proved our first identity:

$$\prod_{k \geq 1} \frac{1}{1-x^k} = \sum_{n \geq 0} p(n)\, x^n. \tag{4}$$

What's more, we see from our analysis that the factor $\frac{1}{1-x^k}$ accounts for the contribution of k to a partition of n. Thus, if we leave out $\frac{1}{1-x^k}$ from the product on the left side of (4), then k does not appear in any partition on the right side. As an example we immediately obtain

$6 = 5 + 1$
$6 = 3 + 3$
$6 = 3 + 1 + 1 + 1$
$6 = 1 + 1 + 1 + 1 + 1 + 1$

Partitions of 6 into odd parts: $p_o(6) = 4$

$$\prod_{i \geq 1} \frac{1}{1-x^{2i-1}} = \sum_{n \geq 0} p_o(n)\, x^n, \tag{5}$$

where $p_o(n)$ is the number of partitions of n all of whose summands are *odd*, and the analogous statement holds when all summands are *even*.

By now it should be clear what the n-th coefficient in the infinite product $\prod_{k \geq 1}(1 + x^k)$ will be. Since we take from any factor in (3) either 1 or x^k, this means that we consider only those partitions where any summand k appears at most once. In other words, our original product (1) is expanded into

$$\prod_{k \geq 1} (1 + x^k) = \sum_{n \geq 0} p_d(n)\, x^n, \tag{6}$$

where $p_d(n)$ is the number of partitions of n into *distinct* summands.

$7 = 7$
$7 = 5 + 1 + 1$
$7 = 3 + 3 + 1$
$7 = 3 + 1 + 1 + 1 + 1$
$7 = 1 + 1 + 1 + 1 + 1 + 1 + 1$

$7 = 7$
$7 = 6 + 1$
$7 = 5 + 2$
$7 = 4 + 3$
$7 = 4 + 2 + 1.$

The partitions of 7 into odd resp. distinct parts: $p_o(7) = p_d(7) = 5$.

Now the method of formal series displays its full power. Since $1 - x^2 = (1-x)(1+x)$ we may write

$$\prod_{k \geq 1} (1 + x^k) = \prod_{k \geq 1} \frac{1 - x^{2k}}{1 - x^k} = \prod_{k \geq 1} \frac{1}{1 - x^{2k-1}}$$

since all factors $1 - x^{2i}$ with even exponent cancel out. So, the infinite products in (5) and (6) are the same, and hence also the series, and we obtain the beautiful result

$$p_o(n) = p_d(n) \qquad \text{for all } n \geq 0. \tag{7}$$

Such a striking equality demands a simple proof by bijection — at least that is the point of view of any combinatorialist.

Problem. *Let $P_o(n)$ and $P_d(n)$ be the partitions of n into odd and into distinct summands, respectively: Find a bijection from $P_o(n)$ onto $P_d(n)$!*

Several bijections are known, but the following one due to J. W. L. Glaisher (1907) is perhaps the neatest. Let λ be a partition of n into odd parts. We collect equal summands and have

$$\begin{aligned} n &= \underbrace{\lambda_1 + \ldots + \lambda_1}_{n_1} + \underbrace{\lambda_2 + \ldots + \lambda_2}_{n_2} + \ldots + \underbrace{\lambda_t + \ldots + \lambda_t}_{n_t} \\ &= n_1 \cdot \lambda_1 + n_2 \cdot \lambda_2 + \ldots + n_t \cdot \lambda_t. \end{aligned}$$

Now we write $n_1 = 2^{m_1} + 2^{m_2} + \ldots + 2^{m_r}$ in its binary representation and similarly for the other n_i. The new partition λ' of n is then

$$\lambda' : \; n = 2^{m_1}\lambda_1 + 2^{m_2}\lambda_1 + \ldots + 2^{m_r}\lambda_1 + 2^{k_1}\lambda_2 + \ldots .$$

We have to check that λ' is in $P_d(n)$, and that $\phi : \lambda \longmapsto \lambda'$ is indeed a bijection. Both claims are easy to verify: If $2^a\lambda_i = 2^b\lambda_j$ then $2^a = 2^b$ since λ_i and λ_j are odd, and so $\lambda_i = \lambda_j$. Hence λ' is in $P_d(n)$. Conversely, when $n = \mu_1 + \mu_2 + \ldots + \mu_s$ is a partition into distinct summands, then we reverse the bijection by collecting all μ_i with the same highest power of 2, and write down the odd parts with the proper multiplicity. The margin displays an example.

For example,
$\lambda : 25 = 5{+}5{+}5{+}3{+}3{+}1{+}1{+}1{+}1$
is mapped by ϕ to
$\lambda' : 25 = (2{+}1)5 + (2)3 + (4)1$
$= 10 + 5 + 6 + 4$
$= 10 + 6 + 5 + 4\,.$

We write
$\lambda' : 30 = 12 + 6 + 5 + 4 + 3$
as $30 = 4(3{+}1) + 2(3) + 1(5{+}3)$
$= (1)5 + (4{+}2{+}1)3 + (4)1$
and obtain as $\phi^{-1}(\lambda')$ the partition
$\lambda : 30 = 5 + 3 + 3 + 3 + 3 + 3 + 3 + 3 + 1 + 1 + 1 + 1$
into odd summands.

Manipulating formal products has thus led to the equality $p_o(n) = p_d(n)$ for partitions which we then verified via a bijection. Now we turn this around, give a bijection proof for partitions and deduce an identity. This time our goal is to identify the pattern in the expansion (2).
Look at

$$1 - x - x^2 + x^5 + x^7 - x^{12} - x^{15} + x^{22} + x^{26} - x^{35} - x^{40} + \ldots .$$

The exponents (apart from 0) seem to come in pairs, and taking the exponents of the first power in each pair gives the sequence

$$1 \quad 5 \quad 12 \quad 22 \quad 35 \quad 51 \quad 70 \quad \ldots$$

well-known to Euler. These are the *pentagonal numbers* $f(j)$, whose name is suggested by the figure in the margin.

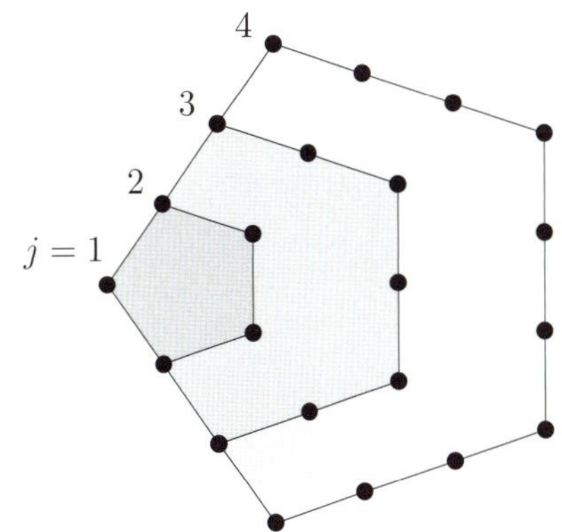

Pentagonal numbers

We easily compute $f(j) = \frac{3j^2-j}{2}$ and $\bar{f}(j) = \frac{3j^2+j}{2}$ for the other number of each pair. In summary, we conjecture, as Euler has done, that the following formula should hold.

Theorem.

$$\prod_{k\geq 1}(1 - x^k) \;=\; 1 + \sum_{j\geq 1}(-1)^j\Big(x^{\frac{3j^2-j}{2}} + x^{\frac{3j^2+j}{2}}\Big). \qquad (8)$$

Euler proved this remarkable theorem by calculations with formal series, but we give a bijection proof from The Book. First of all, we notice by (4) that the product $\prod_{k\geq 1}(1-x^k)$ is precisely the inverse of our partition series $\sum_{n\geq 0} p(n)x^n$. Hence setting $\prod_{k\geq 1}(1-x^k) =: \sum_{n\geq 0} c(n)x^n$, we find

$$\Big(\sum_{n\geq 0} c(n)x^n\Big) \cdot \Big(\sum_{n\geq 0} p(n)x^n\Big) = 1.$$

Comparing coefficients this means that $c(n)$ is the *unique* sequence with $c(0) = 1$ and

$$\sum_{k=0}^{n} c(k)p(n-k) = 0 \qquad \text{for all } n \geq 1. \tag{9}$$

Writing the right-hand of (8) as $\sum_{j=-\infty}^{\infty} (-1)^j x^{\frac{3j^2+j}{2}}$, we have to show that

$$c(k) = \begin{cases} 1 & \text{for } k = \frac{3j^2+j}{2}, \text{ when } j \in \mathbb{Z} \text{ is even,} \\ -1 & \text{for } k = \frac{3j^2+j}{2}, \text{ when } j \in \mathbb{Z} \text{ is odd,} \\ 0 & \text{otherwise} \end{cases}$$

gives this unique sequence. Setting $b(j) = \frac{3j^2+j}{2}$ for $j \in \mathbb{Z}$ and substituting these values into (9), our conjecture takes on the simple form

$$\sum_{j \text{ even}} p(n-b(j)) = \sum_{j \text{ odd}} p(n-b(j)) \qquad \text{for all } n,$$

where of course we only consider j with $b(j) \leq n$. So the stage is set: We have to find a bijection

$$\phi: \bigcup_{j \text{ even}} P(n-b(j)) \longrightarrow \bigcup_{j \text{ odd}} P(n-b(j)).$$

Again several bijections have been suggested, but the following construction by David Bressoud and Doron Zeilberger is astonishingly simple. We just give the definition of ϕ (which is, in fact, an involution), and invite the reader to verify the easy details.

For $\lambda: \lambda_1 + \ldots + \lambda_t \in P(n-b(j))$ set

$$\phi(\lambda) := \begin{cases} (t+3j-1) + (\lambda_1 - 1) + \ldots + (\lambda_t - 1) & \text{if } t+3j \geq \lambda_1, \\ (\lambda_2+1) + \ldots + (\lambda_t+1) + \underbrace{1 + \ldots + 1}_{\lambda_1 - t - 3j - 1} & \text{if } t+3j < \lambda_1, \end{cases}$$

As an example consider $n = 15$, $j = 2$, so $b(2) = 7$. The partition $3+2+2+1$ in $P(15-b(2)) = P(8)$ is mapped to $9+2+1+1$, which is in $P(15-b(1)) = P(13)$.

where we leave out possible 0's. One finds that in the first case $\phi(\lambda)$ is in $P(n-b(j-1))$, and in the second case in $P(n-b(j+1))$.

This was beautiful, and we can get even more out of it. We already know that

$$\prod_{k\geq 1}(1+x^k) = \sum_{n\geq 0} p_d(n)\, x^n.$$

As experienced formal series manipulators we notice that the introduction of the new variable y yields

$$\prod_{k\geq 1}(1+yx^k) \;=\; \sum_{n\geq 0} p_{d,m}(n)\,x^n y^m,$$

where $p_{d,m}(n)$ counts the partitions of n into precisely m distinct summands. With $y=-1$ this yields

$$\prod_{k\geq 1}(1-x^k) \;=\; \sum_{n\geq 0}(E_d(n)-O_d(n))\,x^n, \qquad (10)$$

where $E_d(n)$ is the number of partitions of n into an *even* number of distinct parts, and $O_d(n)$ is the number of partitions into an *odd* number. And here is the punchline. Comparing (10) to Euler's expansion in (8) we infer the beautiful result

$$E_d(n)-O_d(n) \;=\; \begin{cases} 1 & \text{for } n=\frac{3j^2\pm j}{2} \text{ when } j\geq 0 \text{ is even,} \\ -1 & \text{for } n=\frac{3j^2\pm j}{2} \text{ when } j\geq 1 \text{ is odd,} \\ 0 & \text{otherwise.} \end{cases}$$

An example for $n=10$:
$10=9+1$
$10=8+2$
$10=7+3$
$10=6+4$
$10=4+3+2+1$
and
$10=10$
$10=7+2+1$
$10=6+3+1$
$10=5+4+1$
$10=5+3+2$,
so $E_d(10)=O_d(10)=5$.

This is, of course, just the beginning of a longer and still ongoing story. The theory of infinite products is replete with unexpected indentities, and with their bijective counterparts. The most famous examples are the so-called Rogers-Ramanujan identities, named after Leonard Rogers and Srinivasa Ramanujan, in which the number 5 plays a mysterious role:

$$\prod_{k\geq 1}\frac{1}{(1-x^{5k-4})(1-x^{5k-1})} \;=\; \sum_{n\geq 0}\frac{x^{n^2}}{(1-x)(1-x^2)\cdots(1-x^n)},$$

$$\prod_{k\geq 1}\frac{1}{(1-x^{5k-3})(1-x^{5k-2})} \;=\; \sum_{n\geq 0}\frac{x^{n^2+n}}{(1-x)(1-x^2)\cdots(1-x^n)}.$$

Srinivasa Ramanujan

The reader is invited to translate them into the following partition identities first noted by Percy MacMahon:

- Let $f(n)$ be the number of partitions of n all of whose summands are of the form $5k+1$ or $5k+4$, and $g(n)$ the number of partitions whose summands differ by at least 2. Then $f(n)=g(n)$.

- Let $r(n)$ be the number of partitions of n all of whose summands are of the form $5k+2$ or $5k+3$, and $s(n)$ the number of partitions whose parts differ by at least 2 and which do not contain 1. Then $r(n)=s(n)$.

All known formal series proofs of the Rogers-Ramanujan identities are quite involved, and for a long time bijection proofs of $f(n)=g(n)$ and of $r(n)=s(n)$ seemed elusive. Such proofs were eventually given 1981 by Adriano Garsia and Stephen Milne. Their bijections are, however, very complicated — Book proofs are not yet in sight.

References

[1] G. E. ANDREWS: *The Theory of Partitions,* Encyclopedia of Mathematics and its Applications, Vol. 2, Addison-Wesley, Reading MA 1976.

[2] D. BRESSOUD & D. ZEILBERGER: *Bijecting Euler's partitions-recurrence,* Amer. Math. Monthly **92** (1985), 54-55.

[3] A. GARSIA & S. MILNE: *A Rogers-Ramanujan bijection,* J. Combinatorial Theory, Ser. A **31** (1981), 289-339.

[4] S. RAMANUJAN: *Proof of certain identities in combinatory analysis,* Proc. Cambridge Phil. Soc. **19** (1919), 214-216.

[5] L. J. ROGERS: *Second memoir on the expansion of certain infinite products,* Proc. London Math. Soc. **25** (1894), 318-343.

Graph Theory

"The four-colorist geographer"

Five-coloring plane graphs

Chapter 30

Plane graphs and their colorings have been the subject of intensive research since the beginnings of graph theory because of their connection to the four-color problem. As stated originally the four-color problem asked whether it is always possible to color the regions of a plane map with four colors such that regions which share a common boundary (and not just a point) receive different colors. The figure on the right shows that coloring the regions of a map is really the same task as coloring the vertices of a plane graph. As in Chapter 11 (page 65) place a vertex in the interior of each region (including the outer region) and connect two such vertices belonging to neighboring regions by an edge through the common boundary.

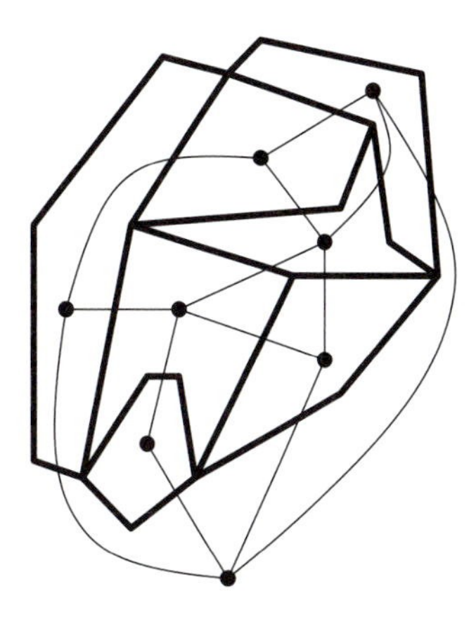

The dual graph of a map

The resulting graph G, the *dual graph* of the map M, is then a plane graph, and coloring the vertices of G in the usual sense is the same as coloring the regions of M. So we may as well concentrate on vertex-coloring plane graphs and will do so from now on. Note that we may assume that G has no loops or multiple edges, since these are irrelevant for coloring.

In the long and arduous history of attacks to prove the four-color theorem many attempts came close, but what finally succeeded in the Appel-Haken proof of 1976 and also in the recent proof of Robertson, Sanders, Seymour and Thomas 1997 was a combination of very old ideas (dating back to the 19th century) and the very new calculating powers of modern-day computers. Twenty-five years after the original proof, the situation is still basically the same, no proof from The Book is in sight.

So let us be more modest and ask whether there is a neat proof that every plane graph can be 5-colored. A proof of this five-color theorem had already been given by Heawood at the turn of the century. The basic tool for his proof (and indeed also for the four-color theorem) was Euler's formula (see Chapter 11). Clearly, when coloring a graph G we may assume that G is connected since we may color the connected pieces separately. A plane graph divides the plane into a set R of regions (including the exterior region). Euler's formula states that for plane connected graphs $G = (V, E)$ we always have

$$|V| - |E| + |R| = 2.$$

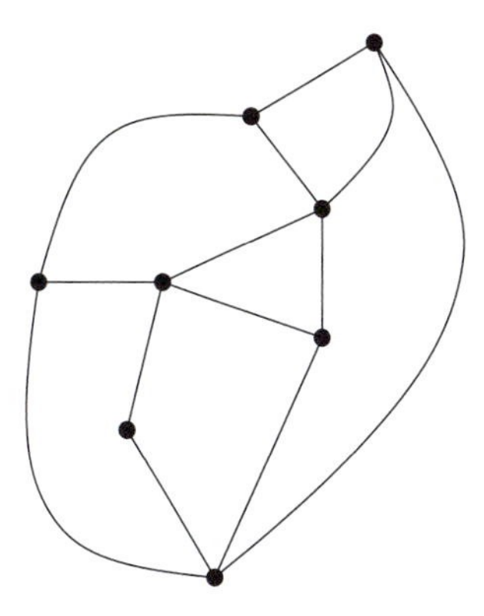

This plane graph has 8 vertices, 13 edges and 7 regions.

As a warm-up, let us see how Euler's formula may be applied to prove that every plane graph G is 6-colorable. We proceed by induction on the number n of vertices. For small values of n (in particular, for $n \leq 6$) this is obvious.

From part (A) of the proposition on page 67 we know that G has a vertex v of degree at most 5. Delete v and all edges incident with v. The resulting graph $G' = G \backslash v$ is a plane graph on $n-1$ vertices. By induction, it can be 6-colored. Since v has at most 5 neighbors in G, at most 5 colors are used for these neighbors in the coloring of G'. So we can extend any 6-coloring of G' to a 6-coloring of G by assigning a color to v which is not used for any of its neighbors in the coloring of G'. Thus G is indeed 6-colorable.

Now let us look at the list chromatic number of plane graphs, as discussed in the previous chapter on the Dinitz problem. Clearly, our 6-coloring method works for lists of colors as well (again we never run out of colors), so $\chi_\ell(G) \leq 6$ holds for any plane graph G. Erdős, Rubin and Taylor conjectured in 1979 that every plane graph has list chromatic number at most 5, and further that there are plane graphs G with $\chi_\ell(G) > 4$. They were right on both counts. Margit Voigt was the first to construct an example of a plane graph G with $\chi_\ell(G) = 5$ (her example had 238 vertices) and around the same time Carsten Thomassen gave a truly stunning proof of the 5-list coloring conjecture. His proof is a telling example of what you can do when you find the right induction hypothesis. It does not use Euler's formula at all!

Theorem. *All planar graphs G can be 5-list colored:*

$$\chi_\ell(G) \leq 5.$$

■ **Proof.** First note that adding edges can only increase the chromatic number. In other words, when H is a subgraph of G, then $\chi_\ell(H) \leq \chi_\ell(G)$ certainly holds. Hence we may assume that G is connected and that all the bounded faces of an embedding have triangles as boundaries. Let us call such a graph *near-triangulated*. The validity of the theorem for near-triangulated graphs will establish the statement for all plane graphs.

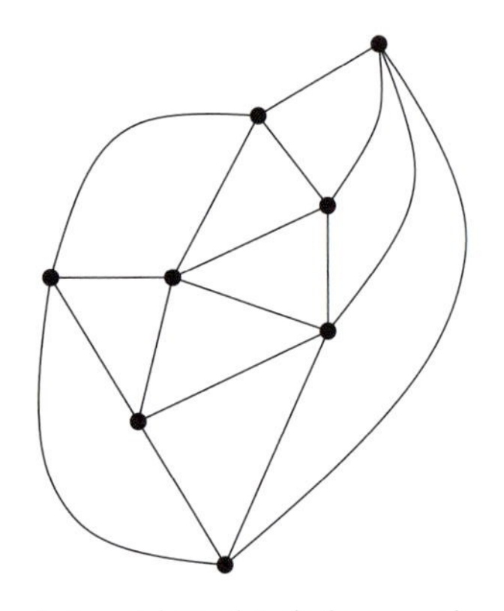

A near-triangulated plane graph

The trick of the proof is to show the following stronger statement (which allows us to use induction):

> *Let $G = (V, E)$ be a near-triangulated graph, and let B be the cycle bounding the outer region. We make the following assumptions on the color sets $C(v)$, $v \in V$:*
>
> (1) *Two adjacent vertices x, y of B are already colored with (different) colors α and β.*
>
> (2) $|C(v)| \geq 3$ *for all other vertices v of B.*
>
> (3) $|C(v)| \geq 5$ *for all vertices v in the interior.*
>
> *Then the coloring of x, y can be extended to a proper coloring of G by choosing colors from the lists. In particular, $\chi_\ell(G) \leq 5$.*

For $|V| = 3$ this is obvious, since for the only uncolored vertex v we have $|C(v)| \geq 3$, so there is a color available. Now we proceed by induction.

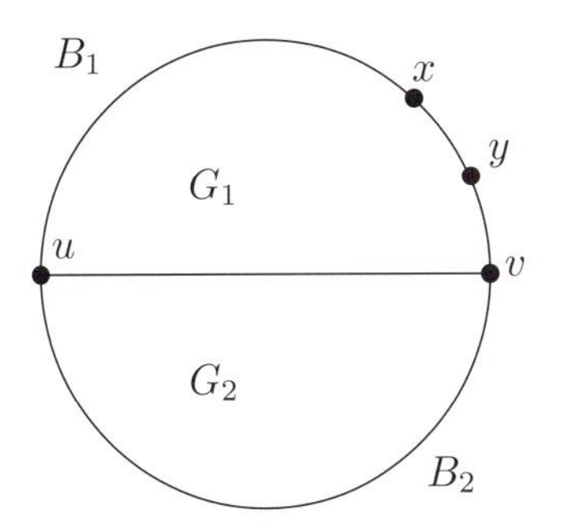

Case 1: Suppose B has a chord, that is, an edge not in B that joins two vertices $u, v \in B$. The subgraph G_1 which is bounded by $B_1 \cup \{uv\}$ and contains x, y, u and v is near-triangulated and therefore has a 5-list coloring by induction. Suppose in this coloring the vertices u and v receive the colors γ and δ. Now we look at the bottom part G_2 bounded by B_2 and uv. Regarding u, v as pre-colored, we see that the induction hypotheses are also satisfied for G_2. Hence G_2 can be 5-list colored with the available colors, and thus the same is true for G.

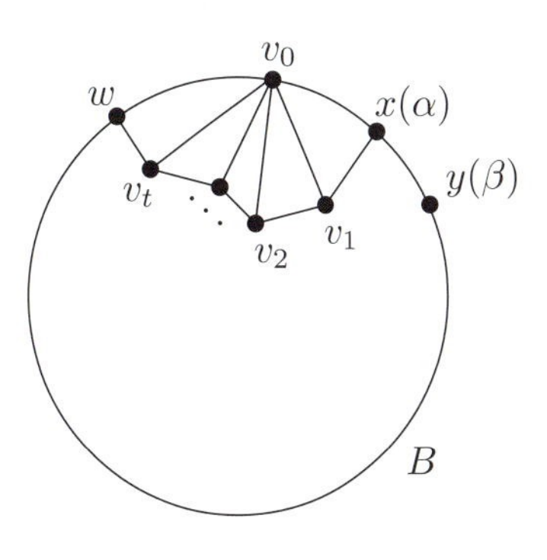

Case 2: Suppose B has no chord. Let v_0 be the vertex on the other side of the α-colored vertex x on B, and let $x, v_1, \ldots, v_t, w$ be the neighbors of v_0. Since G is near-triangulated we have the situation shown in the figure.

Construct the near-triangulated graph $G' = G \backslash v_0$ by deleting from G the vertex v_0 and all edges emanating from v_0. This G' has as outer boundary $B' = (B \backslash v_0) \cup \{v_1, \ldots, v_t\}$. Since $|C(v_0)| \geq 3$ by assumption (2) there exist two colors γ, δ in $C(v_0)$ different from α. Now we replace every color set $C(v_i)$ by $C(v_i) \backslash \{\gamma, \delta\}$, keeping the original color sets for all other vertices in G'. Then G' clearly satisfies all assumptions and is thus 5-list colorable by induction. Choosing γ or δ for v_0, different from the color of w, we can extend the list coloring of G' to all of G. □

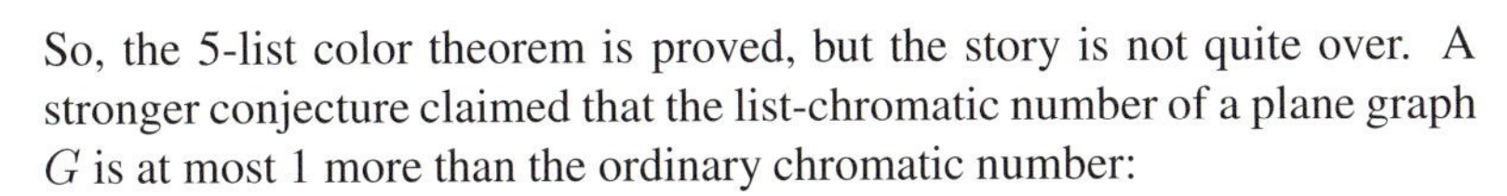

So, the 5-list color theorem is proved, but the story is not quite over. A stronger conjecture claimed that the list-chromatic number of a plane graph G is at most 1 more than the ordinary chromatic number:

Is $\chi_\ell(G) \leq \chi(G) + 1$ *for every plane graph G ?*

Since $\chi(G) \leq 4$ by the four-color theorem, we have three cases:

Case I: $\chi(G) = 2 \Longrightarrow \chi_\ell(G) \leq 3$
Case II: $\chi(G) = 3 \Longrightarrow \chi_\ell(G) \leq 4$
Case III: $\chi(G) = 4 \Longrightarrow \chi_\ell(G) \leq 5$.

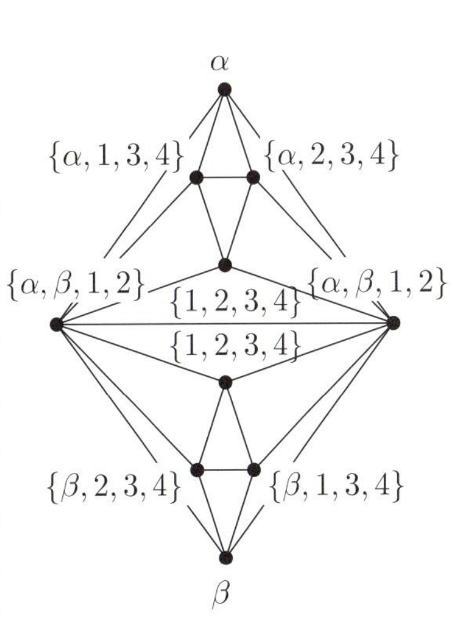

Thomassen's result settles Case III, and Case I was proved by an ingenious (and much more sophisticated) argument by Alon and Tarsi. Furthermore, there are plane graphs G with $\chi(G) = 2$ and $\chi_\ell(G) = 3$, for example the graph $K_{2,4}$ that we considered in the preceding chapter on the Dinitz problem.

But what about Case II? Here the conjecture fails: this was first shown by Margit Voigt for a graph that was earlier constructed by Shai Gutner. His graph on 130 vertices can be obtained as follows. First we look at the "double octahedron" (see the figure), which is clearly 3-colorable. Let $\alpha \in \{5, 6, 7, 8\}$ and $\beta \in \{9, 10, 11, 12\}$, and consider the lists that are given in the figure. You are invited to check that with these lists a coloring is not possible. Now take 16 copies of this graph, and identify all top vertices and all bottom vertices. This yields a graph on $16 \cdot 8 + 2 = 130$ vertices which is still plane and 3-colorable. We assign $\{5,6,7,8\}$ to the top vertex and $\{9,10,11,12\}$ to the bottom vertex, with the inner lists corresponding to the 16 pairs (α,β), $\alpha \in \{5,6,7,8\}$, $\beta \in \{9,10,11,12\}$. For every choice of α and β we thus obtain a subgraph as in the figure, and so a list coloring of the big graph is not possible.

By modifying another one of Gutner's examples, Voigt and Wirth came up with an even smaller plane graph with 75 vertices and $\chi = 3$, $\chi_\ell = 5$, which in addition uses only the minimal number of 5 colors in the combined lists. The current record is 63 vertices.

References

[1] N. ALON & M. TARSI: *Colorings and orientations of graphs,* Combinatorica **12** (1992), 125-134.

[2] P. ERDŐS, A. L. RUBIN & H. TAYLOR: *Choosability in graphs,* Proc. West Coast Conference on Combinatorics, Graph Theory and Computing, Congressus Numerantium **26** (1979), 125-157.

[3] S. GUTNER: *The complexity of planar graph choosability,* Discrete Math. **159** (1996), 119-130.

[4] N. ROBERTSON, D. P. SANDERS, P. SEYMOUR & R. THOMAS: *The four-colour theorem,* J. Combinatorial Theory, Ser. B **70** (1997), 2-44.

[5] C. THOMASSEN: *Every planar graph is* 5*-choosable,* J. Combinatorial Theory, Ser. B **62** (1994), 180-181.

[6] M. VOIGT: *List colorings of planar graphs,* Discrete Math. **120** (1993), 215-219.

[7] M. VOIGT & B. WIRTH: *On* 3*-colorable non-*4*-choosable planar graphs,* J. Graph Theory **24** (1997), 233-235.

How to guard a museum

Chapter 31

Here is an appealing problem which was raised by Victor Klee in 1973. Suppose the manager of a museum wants to make sure that at all times every point of the museum is watched by a guard. The guards are stationed at fixed posts, but they are able to turn around. How many guards are needed?

We picture the walls of the museum as a polygon consisting of n sides. Of course, if the polygon is *convex*, then one guard is enough. In fact, the guard may be stationed at any point of the museum. But, in general, the walls of the museum may have the shape of any closed polygon.

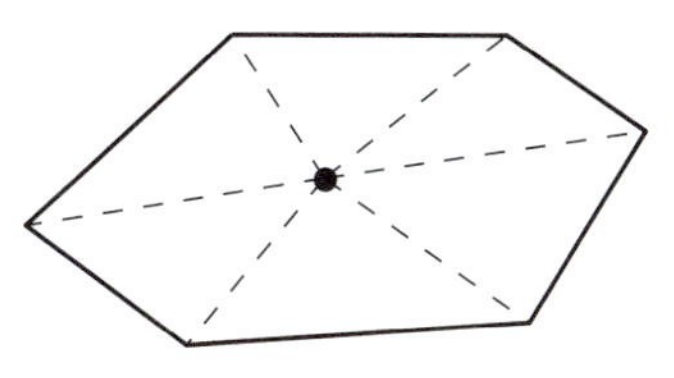

A convex exhibition hall

Consider a comb-shaped museum with $n = 3m$ walls, as depicted on the right. It is easy to see that this requires at least $m = \frac{n}{3}$ guards. In fact, there are n walls. Now notice that the point 1 can only be observed by a guard stationed in the shaded triangle containing 1, and similarly for the other points $2, 3, \ldots, m$. Since all these triangles are disjoint we conclude that at least m guards are needed. But m guards are also enough, since they can be placed at the top lines of the triangles. By cutting off one or two walls at the end, we conclude that for any n there is an n-walled museum which requires $\lfloor \frac{n}{3} \rfloor$ guards.

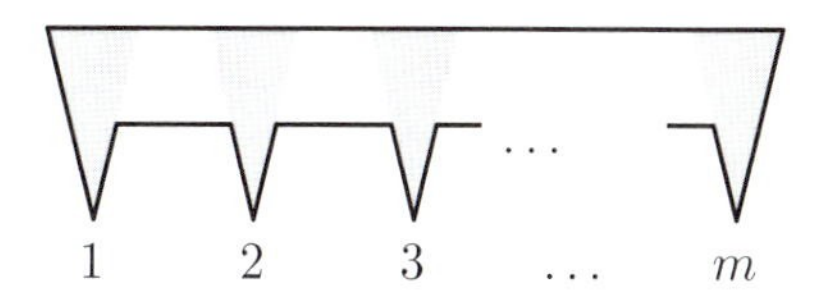

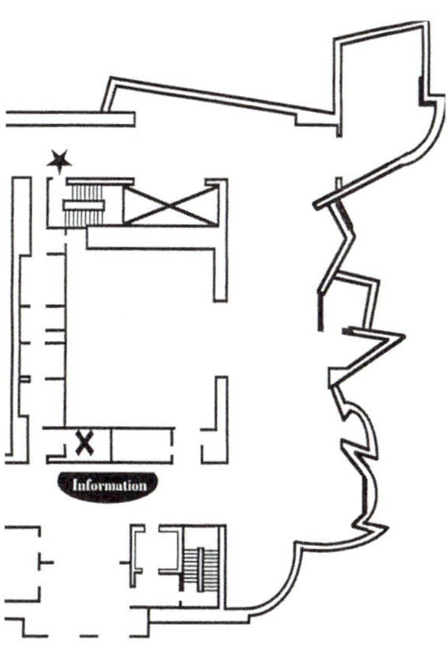

A real life art gallery...

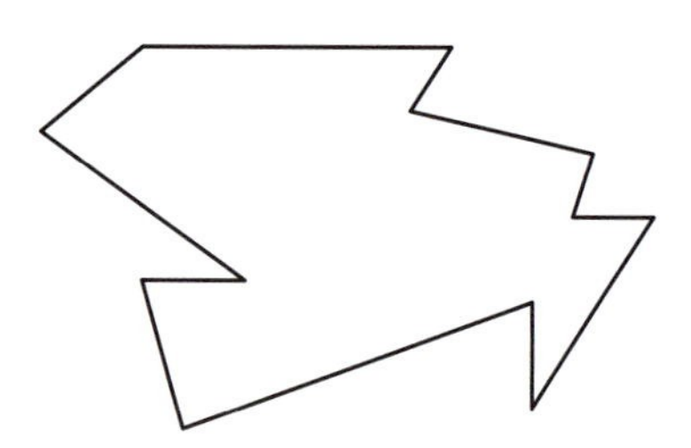

A museum with $n = 12$ walls

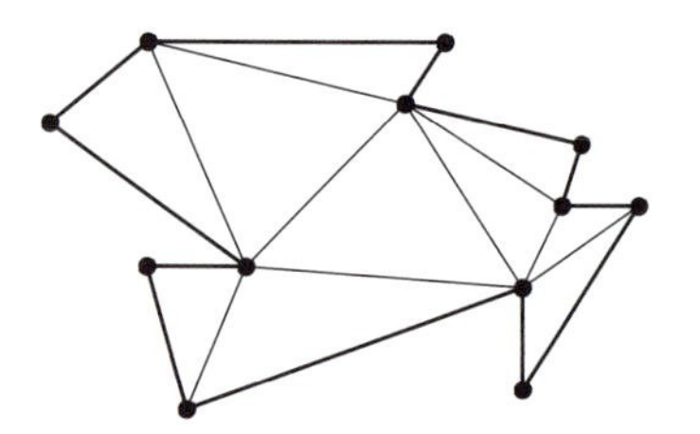

A triangulation of the museum

The following result states that this is the worst case.

> **Theorem.** *For any museum with n walls, $\lfloor \frac{n}{3} \rfloor$ guards suffice.*

This "art gallery theorem" was first proved by Vašek Chvátal by a clever argument, but here is a proof due to Steve Fisk that is truly beautiful.

■ **Proof.** First of all, let us draw $n - 3$ non-crossing diagonals between corners of the walls until the interior is triangulated. For example, we can draw 9 diagonals in the museum depicted in the margin to produce a triangulation. It does not matter which triangulation we choose, any one will do. Now think of the new figure as a plane graph with the corners as vertices and the walls and diagonals as edges.

Claim. *This graph is 3-colorable.*

For $n = 3$ there is nothing to prove. Now for $n > 3$ pick any two vertices u and v which are connected by a diagonal. This diagonal will split the graph into two smaller triangulated graphs both containing the edge uv. By induction we may color each part with 3 colors where we may choose color 1 for u and color 2 for v in each coloring. Pasting the colorings together yields a 3-coloring of the whole graph.

The rest is easy. Since there are n vertices, at least one of the color classes, say the vertices colored 1, contains at most $\lfloor \frac{n}{3} \rfloor$ vertices, and this is where we place the guards. Since every triangle contains a vertex of color 1 we infer that every triangle is guarded, and hence so is the whole museum. □

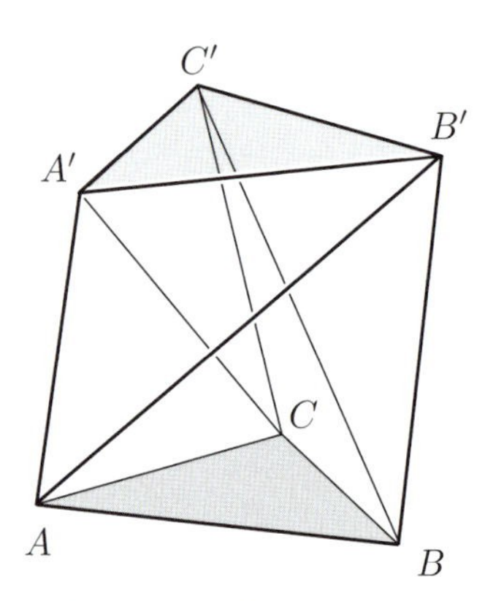

Schönhardt's polyhedron: The interior dihedral angles at the edges AB', BC' and CA' are greater than $180°$.

The astute reader may have noticed a subtle point in our reasoning. Does a triangulation always exist? Probably everybody's first reaction is: Obviously, yes! Well, it does exist, but this is not completely obvious, and, in fact, the natural generalization to three dimensions (partitioning into tetrahedra) is false! This may be seen from *Schönhardt's polyhedron*, depicted on the left. It is obtained from a triangular prism by rotating the top triangle, so that each of the quadrilateral faces breaks into two triangles with a non-convex edge. Try to triangulate this polyhedron! You will notice that any tetrahedron that contains the bottom triangle must contain one of the three top vertices: but the resulting tetrahedron will not be contained in Schönhardt's polyhedron. So there is no triangulation without an additional vertex.

To prove that a triangulation exists in the case of a planar non-convex polygon, we proceed by induction on the number n of vertices. For $n = 3$ the polygon is a triangle, and there is nothing to prove. Let $n \geq 4$. To use induction, all we have to produce is *one* diagonal which will split the polygon P into two smaller parts, such that a triangulation of the polygon can be pasted together from triangulations of the parts.

Call a vertex A *convex* if the interior angle at the vertex is less than $180°$. Since the sum of the interior angles of P is $(n - 2)180°$, there must be a convex vertex A. In fact, there must be at least three of them: In essence this is an application of the pigeonhole principle! Or you may consider the convex hull of the polygon, and note that all its vertices are convex also for the original polygon.

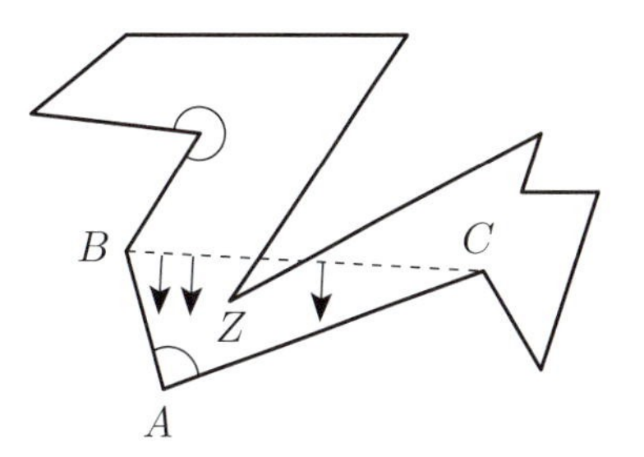

Now look at the two neighboring vertices B and C of A. If the segment BC lies entirely in P, then this is our diagonal. If not, the triangle ABC contains other vertices. Slide BC towards A until it hits the last vertex Z in ABC. Now AZ is within P, and we have a diagonal.

There are many variants to the art gallery theorem. For example, we may only want to guard the walls (which is, after all, where the paintings hang), or the guards are all stationed at vertices. A particularly nice (unsolved) variant goes as follows:

> *Suppose each guard may patrol one wall of the museum, so he walks along his wall and sees anything that can be seen from any point along this wall.*
> *How many "wall guards" do we then need to keep control?*

Godfried Toussaint constructed the example of a museum displayed here which shows that $\lfloor \frac{n}{4} \rfloor$ guards may be necessary.

This polygon has 28 sides (and, in general, $4m$ sides), and the reader is invited to check that m wall-guards are needed. It is conjectured that, except for some small values of n, this number is also sufficient, but a proof, let alone a Book Proof, is still missing.

References

[1] V. Chvátal: *A combinatorial theorem in plane geometry,* J. Combinatorial Theory, Ser. B **18** (1975), 39-41.

[2] S. Fisk: *A short proof of Chvátal's watchman theorem,* J. Combinatorial Theory, Ser. B **24** (1978), 374.

[3] J. O'Rourke: *Art Gallery Theorems and Algorithms,* Oxford University Press 1987.

[4] E. Schönhardt: *Über die Zerlegung von Dreieckspolyedern in Tetraeder,* Math. Annalen **98** (1928), 309-312.

"Museum guards"
(A 3-dimensional art-gallery problem)

Turán's graph theorem

Chapter 32

One of the fundamental results in graph theory is the theorem of Turán from 1941, which initiated extremal graph theory. Turán's theorem was rediscovered many times with various different proofs. We will discuss five of them and let the reader decide which one belongs in The Book.

Paul Turán

Let us fix some notation. We consider simple graphs G on the vertex set $V = \{v_1, \dots, v_n\}$ and edge set E. If v_i and v_j are neighbors, then we write $v_iv_j \in E$. A *p-clique* in G is a complete subgraph of G on p vertices, denoted by K_p. Paul Turán posed the following question:

> *Suppose G is a simple graph that does not contain a p-clique. What is the largest number of edges that G can have?*

We readily obtain examples of such graphs by dividing V into $p-1$ pairwise disjoint subsets $V = V_1 \cup \ldots \cup V_{p-1}$, $|V_i| = n_i$, $n = n_1 + \ldots + n_{p-1}$, joining two vertices if and only if they lie in distinct sets V_i, V_j. We denote the resulting graph by $K_{n_1,\dots,n_{p-1}}$; it has $\sum_{i<j} n_in_j$ edges. We obtain a maximal number of edges among such graphs with given n if we divide the numbers n_i as evenly as possible, that is, if $|n_i - n_j| \le 1$ for all i, j. Indeed, suppose $n_1 \ge n_2 + 2$. By shifting one vertex from V_1 to V_2, we obtain $K_{n_1-1,n_2+1,\dots,n_{p-1}}$ which contains $(n_1 - 1)(n_2 + 1) - n_1n_2 = n_1 - n_2 - 1 \ge 1$ more edges than $K_{n_1,n_2,\dots,n_{p-1}}$. Let us call the graphs $K_{n_1,\dots,n_{p-1}}$ with $|n_i - n_j| \le 1$ the *Turán graphs*. In particular, if $p - 1$ divides n, then we may choose $n_i = \frac{n}{p-1}$ for all i, obtaining

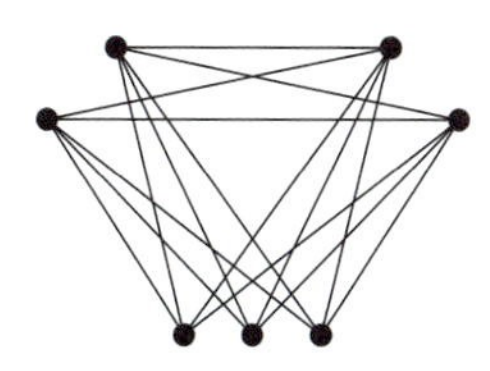

The graph $K_{2,2,3}$

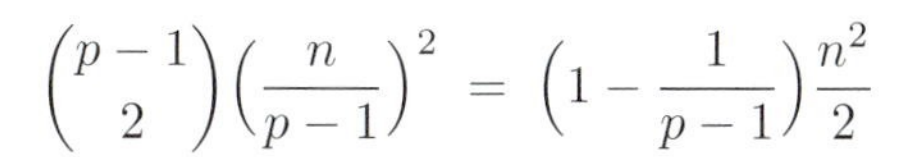

$$\binom{p-1}{2}\left(\frac{n}{p-1}\right)^2 = \left(1 - \frac{1}{p-1}\right)\frac{n^2}{2}$$

edges. Turán's theorem now states that this number is an upper bound for the edge-number of *any* graph on n vertices without a p-clique.

Theorem. *If a graph $G = (V, E)$ on n vertices has no p-clique, $p \ge 2$, then*

$$|E| \le \left(1 - \frac{1}{p-1}\right)\frac{n^2}{2}. \qquad (1)$$

For $p = 2$ this is trivial. In the first interesting case $p = 3$ the theorem states that a triangle-free graph on n vertices contains at most $\frac{n^2}{4}$ edges. Proofs of this special case were known prior to Turán's result. Two elegant proofs using inequalities are contained in Chapter 17.

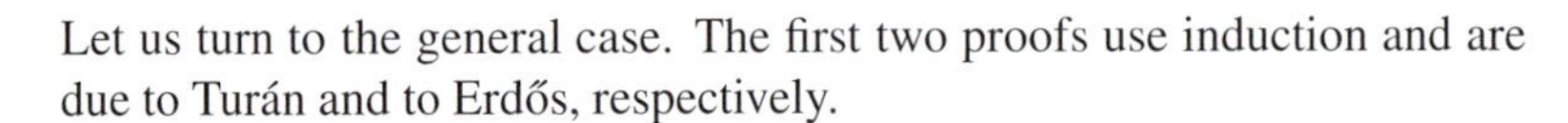

Let us turn to the general case. The first two proofs use induction and are due to Turán and to Erdős, respectively.

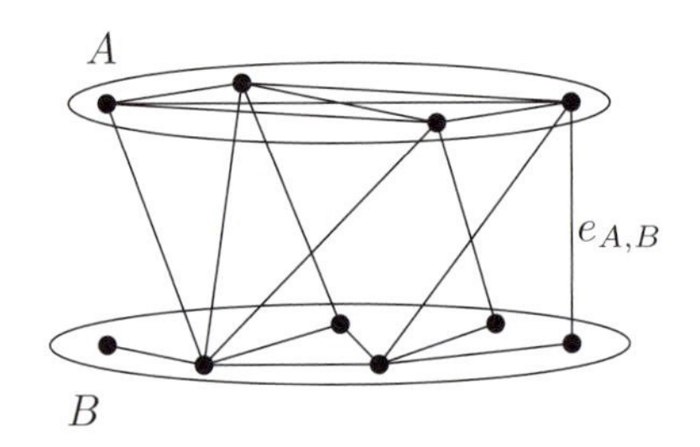

■ **First proof.** We use induction on n. One easily computes that (1) is true for $n < p$. Let G be a graph on $V = \{v_1, \ldots, v_n\}$ without p-cliques with a maximal number of edges, where $n \geq p$. G certainly contains $(p-1)$-cliques, since otherwise we could add edges. Let A be a $(p-1)$-clique, and set $B := V \backslash A$.

A contains $\binom{p-1}{2}$ edges, and we now estimate the edge-number e_B in B and the edge-number $e_{A,B}$ between A and B. By induction, we have $e_B \leq \frac{1}{2}(1-\frac{1}{p-1})(n-p+1)^2$. Since G has no p-clique, every $v_j \in B$ is adjacent to at most $p-2$ vertices in A, and we obtain $e_{A,B} \leq (p-2)(n-p+1)$. Altogether, this yields

$$|E| \leq \binom{p-1}{2} + \frac{1}{2}\Big(1 - \frac{1}{p-1}\Big)(n-p+1)^2 + (p-2)(n-p+1),$$

which is precisely $(1-\frac{1}{p-1})\frac{n^2}{2}$. □

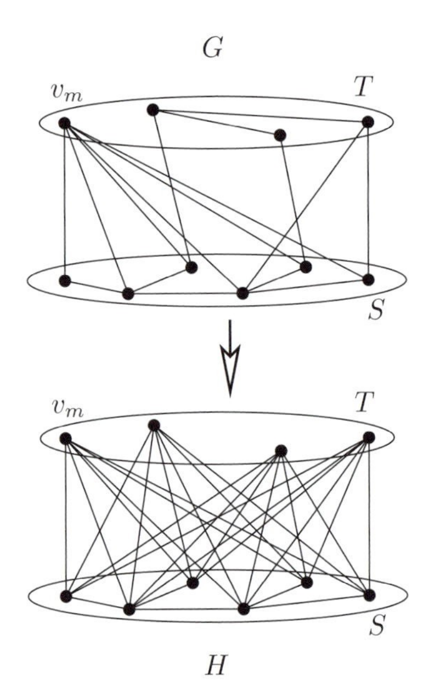

■ **Second proof.** This proof makes use of the structure of the Turán graphs. Let $v_m \in V$ be a vertex of maximal degree $d_m = \max_{1 \leq j \leq n} d_j$. Denote by S the set of neighbors of v_m, $|S| = d_m$, and set $T := V \backslash S$. As G contains no p-clique, and v_m is adjacent to all vertices of S, we note that S contains no $(p-1)$-clique.

We now construct the following graph H on V (see the figure). H corresponds to G on S and contains all edges between S and T, but no edges within T. In other words, T is an independent set in H, and we conclude that H has again no p-cliques. Let d'_j be the degree of v_j in H. If $v_j \in S$, then we certainly have $d'_j \geq d_j$ by the construction of H, and for $v_j \in T$, we see $d'_j = |S| = d_m \geq d_j$ by the choice of v_m. We infer $|E(H)| \geq |E|$, and find that among all graphs with a maximal number of edges, there must be one of the form of H. By induction, the graph induced by S has at most as many edges as a suitable graph $K_{n_1,\ldots,n_{p-2}}$ on S. So $|E| \leq |E(H)| \leq E(K_{n_1,\ldots,n_{p-1}})$ with $n_{p-1} = |T|$, which implies (1). □

The next two proofs are of a totally different nature, using a maximizing argument and ideas from probability theory. They are due to Motzkin and Straus and to Alon and Spencer, respectively.

■ **Third proof.** Consider a *probability distribution* $\boldsymbol{w} = (w_1, \ldots, w_n)$ on the vertices, that is, an assignment of values $w_i \geq 0$ to the vertices with $\sum_{i=1}^{n} w_i = 1$. Our goal is to maximize the function

$$f(\boldsymbol{w}) = \sum_{v_i v_j \in E} w_i w_j .$$

Suppose $\boldsymbol{w}$ is any distribution, and let v_i and v_j be a pair of non-adjacent vertices with positive weights w_i, w_j. Let s_i be the sum of the weights of all vertices adjacent to v_i, and define s_j similarly for v_j, where we may assume that $s_i \geq s_j$. Now we move the weight from v_j to v_i, that is, the new weight of v_i is $w_i + w_j$, while the weight of v_j drops to 0. For the new new distribution $\boldsymbol{w}'$ we find

$$f(\boldsymbol{w}') = f(\boldsymbol{w}) + w_j s_i - w_j s_j \geq f(\boldsymbol{w}).$$

"Moving weights"

We repeat this (reducing the number of vertices with a positive weight by one in each step) until there are no non-adjacent vertices of positive weight anymore. Thus we conclude that there is an optimal distribution whose nonzero weights are concentrated on a clique, say on a k-clique. Now if, say, $w_1 > w_2 > 0$, then choose ε with $0 < \varepsilon < w_1 - w_2$ and change w_1 to $w_1 - \varepsilon$ and w_2 to $w_2 + \varepsilon$. The new distribution $\boldsymbol{w}'$ satisfies $f(\boldsymbol{w}') = f(\boldsymbol{w}) + \varepsilon(w_1 - w_2) - \varepsilon^2 > f(\boldsymbol{w})$, and we infer that the maximal value of $f(\boldsymbol{w})$ is attained for $w_i = \frac{1}{k}$ on a k-clique and $w_i = 0$ otherwise. Since a k-clique contains $\frac{k(k-1)}{2}$ edges, we obtain

$$f(\boldsymbol{w}) = \frac{k(k-1)}{2}\frac{1}{k^2} = \frac{1}{2}\Big(1 - \frac{1}{k}\Big).$$

Since this expression is increasing in k, the best we can do is to set $k = p-1$ (since G has no p-cliques). So we conclude

$$f(\boldsymbol{w}) \leq \frac{1}{2}\Big(1 - \frac{1}{p-1}\Big)$$

for *any* distribution $\boldsymbol{w}$. In particular, this inequality holds for the *uniform* distribution given by $w_i = \frac{1}{n}$ for all i. Thus we find

$$\frac{|E|}{n^2} = f\Big(w_i = \frac{1}{n}\Big) \leq \frac{1}{2}\Big(1 - \frac{1}{p-1}\Big),$$

which is precisely (1). □

■ **Fourth proof.** This time we use some concepts from probability theory. Let G be an arbitrary graph on the vertex set $V = \{v_1, \ldots, v_n\}$. Denote the degree of v_i by d_i, and write $\omega(G)$ for the number of vertices in a largest clique, called the *clique number* of G.

Claim. *We have* $\omega(G) \geq \sum_{i=1}^{n} \frac{1}{n - d_i}$.

We choose a random permutation $\pi = v_1 v_2 \ldots v_n$ of the vertex set V, where each permutation is supposed to appear with the same probability $\frac{1}{n!}$, and then consider the following set C_π. We put v_i into C_π if and only if v_i is adjacent to all v_j $(j < i)$ preceding v_i. By definition, C_π is a clique in G. Let $X = |C_\pi|$ be the corresponding random variable. We have $X = \sum_{i=1}^{n} X_i$, where X_i is the indicator random variable of the vertex v_i, that is, $X_i = 1$ or $X_i = 0$ depending on whether $v_i \in C_\pi$ or $v_i \notin C_\pi$. Note that v_i belongs to C_π with respect to the permutation $v_1 v_2 \ldots v_n$ if and only if v_i appears *before* all $n - 1 - d_i$ vertices which are not adjacent to v_i, or in other words, if v_i is the *first* among v_i and its $n - 1 - d_i$ non-neighbors. The probability that this happens is $\frac{1}{n-d_i}$, hence $EX_i = \frac{1}{n-d_i}$.

Thus by linearity of expectation (see page 84) we obtain

$$E(|C_\pi|) \;=\; EX \;=\; \sum_{i=1}^{n} EX_i \;=\; \sum_{i=1}^{n} \frac{1}{n-d_i}.$$

Consequently, there must be a clique of at least that size, and this was our claim. To deduce Turán's theorem from the claim we use the Cauchy-Schwarz inequality from Chapter 17,

$$\Big(\sum_{i=1}^{n} a_i b_i\Big)^2 \;\le\; \Big(\sum_{i=1}^{n} a_i^2\Big)\Big(\sum_{n=1}^{n} b_i^2\Big).$$

Set $a_i = \sqrt{n-d_i}$, $b_i = \frac{1}{\sqrt{n-d_i}}$, then $a_i b_i = 1$, and so

$$n^2 \;\le\; \Big(\sum_{i=1}^{n}(n-d_i)\Big)\Big(\sum_{i=1}^{n}\frac{1}{n-d_i}\Big) \;\le\; \omega(G)\sum_{i=1}^{n}(n-d_i). \qquad (2)$$

At this point we apply the hypothesis $\omega(G) \le p-1$ of Turán's theorem. Using also $\sum_{i=1}^{n} d_i = 2|E|$ from the chapter on double counting, inequality (2) leads to

$$n^2 \;\le\; (p-1)(n^2 - 2|E|),$$

and this is equivalent to Turán's inequality. □

Now we are ready for the last proof, which may be the most beautiful of them all. Its origin is not clear; we got it from Stephan Brandt, who heard it in Oberwolfach. It may be "folklore" graph theory. It yields in one stroke that the Turán graph is in fact the unique example with a maximal number of edges. It may be noted that both proofs 1 and 2 also imply this stronger result.

■ **Fifth proof.** Let G be a graph on n vertices without a p-clique and with a maximal number of edges.

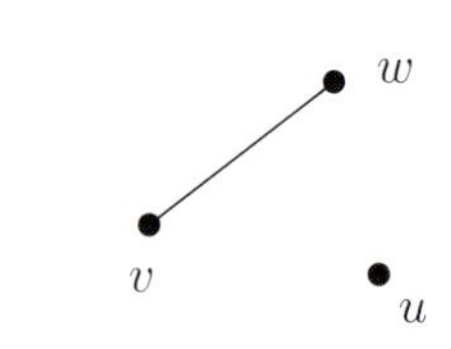

> ***Claim.*** *G does not contain three vertices u, v, w such that $vw \in E$, but $uv \notin E$, $uw \notin E$.*

Suppose otherwise, and consider the following cases.

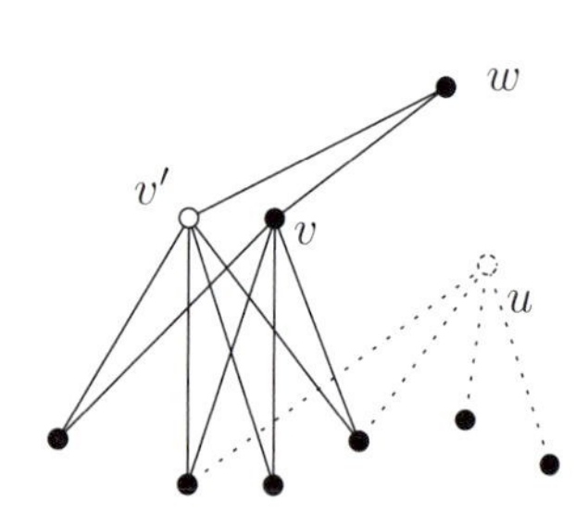

Case 1: $d(u) < d(v)$ or $d(u) < d(w)$.
We may suppose that $d(u) < d(v)$. Then we duplicate v, that is, we create a new vertex v' which has exactly the same neighbors as v (but vv' is not an edge), delete u, and keep the rest unchanged.

The new graph G' has again no p-clique, and for the number of edges we find

$$|E(G')| \;=\; |E(G)| + d(v) - d(u) \;>\; |E(G)|,$$

a contradiction.

Case 2: $d(u) \geq d(v)$ and $d(u) \geq d(w)$.
Duplicate u twice and delete v and w (as illustrated in the margin). Again, the new graph G' has no p-clique, and we compute (the -1 results from the edge vw):

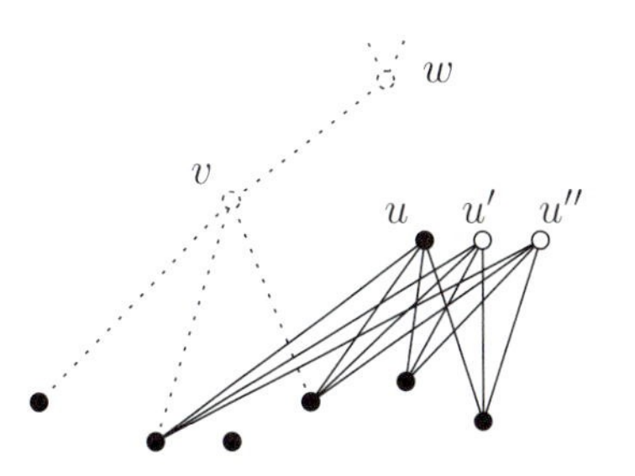

$$|E(G')| \;=\; |E(G)| + 2d(u) - (d(v) + d(w) - 1) \;>\; |E(G)|.$$

So we have a contradiction once more.

A moment's thought shows that the claim we have proved is equivalent to the statement that

$$u \sim v \;:\Longleftrightarrow\; uv \notin E(G)$$

defines an equivalence relation. Thus G is a complete multipartite graph, $G = K_{n_1,\ldots,n_{p-1}}$, and we are finished. □

References

[1] M. AIGNER: *Turán's graph theorem,* Amer. Math. Monthly **102** (1995), 808-816.

[2] N. ALON & J. SPENCER: *The Probabilistic Method,* Wiley Interscience 1992.

[3] P. ERDŐS: *On the graph theorem of Turán (in Hungarian),* Math. Fiz. Lapok **21** (1970), 249-251.

[4] T. S. MOTZKIN & E. G. STRAUS: *Maxima for graphs and a new proof of a theorem of Turán,* Canad. J. Math. **17** (1965), 533-540.

[5] P. TURÁN: *On an extremal problem in graph theory,* Math. Fiz. Lapok **48** (1941), 436-452.

"Larger weights to move"

Communicating without errors

Chapter 33

In 1956, Claude Shannon, the founder of information theory, posed the following very interesting question:

> *Suppose we want to transmit messages across a channel (where some symbols may be distorted) to a receiver. What is the maximum rate of transmission such that the receiver may recover the original message without errors?*

Claude Shannon

Let us see what Shannon meant by "channel" and "rate of transmission." We are given a set V of symbols, and a message is just a string of symbols from V. We model the channel as a graph $G = (V, E)$, where V is the set of symbols, and E the set of edges between unreliable pairs of symbols, that is, symbols which may be confused during transmission. For example, communicating over a phone in everyday language, we connnect the symbols B and P by an edge since the receiver may not be able to distinguish them. Let us call G the *confusion graph*.

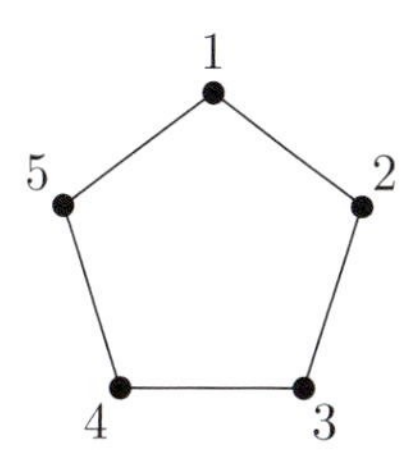

The 5-cycle C_5 will play a prominent role in our discussion. In this example, 1 and 2 may be confused, but not 1 and 3, etc. Ideally we would like to use all 5 symbols for transmission, but since we want to communicate error-free we can — if we only send single symbols — use only one letter from each pair that might be confused. Thus for the 5-cycle we can use only two different letters (any two that are not connected by an edge). In the language of information theory, this means that for the 5-cycle we achieve an information rate of $\log_2 2 = 1$ (instead of the maximal $\log_2 5 \approx 2.32$).

It is clear that in this model, for an arbitrary graph $G = (V, E)$, the best we can do is to transmit symbols from a largest independent set. Thus the information rate, when sending single symbols, is $\log_2 \alpha(G)$, where $\alpha(G)$ is the *independence number* of G.

Let us see whether we can increase the information rate by using larger strings in place of single symbols. Suppose we want to transmit strings of length 2. The strings u_1u_2 and v_1v_2 can only be confused if one of the following three cases holds:

- $u_1 = v_1$ and u_2 can be confused with v_2,
- $u_2 = v_2$ and u_1 can be confused with v_1, or
- $u_1 \neq v_1$ can be confused and $u_2 \neq v_2$ can be confused.

In graph-theoretic terms this amounts to considering the *product* $G_1 \times G_2$ of two graphs $G_1 = (V_1, E_1)$ and $G_2 = (V_2, E_2)$. $G_1 \times G_2$ has the vertex set $V_1 \times V_2 = \{(u_1, u_2) : u_1 \in V_1, u_2 \in V_2\}$, with $(u_1, u_2) \neq (v_1, v_2)$ connected by an edge if and only if $u_i = v_i$ or $u_i v_i \in E$ for $i = 1, 2$. The confusion graph for strings of length 2 is thus $G^2 = G \times G$, the product of the confusion graph G for single symbols with itself. The information rate of strings of length 2 *per symbol* is then given by

$$\frac{\log_2 \alpha(G^2)}{2} = \log_2 \sqrt{\alpha(G^2)}.$$

Now, of course, we may use strings of any length n. The n-th confusion graph $G^n = G \times G \times \ldots \times G$ has vertex set $V^n = \{(u_1, \ldots, u_n) : u_i \in V\}$ with $(u_1, \ldots, u_n) \neq (v_1, \ldots v_n)$ being connected by an edge if $u_i = v_i$ or $u_i v_i \in E$ for all i. The rate of information per symbol determined by strings of length n is

$$\frac{\log_2 \alpha(G^n)}{n} = \log_2 \sqrt[n]{\alpha(G^n)}.$$

What can we say about $\alpha(G^n)$? Here is a first observation. Let $U \subseteq V$ be a largest independent set in G, $|U| = \alpha$. The α^n vertices in G^n of the form $(u_1, \ldots, u_n)$, $u_i \in U$ for all i, clearly form an independent set in G^n. Hence

$$\alpha(G^n) \geq \alpha(G)^n$$

and therefore

$$\sqrt[n]{\alpha(G^n)} \geq \alpha(G),$$

meaning that we never decrease the information rate by using longer strings instead of single symbols. This, by the way, is a basic idea of coding theory: By encoding symbols into longer strings we can make error-free communication more efficient.

Disregarding the logarithm we thus arrive at Shannon's fundamental definition: The *zero-error capacity* of a graph G is given by

$$\Theta(G) := \sup_{n \geq 1} \sqrt[n]{\alpha(G^n)},$$

and Shannon's problem was to compute $\Theta(G)$, and in particular $\Theta(C_5)$.

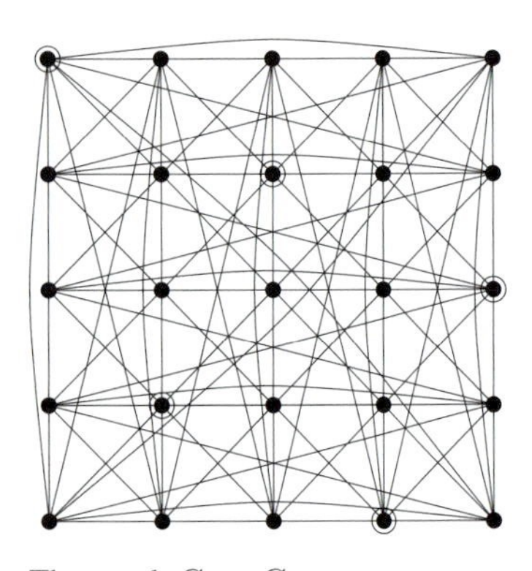
The graph $C_5 \times C_5$

Let us look at C_5. So far we know $\alpha(C_5) = 2 \leq \Theta(C_5)$. Looking at the 5-cycle as depicted earlier, or at the product $C_5 \times C_5$ as drawn on the left, we see that the set $\{(1,1), (2,3), (3,5), (4,2), (5,4)\}$ is independent in C_5^2. Thus we have $\alpha(C_5^2) \geq 5$. Since an independent set can contain only two vertices from any two consecutive rows we see that $\alpha(C_5^2) = 5$. Hence, by using strings of length 2 we have increased the lower bound for the capacity to $\Theta(C_5) \geq \sqrt{5}$.

So far we have no upper bounds for the capacity. To obtain such bounds we again follow Shannon's original ideas. First we need the dual definition of an independent set. We recall that a subset $C \subseteq V$ is a *clique* if any two vertices of C are joined by an edge. Thus the vertices form trivial cliques of size 1, the edges are the cliques of size 2, the triangles are cliques of size 3, and so on. Let $\mathcal{C}$ be the set of cliques in G. Consider an arbitrary probability distribution $\boldsymbol{x} = (x_v : v \in V)$ on the set of vertices, that is, $x_v \geq 0$ and $\sum_{v\in V} x_v = 1$. To every distribution $\boldsymbol{x}$ we associate the "maximal value of a clique"

$$\lambda(\boldsymbol{x}) = \max_{C\in\mathcal{C}} \sum_{v\in C} x_v,$$

and finally we set

$$\lambda(G) = \min_{\boldsymbol{x}} \lambda(\boldsymbol{x}) = \min_{\boldsymbol{x}} \max_{C\in\mathcal{C}} \sum_{v\in C} x_v.$$

To be precise we should use inf instead of min, but the minimum exists because $\lambda(\boldsymbol{x})$ is continuous on the compact set of all distributions.

Consider now an independent set $U \subseteq V$ of maximal size $\alpha(G) = \alpha$. Associated to U we define the distribution $\boldsymbol{x}_U = (x_v : v \in V)$ by setting $x_v = \frac{1}{\alpha}$ if $v \in U$ and $x_v = 0$ otherwise. Since any clique contains at most one vertex from U, we infer $\lambda(\boldsymbol{x}_U) = \frac{1}{\alpha}$, and thus by the definition of $\lambda(G)$

$$\lambda(G) \leq \frac{1}{\alpha(G)} \quad \text{or} \quad \alpha(G) \leq \lambda(G)^{-1}.$$

What Shannon observed is that $\lambda(G)^{-1}$ is, in fact, an upper bound for all $\sqrt[n]{\alpha(G^n)}$, and hence also for $\Theta(G)$. In order to prove this it suffices to show that for graphs G, H

$$\lambda(G \times H) = \lambda(G)\lambda(H) \tag{1}$$

holds, since this will imply $\lambda(G^n) = \lambda(G)^n$ and hence

$$\begin{aligned} \alpha(G^n) &\leq \lambda(G^n)^{-1} = \lambda(G)^{-n} \\ \sqrt[n]{\alpha(G^n)} &\leq \lambda(G)^{-1}. \end{aligned}$$

To prove (1) we make use of the duality theorem of linear programming (see [1]) and get

$$\lambda(G) = \min_{\boldsymbol{x}} \max_{C\in\mathcal{C}} \sum_{v\in C} x_v = \max_{\boldsymbol{y}} \min_{v\in V} \sum_{C\ni v} y_C, \tag{2}$$

where the right-hand side runs through all probability distributions $\boldsymbol{y} = (y_C : C \in \mathcal{C})$ on $\mathcal{C}$.

Consider $G \times H$, and let $\boldsymbol{x}$ and $\boldsymbol{x}'$ be distributions which achieve the minima, $\lambda(\boldsymbol{x}) = \lambda(G)$, $\lambda(\boldsymbol{x}') = \lambda(H)$. In the vertex set of $G \times H$ we assign the value $z_{(u,v)} = x_u x'_v$ to the vertex (u, v). Since $\sum_{(u,v)} z_{(u,v)} = \sum_u x_u \sum_v x'_v = 1$, we obtain a distribution. Next we observe that the maximal cliques in $G \times H$ are of the form $C \times D = \{(u, v) : u \in C, v \in D\}$ where C and D are cliques in G and H, respectively. Hence we obtain

$$\begin{aligned} \lambda(G \times H) \leq \lambda(\boldsymbol{z}) &= \max_{C\times D} \sum_{(u,v)\in C\times D} z_{(u,v)} \\ &= \max_{C\times D} \sum_{u\in C} x_u \sum_{v\in D} x'_v = \lambda(G)\lambda(H) \end{aligned}$$

by the definition of $\lambda(G \times H)$. In the same way the converse inequality $\lambda(G \times H) \geq \lambda(G)\lambda(H)$ is shown by using the dual expression for $\lambda(G)$ in (2). In summary we can state:

$$\Theta(G) \leq \lambda(G)^{-1},$$

for any graph G.

Let us apply our findings to the 5-cycle and, more generally, to the m-cycle C_m. By using the uniform distribution $(\frac{1}{m}, \dots, \frac{1}{m})$ on the vertices, we obtain $\lambda(C_m) \leq \frac{2}{m}$, since any clique contains at most two vertices. Similarly, choosing $\frac{1}{m}$ for the edges and 0 for the vertices, we have $\lambda(C_m) \geq \frac{2}{m}$ by the dual expression in (2). We conclude that $\lambda(C_m) = \frac{2}{m}$ and therefore

$$\Theta(C_m) \leq \frac{m}{2}$$

for all m. Now, if m is even, then clearly $\alpha(C_m) = \frac{m}{2}$ and thus also $\Theta(C_m) = \frac{m}{2}$. For odd m, however, we have $\alpha(C_m) = \frac{m-1}{2}$. For $m = 3$, C_3 is a clique, and so is every product C_3^n, implying $\alpha(C_3) = \Theta(C_3) = 1$. So, the first interesting case is the 5-cycle, where we know up to now

$$\sqrt{5} \leq \Theta(C_5) \leq \frac{5}{2}. \tag{3}$$

Using his linear programming approach (and some other ideas) Shannon was able to compute the capacity of many graphs and, in particular, of all graphs with five or fewer vertices — with the single exception of C_5, where he could not go beyond the bounds in (3). This is where things stood for more than 20 years until László Lovász showed by an astonishingly simple argument that indeed $\Theta(C_5) = \sqrt{5}$. A seemingly very difficult combinatorial problem was provided with an unexpected and elegant solution.

Lovász' main new idea was to represent the vertices v of the graph by real vectors of length 1 such that any two vectors which belong to non-adjacent vertices in G are orthogonal. Let us call such a set of vectors an *orthonormal representation* of G. Clearly, such a representation always exists: just take the unit vectors $(1, 0, \dots, 0)^T$, $(0, 1, 0, \dots, 0)^T$, $\dots$, $(0, 0, \dots, 1)^T$ of dimension $m = |V|$.

For the graph C_5 we may obtain an orthonormal representation in $\mathbb{R}^3$ by considering an "umbrella" with five ribs $\boldsymbol{v}_1, \dots, \boldsymbol{v}_5$ of unit length. Now open the umbrella (with tip at the origin) to the point where the angles between alternate ribs are $90°$.

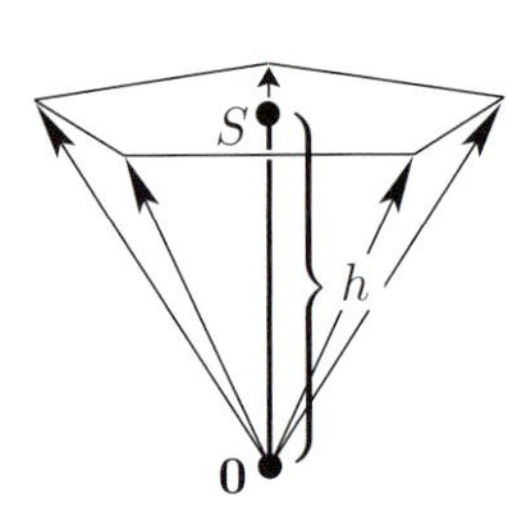

The Lovász umbrella

Lovász then went on to show that the height h of the umbrella, that is, the distance between $\mathbf{0}$ and S, provides the bound

$$\Theta(C_5) \leq \frac{1}{h^2}. \tag{4}$$

A simple calculation yields $h^2 = \frac{1}{\sqrt{5}}$; see the box on the next page. From this $\Theta(C_5) \leq \sqrt{5}$ follows, and therefore $\Theta(C_5) = \sqrt{5}$.

Let us see how Lovász proceeded to prove the inequality (4). (His results were, in fact, much more general.) Consider the usual inner product

$$\langle \boldsymbol{x}, \boldsymbol{y} \rangle = x_1 y_1 + \ldots + x_s y_s$$

of two vectors $\boldsymbol{x} = (x_1, \ldots, x_s)$, $\boldsymbol{y} = (y_1, \ldots, y_s)$ in $\mathbb{R}^s$. Then $|\boldsymbol{x}|^2 = \langle \boldsymbol{x}, \boldsymbol{x} \rangle = x_1^2 + \ldots + x_s^2$ is the square of the length $|\boldsymbol{x}|$ of $\boldsymbol{x}$, and the angle γ between $\boldsymbol{x}$ and $\boldsymbol{y}$ is given by

$$\cos\gamma = \frac{\langle \boldsymbol{x}, \boldsymbol{y} \rangle}{|\boldsymbol{x}||\boldsymbol{y}|}.$$

Thus $\langle \boldsymbol{x}, \boldsymbol{y} \rangle = 0$ if and only if $\boldsymbol{x}$ and $\boldsymbol{y}$ are orthogonal.

Pentagons and the golden section

Tradition has it that a rectangle was considered aesthetically pleasing if, after cutting off a square of length a, the remaining rectangle had the same shape as the original one. The side lengths a, b of such a rectangle must satisfy $\frac{b}{a} = \frac{a}{b-a}$. Setting $\tau := \frac{b}{a}$ for the ratio, we obtain $\tau = \frac{1}{\tau - 1}$ or $\tau^2 - \tau - 1 = 0$. Solving the quadratic equation yields the *golden section* $\tau = \frac{1+\sqrt{5}}{2} \approx 1.6180$.

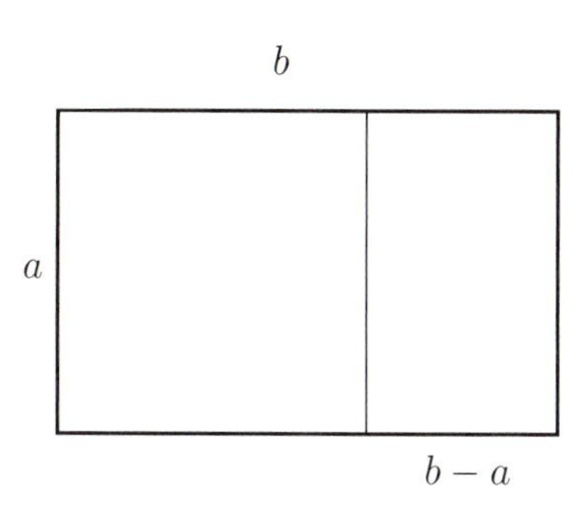

Consider now a regular pentagon of side length a, and let d be the length of its diagonals. It was already known to Euclid (Book XIII,8) that $\frac{d}{a} = \tau$, and that the intersection point of two diagonals divides the diagonals in the golden section.

Here is Euclid's Book Proof. Since the total angle sum of the pentagon is 3π, the angle at any vertex equals $\frac{3\pi}{5}$. It follows that $\sphericalangle ABE = \frac{\pi}{5}$, since ABE is an isosceles triangle. This, in turn, implies $\sphericalangle AMB = \frac{3\pi}{5}$, and we conclude that the triangles ABC and AMB are similar. The quadrilateral $CMED$ is a rhombus since opposing sides are parallel (look at the angles), and so $|MC| = a$ and thus $|AM| = d - a$. By the similarity of ABC and AMB we conclude

$$\frac{d}{a} = \frac{|AC|}{|AB|} = \frac{|AB|}{|AM|} = \frac{a}{d-a} = \frac{|MC|}{|MA|} = \tau.$$

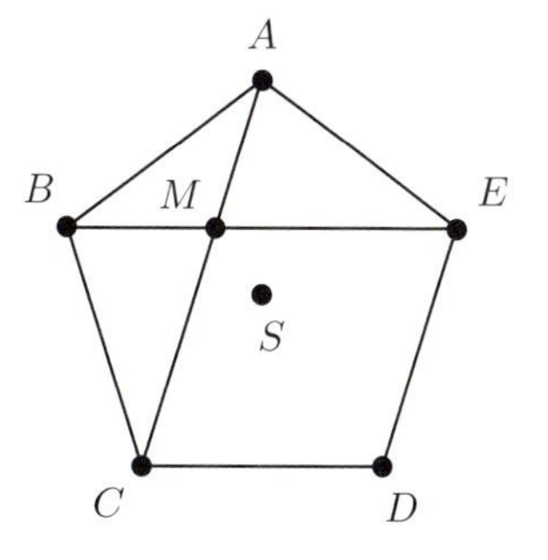

There is more to come. For the distance s of a vertex to the center of the pentagon S, the reader is invited to prove the relation $s^2 = \frac{d^2}{\tau+2}$ (note that BS cuts the diagonal AC at a right angle and halves it).

To finish our excursion into geometry, consider now the umbrella with the regular pentagon on top. Since alternate ribs (of length 1) form a right angle, the theorem of Pythagoras gives us $d = \sqrt{2}$, and hence $s^2 = \frac{2}{\tau+2} = \frac{4}{\sqrt{5}+5}$. So, with Pythagoras again, we find for the height $h = |OS|$ our promised result

$$h^2 = 1 - s^2 = \frac{1+\sqrt{5}}{\sqrt{5}+5} = \frac{1}{\sqrt{5}}.$$

Now we head for an upper bound "$\Theta(G) \leq \sigma_T^{-1}$" for the Shannon capacity of any graph G that has an especially "nice" orthonormal representation. For this let $T = \{\boldsymbol{v}^{(1)}, \ldots, \boldsymbol{v}^{(m)}\}$ be an orthonormal representation of G in $\mathbb{R}^s$, where $\boldsymbol{v}^{(i)}$ corresponds to the vertex v_i. We assume in addition that all the vectors $\boldsymbol{v}^{(i)}$ have the *same* angle ($\neq 90°$) with the vector $\boldsymbol{u} := \frac{1}{m}(\boldsymbol{v}^{(1)} + \ldots + \boldsymbol{v}^{(m)})$, or equivalently that the inner product

$$\langle \boldsymbol{v}^{(i)}, \boldsymbol{u} \rangle = \sigma_T$$

has the same value $\sigma_T \neq 0$ for all i. Let us call this value σ_T the *constant* of the representation T. For the Lovász umbrella that represents C_5 the condition $\langle \boldsymbol{v}^{(i)}, \boldsymbol{u} \rangle = \sigma_T$ certainly holds, for $\boldsymbol{u} = \vec{OS}$.

Now we proceed in the following three steps.

(A) Consider a probability distribution $\boldsymbol{x} = (x_1, \ldots, x_m)$ on V and set

$$\mu(\boldsymbol{x}) := |x_1 \boldsymbol{v}^{(1)} + \ldots + x_m \boldsymbol{v}^{(m)}|^2,$$

and

$$\mu_T(G) := \inf_{\boldsymbol{x}} \mu(\boldsymbol{x}).$$

Let U be a largest independent set in G with $|U| = \alpha$, and define $\boldsymbol{x}_U = (x_1, \ldots, x_m)$ with $x_i = \frac{1}{\alpha}$ if $v_i \in U$ and $x_i = 0$ otherwise. Since all vectors $\boldsymbol{v}^{(i)}$ have unit length and $\langle \boldsymbol{v}^{(i)}, \boldsymbol{v}^{(j)} \rangle = 0$ for any two non-adjacent vertices, we infer

$$\mu_T(G) \leq \mu(\boldsymbol{x}_U) = \Big| \sum_{i=1}^{m} x_i \boldsymbol{v}^{(i)} \Big|^2 = \sum_{i=1}^{m} x_i^2 = \alpha \frac{1}{\alpha^2} = \frac{1}{\alpha}.$$

Thus we have $\mu_T(G) \leq \alpha^{-1}$, and therefore

$$\alpha(G) \leq \frac{1}{\mu_T(G)}.$$

(B) Next we compute $\mu_T(G)$. We need the Cauchy-Schwarz inequality

$$\langle \boldsymbol{a}, \boldsymbol{b} \rangle^2 \leq |\boldsymbol{a}|^2 \, |\boldsymbol{b}|^2$$

for vectors $\boldsymbol{a}, \boldsymbol{b} \in \mathbb{R}^s$. Applied to $\boldsymbol{a} = x_1 \boldsymbol{v}^{(1)} + \ldots + x_m \boldsymbol{v}^{(m)}$ and $\boldsymbol{b} = \boldsymbol{u}$, the inequality yields

$$\langle x_1 \boldsymbol{v}^{(1)} + \ldots + x_m \boldsymbol{v}^{(m)}, \boldsymbol{u} \rangle^2 \leq \mu(\boldsymbol{x}) \, |\boldsymbol{u}|^2. \tag{5}$$

By our assumption that $\langle \boldsymbol{v}^{(i)}, \boldsymbol{u} \rangle = \sigma_T$ for all i, we have

$$\langle x_1 \boldsymbol{v}^{(1)} + \ldots + x_m \boldsymbol{v}^{(m)}, \boldsymbol{u} \rangle = (x_1 + \ldots + x_m) \, \sigma_T = \sigma_T$$

for *any* distribution $\boldsymbol{x}$. Thus, in particular, this has to hold for the uniform distribution $(\frac{1}{m}, \ldots, \frac{1}{m})$, which implies $|\boldsymbol{u}|^2 = \sigma_T$. Hence (5) reduces to

$$\sigma_T^2 \leq \mu(\boldsymbol{x}) \, \sigma_T \qquad \text{or} \qquad \mu_T(G) \geq \sigma_T.$$

On the other hand, for $\boldsymbol{x} = (\frac{1}{m}, \dots, \frac{1}{m})$ we obtain

$$\mu_T(G) \le \mu(\boldsymbol{x}) = |\tfrac{1}{m}(\boldsymbol{v}^{(1)} + \dots + \boldsymbol{v}^{(m)})|^2 = |\boldsymbol{u}|^2 = \sigma_T,$$

and so we have proved

$$\mu_T(G) = \sigma_T. \tag{6}$$

In summary, we have established the inequality

$$\alpha(G) \le \frac{1}{\sigma_T} \tag{7}$$

for *any* orthonormal respresentation T with constant σ_T.

(C) To extend this inequality to $\Theta(G)$, we proceed as before. Consider again the product $G \times H$ of two graphs. Let G and H have orthonormal representations R and S in $\mathbb{R}^r$ and $\mathbb{R}^s$, respectively, with constants σ_R and σ_S. Let $\boldsymbol{v} = (v_1, \dots, v_r)$ be a vector in R and $\boldsymbol{w} = (w_1, \dots, w_s)$ be a vector in S. To the vertex in $G \times H$ corresponding to the pair $(\boldsymbol{v}, \boldsymbol{w})$ we associate the vector

$$\boldsymbol{v}\boldsymbol{w}^T := (v_1w_1, \dots, v_1w_s, v_2w_1, \dots, v_2w_s, \dots, v_rw_1, \dots, v_rw_s) \in \mathbb{R}^{rs}.$$

It is immediately checked that $R \times S := \{\boldsymbol{v}\boldsymbol{w}^T : \boldsymbol{v} \in R, \boldsymbol{w} \in S\}$ is an orthonormal representation of $G \times H$ with constant $\sigma_R\sigma_S$. Hence by (6) we obtain

$$\mu_{R\times S}(G \times H) = \mu_R(G)\mu_S(H).$$

For $G^n = G \times \dots \times G$ and the representation T with constant σ_T this means

$$\mu_{T^n}(G^n) = \mu_T(G)^n = \sigma_T^n$$

and by (7) we obtain

$$\alpha(G^n) \le \sigma_T^{-n}, \qquad \sqrt[n]{\alpha(G^n)} \le \sigma_T^{-1}.$$

Taking all things together we have thus completed Lovász' argument:

Theorem. *Whenever $T = \{\boldsymbol{v}^{(1)}, \dots, \boldsymbol{v}^{(m)}\}$ is an orthonormal representation of G with constant σ_T, then*

$$\Theta(G) \le \frac{1}{\sigma_T}. \tag{8}$$

Looking at the Lovász umbrella, we have $\boldsymbol{u} = (0, 0, h{=}\frac{1}{\sqrt[4]{5}})^T$ and hence $\sigma = \langle \boldsymbol{v}^{(i)}, \boldsymbol{u}\rangle = h^2 = \frac{1}{\sqrt{5}}$, which yields $\Theta(C_5) \le \sqrt{5}$. Thus Shannon's problem is solved.

"Umbrellas with five ribs"

Let us carry our discussion a little further. We see from (8) that the larger σ_T is for a representation of G, the better a bound for $\Theta(G)$ we will get. Here is a method that gives us an orthonormal representation for *any* graph G. To $G = (V, E)$ we associate the *adjacency matrix* $A = (a_{ij})$, which is defined as follows: Let $V = \{v_1, \ldots, v_m\}$, then we set

$$a_{ij} := \begin{cases} 1 & \text{if } v_i v_j \in E \\ 0 & \text{otherwise.} \end{cases}$$

$$A = \begin{pmatrix} 0 & 1 & 0 & 0 & 1 \\ 1 & 0 & 1 & 0 & 0 \\ 0 & 1 & 0 & 1 & 0 \\ 0 & 0 & 1 & 0 & 1 \\ 1 & 0 & 0 & 1 & 0 \end{pmatrix}$$

The adjacency matrix for the 5-cycle C_5

A is a real symmetric matrix with 0's in the main diagonal.

Now we need two facts from linear algebra. First, as a symmetric matrix, A has m real eigenvalues $\lambda_1 \geq \lambda_2 \geq \ldots \geq \lambda_m$ (some of which may be equal), and the sum of the eigenvalues equals the sum of the diagonal entries of A, that is, 0. Hence the smallest eigenvalue must be negative (except in the trivial case when G has no edges). Let $p = |\lambda_m| = -\lambda_m$ be the absolute value of the smallest eigenvalue, and consider the matrix

$$M := I + \frac{1}{p} A,$$

where I denotes the $(m \times m)$-identity matrix. This M has the eigenvalues $1+\frac{\lambda_1}{p} \geq 1+\frac{\lambda_2}{p} \geq \ldots \geq 1+\frac{\lambda_m}{p} = 0$. Now we quote the second result (the principal axis theorem of linear algebra): If $M = (m_{ij})$ is a real symmetric matrix with all eigenvalues ≥ 0, then there are vectors $\boldsymbol{v}^{(1)}, \ldots, \boldsymbol{v}^{(m)} \in \mathbb{R}^s$ for $s = \text{rank}(M)$, such that

$$m_{ij} = \langle \boldsymbol{v}^{(i)}, \boldsymbol{v}^{(j)} \rangle \qquad (1 \leq i, j \leq m).$$

In particular, for $M = I + \frac{1}{p} A$ we obtain

$$\langle \boldsymbol{v}^{(i)}, \boldsymbol{v}^{(i)} \rangle = m_{ii} = 1 \qquad \text{for all } i$$

and

$$\langle \boldsymbol{v}^{(i)}, \boldsymbol{v}^{(j)} \rangle = \frac{1}{p} a_{ij} \qquad \text{for } i \neq j.$$

Since $a_{ij} = 0$ whenever $v_i v_j \notin E$, we see that the vectors $\boldsymbol{v}^{(1)}, \ldots, \boldsymbol{v}^{(m)}$ form indeed an orthonormal representation of G.

Let us, finally, apply this construction to the m-cycles C_m for odd $m \geq 5$. Here one easily computes $p = |\lambda_{\min}| = 2\cos\frac{\pi}{m}$ (see the box). Every row of the adjacency matrix contains two 1's, implying that every row of the matrix M sums to $1 + \frac{2}{p}$. For the representation $\{\boldsymbol{v}^{(1)}, \ldots, \boldsymbol{v}^{(m)}\}$ this means

$$\langle \boldsymbol{v}^{(i)}, \boldsymbol{v}^{(1)} + \ldots + \boldsymbol{v}^{(m)} \rangle = 1 + \frac{2}{p} = 1 + \frac{1}{\cos\frac{\pi}{m}}$$

and hence

$$\langle \boldsymbol{v}^{(i)}, \boldsymbol{u} \rangle = \frac{1}{m}(1 + (\cos\tfrac{\pi}{m})^{-1}) = \sigma$$

for all i. We can therefore apply our main result (8) and conclude

$$\Theta(C_m) \leq \frac{m}{1 + (\cos\frac{\pi}{m})^{-1}} \qquad (\text{for } m \geq 5 \text{ odd}). \tag{9}$$

Notice that because of $\cos\frac{\pi}{m} < 1$ the bound (9) is better than the bound $\Theta(C_m) \le \frac{m}{2}$ we found before. Note further $\cos\frac{\pi}{5} = \frac{\tau}{2}$, where $\tau = \frac{\sqrt{5}+1}{2}$ is the golden section. Hence for $m = 5$ we again obtain

$$\Theta(C_5) \le \frac{5}{1+\frac{4}{\sqrt{5}+1}} = \frac{5(\sqrt{5}+1)}{5+\sqrt{5}} = \sqrt{5}.$$

The orthonormal representation given by this construction is, of course, precisely the "Lovász umbrella."

And what about C_7, C_9, and the other odd cycles? By considering $\alpha(C_m^2)$, $\alpha(C_m^3)$ and other small powers the lower bound $\frac{m-1}{2} \le \Theta(C_m)$ can certainly be increased, but for no odd $m \ge 7$ do the best known lower bounds agree with the upper bound given in (8). So, twenty years after Lovász' marvelous proof of $\Theta(C_5) = \sqrt{5}$, these problems remain open and are considered very difficult — but after all we had this situation before.

For example, for $m = 7$ all we know is

$$\sqrt[5]{343} \le \Theta(C_7) \le \frac{7}{1+(\cos\frac{\pi}{7})^{-1}},$$

which is $3.2141 \le \Theta(C_7) \le 3.3177$.

The eigenvalues of C_m

Look at the adjacency matrix A of the cycle C_m. To find the eigenvalues (and eigenvectors) we use the m-th roots of unity. These are given by $1, \zeta, \zeta^2, \ldots, \zeta^{m-1}$ for $\zeta = e^{\frac{2\pi i}{m}}$ — see the box on page 25. Let $\lambda = \zeta^k$ be any of these roots, then we claim that $(1, \lambda, \lambda^2, \ldots, \lambda^{m-1})^T$ is an eigenvector of A to the eigenvalue $\lambda + \lambda^{-1}$. In fact, by the set-up of A we find

$$A\begin{pmatrix} 1 \\ \lambda \\ \lambda^2 \\ \vdots \\ \lambda^{m-1} \end{pmatrix} = \begin{pmatrix} \lambda + \lambda^{m-1} \\ \lambda^2 + 1 \\ \lambda^3 + \lambda \\ \vdots \\ 1 + \lambda^{m-2} \end{pmatrix} = (\lambda+\lambda^{-1})\begin{pmatrix} 1 \\ \lambda \\ \lambda^2 \\ \vdots \\ \lambda^{m-1} \end{pmatrix}.$$

Since the vectors $(1, \lambda, \ldots, \lambda^{m-1})$ are independent (they form a so-called Vandermonde matrix) we conclude that for odd m

$$\begin{aligned} \zeta^k + \zeta^{-k} &= [(\cos(2k\pi/m) + i\sin(2k\pi/m)] \\ &\quad + [\cos(2k\pi/m) - i\sin(2k\pi/m)] \\ &= 2\cos(2k\pi/m) \qquad (0 \le k \le \tfrac{m-1}{2}) \end{aligned}$$

are all the eigenvalues of A. Now the cosine is a decreasing function, and so

$$2\cos\Big(\frac{(m-1)\pi}{m}\Big) = -2\cos\frac{\pi}{m}$$

is the smallest eigenvalue of A.

References

[1] V. CHVÁTAL: *Linear Programming,* Freeman, New York 1983.

[2] W. HAEMERS: *Eigenvalue methods,* in: "Packing and Covering in Combinatorics" (A. Schrijver, ed.), Math. Centre Tracts **106** (1979), 15-38.

[3] L. LOVÁSZ: *On the Shannon capacity of a graph,* IEEE Trans. Information Theory **25** (1979), 1-7.

[4] C. E. SHANNON: *The zero-error capacity of a noisy channel,* IRE Trans. Information Theory **3** (1956), 3-15.

Of friends and politicians

Chapter 34

It is not known who first raised the following problem or who gave it its human touch. Here it is:

> *Suppose in a group of people we have the situation that any pair of persons have precisely one common friend. Then there is always a person (the "politician") who is everybody's friend.*

"A politician's smile"

In the mathematical jargon this is called the *friendship theorem.*

Before tackling the proof let us rephrase the problem in graph-theoretic terms. We interpret the people as the set of vertices V and join two vertices by an edge if the corresponding people are friends. We tacitly assume that friendship is always two-ways, that is, if u is a friend of v, then v is also a friend of u, and further that nobody is his or her own friend. Thus the theorem takes on the following form:

Theorem. *Suppose that G is a finite graph in which any two vertices have precisely one common neighbor. Then there is a vertex which is adjacent to all other vertices.*

Note that there are finite graphs with this property; see the figure, where u is the politician. However, these "windmill graphs" also turn out to be the only graphs with the desired property. Indeed, it is not hard to verify that in the presence of a politician only the windmill graphs are possible.

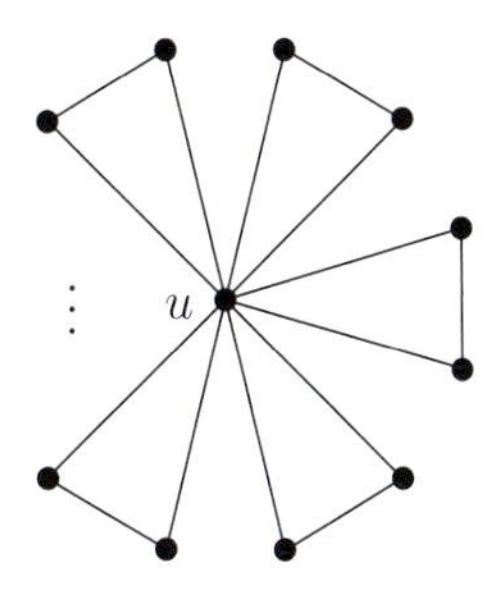

A windmill graph

Surprisingly, the friendship theorem does not hold for infinite graphs! Indeed, for an inductive construction of a counterexample one may start for example with a 5-cycle, and repeatedly add common neighbors for all pairs of vertices in the graph that don't have one, yet. This leads to a (countably) infinite friendship graph without a politician.

Several proofs of the friendship theorem exist, but the first proof, given by Paul Erdős, Alfred Rényi and Vera Sós, is still the most accomplished.

■ **Proof.** Suppose the assertion is false, and G is a counterexample, that is, no vertex of G is adjacent to all other vertices. To derive a contradiction we proceed in two steps. The first part is combinatorics, and the second part is linear algebra.

(1) We claim that G is a regular graph, that is, $d(u) = d(v)$ for any $u, v \in V$. Note first that the condition of the theorem implies that there are no cycles of length 4 in G. Let us call this the *C_4-condition.*

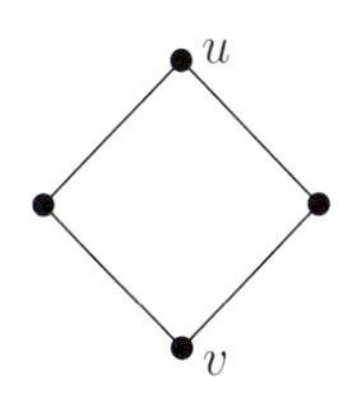

We first prove that any two *non-adjacent* vertices u and v have equal degree $d(u) = d(v)$. Suppose $d(u) = k$, where $w_1, \ldots, w_k$ are the neighbors of u. Exactly one of the w_i, say w_2, is adjacent to v, and w_2 adjacent to exactly one of the other w_i's, say w_1, so that we have the situation of the figure to the left. The vertex v has with w_1 the common neighbor w_2, and with w_i $(i \geq 2)$ a common neighbor z_i $(i \geq 2)$. By the C_4-condition, all these z_i must be distinct. We conclude $d(v) \geq k = d(u)$, and thus $d(u) = d(v) = k$ by symmetry.

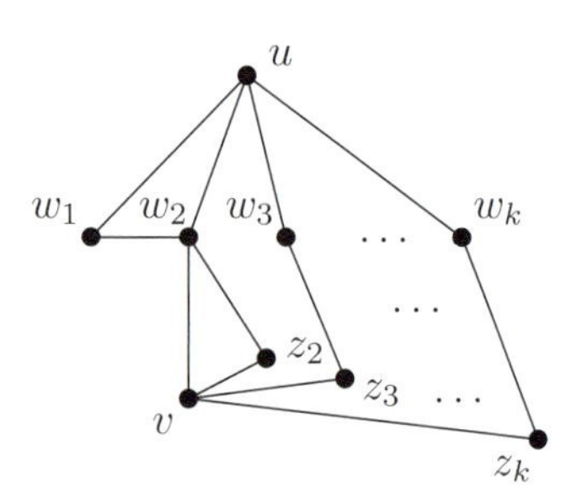

To finish the proof of **(1)**, observe that any vertex different from w_2 is not adjacent to either u or v, and hence has degree k, by what we already proved. But since w_2 also has a non-neighbor, it has degree k as well, and thus G is k-regular.

Summing over the degrees of the k neighbors of u we get k^2. Since every vertex (except u) has exactly one common neighbor with u, we have counted every vertex once, except for u, which was counted k times. So the total number of vertices of G is

$$n = k^2 - k + 1. \tag{1}$$

(2) The rest of the proof is a beautiful application of some standard results of linear algebra. Note first that k must be greater than 2, since for $k \leq 2$ only $G = K_1$ and $G = K_3$ are possible by (1), both of which are trivial windmill graphs. Consider the adjacency matrix $A = (a_{ij})$, as defined on page 220. By part **(1)**, any row has exactly k 1's, and by the condition of the theorem, for any two rows there is exactly one column where they both have a 1. Note further that the main diagonal consists of 0's. Hence we have

$$A^2 = \begin{pmatrix} k & 1 & \ldots & 1 \\ 1 & k & & 1 \\ \vdots & & \ddots & \vdots \\ 1 & \ldots & 1 & k \end{pmatrix} = (k-1)\,I + J\,,$$

where I is the identity matrix, and J the matrix of all 1's. It is immediately checked that J has the eigenvalues n (of multiplicity 1) and 0 (of multiplicity $n-1$). It follows that A^2 has the eigenvalues $k - 1 + n = k^2$ (of multiplicity 1) and $k - 1$ (of multiplicity $n - 1$).

Since A is symmetric and hence diagonalizable, we conclude that A has the eigenvalues k (of multiplicity 1) and $\pm\sqrt{k-1}$. Suppose r of the eigenvalues are equal to $\sqrt{k-1}$ and s of them are equal to $-\sqrt{k-1}$, with $r + s = n - 1$. Now we are almost home. Since the sum of the eigenvalues of A equals the trace (which is 0), we find

$$k + r\sqrt{k-1} - s\sqrt{k-1} = 0,$$

and, in particular, $r \neq s$, and

$$\sqrt{k-1} = \frac{k}{s-r}.$$

Now if the square root $\sqrt{m}$ of a natural number m is rational, then it is an integer! An elegant proof for this was presented by Dedekind in 1858: Let n_0 be the smallest natural number with $n_0\sqrt{m} \in \mathbb{N}$. If $\sqrt{m} \notin \mathbb{N}$, then there exists $\ell \in \mathbb{N}$ with $0 < \sqrt{m} - \ell < 1$. Setting $n_1 := n_0(\sqrt{m} - \ell)$, we find $n_1 \in \mathbb{N}$ and $n_1\sqrt{m} = n_0(\sqrt{m} - \ell)\sqrt{m} = n_0 m - \ell(n_0\sqrt{m}) \in \mathbb{N}$. With $n_1 < n_0$ this yields a contradiction to the choice of n_0.
Returning to our equation, let us set $h = \sqrt{k-1} \in \mathbb{N}$, then

$$h(s-r) \;=\; k \;=\; h^2 + 1.$$

Since h divides $h^2 + 1$ and h^2, we find that h must be equal to 1, and thus $k = 2$, which we have already excluded. So we have arrived at a contradiction, and the proof is complete. □

However, the story is not quite over. Let us rephrase our theorem in the following way: Suppose G is a graph with the property that between any two vertices there is exactly one path of length 2. Clearly, this is an equivalent formulation of the friendship condition. Our theorem then says that the only such graphs are the windmill graphs. But what if we consider paths of length more than 2? A conjecture of Anton Kotzig asserts that the analogous situation is impossible.

Kotzig's Conjecture. *Let $\ell > 2$. Then there are no finite graphs with the property that between any two vertices there is precisely one path of length ℓ.*

Kotzig himself verified his conjecture for $\ell \leq 8$. In [3] his conjecture is proved up to $\ell = 20$, and A. Kostochka has told us recently that it is now verified for all $\ell \leq 33$. A general proof, however, seems to be out of reach ...

References

[1] P. ERDŐS, A. RÉNYI & V. SÓS: *On a problem of graph theory,* Studia Sci. Math. **1** (1966), 215-235.

[2] A. KOTZIG: *Regularly k-path connected graphs,* Congressus Numerantium **40** (1983), 137-141.

[3] A. KOSTOCHKA: *The nonexistence of certain generalized friendship graphs,* in: "Combinatorics" (Eger, 1987), Colloq. Math. Soc. János Bolyai **52**, North-Holland, Amsterdam 1988, 341-356.

Probability makes counting (sometimes) easy

Chapter 35

Just as we started this book with the first papers of Paul Erdős in number theory, we close it by discussing what will possibly be considered his most lasting legacy — the introduction, together with Alfred Rényi, of the *probabilistic method*. Stated in the simplest way it says:

> *If, in a given set of objects, the probability that an object does not have a certain property is less than 1, then there must exist an object with this property.*

Thus we have an *existence* result. It may be (and often is) very difficult to find this object, but we know that it exists. We present here three examples (of increasing sophistication) of this probabilistic method due to Erdős, and end with a particularly elegant recent application.

As a warm-up, consider a family $\mathcal{F}$ of subsets A_i, all of size $d \geq 2$, of a finite ground-set X. We say that $\mathcal{F}$ is 2-*colorable* if there exists a coloring of X with two colors such that in every set A_i both colors appear. It is immediate that not every family can be colored in this way. As an example, take *all* subsets of size d of a $(2d-1)$-set X. Then no matter how we 2-color X, there must be d elements which are colored alike. On the other hand, it is equally clear that every subfamily of a 2-colorable family of d-sets is itself 2-colorable. Hence we are interested in the *smallest* number $m = m(d)$ for which a family with m sets exists which is not 2-colorable. Phrased differently, $m(d)$ is the largest number which guarantees that *every* family with less than $m(d)$ sets is 2-colorable.

Theorem 1. *Every family of at most 2^{d-1} d-sets is 2-colorable, that is, $m(d) > 2^{d-1}$.*

■ **Proof.** Suppose $\mathcal{F}$ is a family of d-sets with at most 2^{d-1} sets. Color X randomly with two colors, all colorings being equally likely. For each set $A \in \mathcal{F}$ let E_A be the event that all elements of A are colored alike. Since there are precisely two such colorings, we have

$$\text{Prob}(E_A) \;=\; \left(\tfrac{1}{2}\right)^{d-1},$$

and hence with $m = |\mathcal{F}| \leq 2^{d-1}$ (note that the events E_A are not disjoint)

$$\text{Prob}\Big(\bigcup_{A\in\mathcal{F}} E_A\Big) \;<\; \sum_{A\in\mathcal{F}} \text{Prob}(E_A) \;=\; m\left(\tfrac{1}{2}\right)^{d-1} \;\leq\; 1.$$

We conclude that there exists some 2-coloring of X without a unicolored d-set from $\mathcal{F}$, and this is just our condition of 2-colorability. □

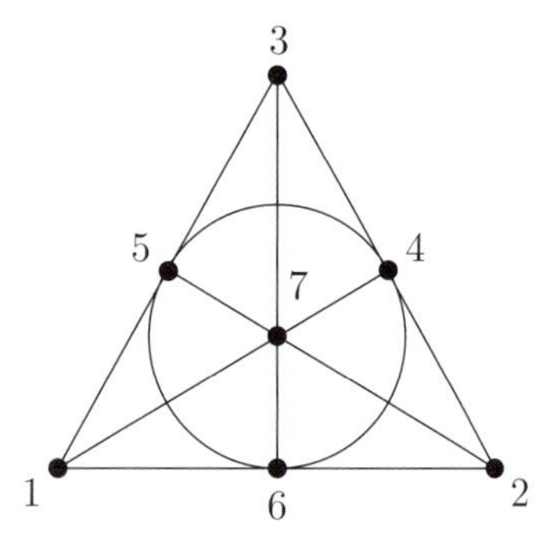

An upper bound for $m(d)$, roughly equal to $d^2 2^d$, was also established by Erdős, again using the probabilistic method, this time taking random sets and a fixed coloring. As for exact values, only the first two $m(2) = 3$, $m(3) = 7$ are known. Of course, $m(2) = 3$ is realized by the graph K_3, while the Fano configuration yields $m(3) \leq 7$. Here $\mathcal{F}$ consists of the seven 3-sets of the figure (including the circle set $\{4, 5, 6\}$). The reader may find it fun to show that $\mathcal{F}$ needs 3 colors. To prove that all families of six 3-sets are 2-colorable, and hence $m(3) = 7$, requires a little more care.

Our next example is the classic in the field — Ramsey numbers. Consider the complete graph K_N on N vertices. We say that K_N has property (m, n) if, no matter how we color the edges of K_N red and blue, there is always a complete subgraph on m vertices with all edges colored red or a complete subgraph on n vertices with all edges colored blue. It is clear that if K_N has property (m, n), then so does every K_s with $s \geq N$. So, as in the first example, we ask for the *smallest* number N (if it exists) with this property — and this is the *Ramsey number* $R(m, n)$.

As a start, we certainly have $R(m, 2) = m$ because either all of the edges of K_m are red or there is a blue edge, resulting in a blue K_2. By symmetry, we have $R(2, n) = n$. Now, suppose $R(m-1, n)$ and $R(m, n-1)$ exist. We then prove that $R(m, n)$ exists and that

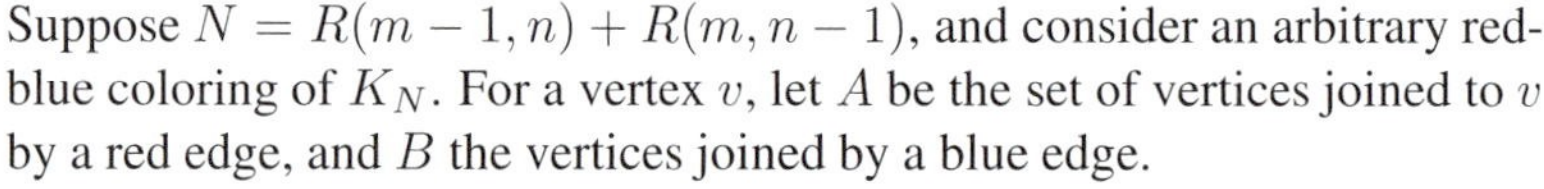

$$R(m,n) \;\leq\; R(m-1,n) \;+\; R(m,n-1). \qquad (1)$$

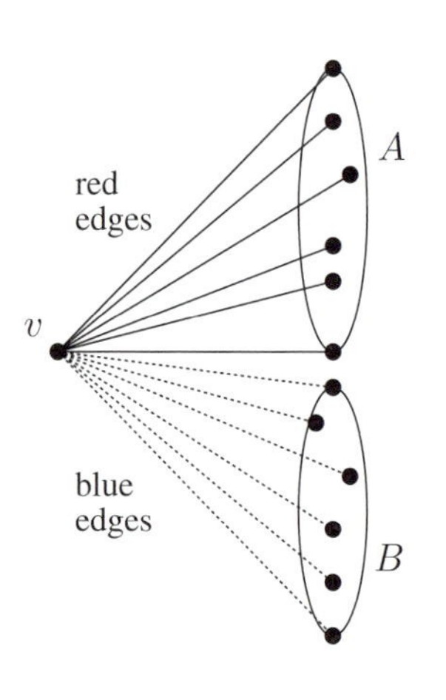

Suppose $N = R(m-1, n) + R(m, n-1)$, and consider an arbitrary red-blue coloring of K_N. For a vertex v, let A be the set of vertices joined to v by a red edge, and B the vertices joined by a blue edge.

Since $|A| + |B| = N - 1$, we find that either $|A| \geq R(m-1, n)$ or $|B| \geq R(m, n-1)$. Suppose $|A| \geq R(m-1, n)$, the other case being analogous. Then by the definition of $R(m-1, n)$, there either exists in A a subset A_R of size $m-1$ all of whose edges are colored red which together with v yields a red K_m, or there is a subset A_B of size n with all edges colored blue. We infer that K_N satisfies the (m, n)-property and Claim (1) follows.

Combining (1) with the starting values $R(m, 2) = m$ and $R(2, n) = n$, we obtain from the familiar recursion for binomial coefficients

$$R(m,n) \;\leq\; \binom{m+n-2}{m-1},$$

and, in particular

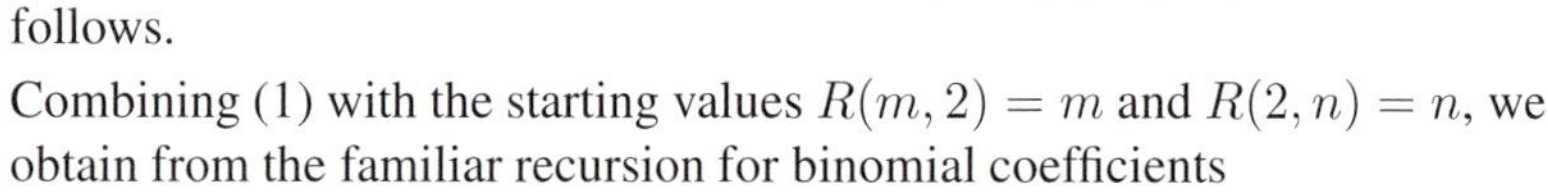

$$R(k,k) \;\leq\; \binom{2k-2}{k-1} \;=\; \binom{2k-3}{k-1} + \binom{2k-3}{k-2} \;\leq\; 2^{2k-3}. \qquad (2)$$

Now what we are really interested in is a lower bound for $R(k, k)$. This amounts to proving for an as-large-as-possible $N < R(k, k)$ that there *exists* a coloring of the edges such that no red or blue K_k results. And this is where the probabilistic method comes into play.

Theorem 2. *For all $k \geq 2$, the following lower bound holds for the Ramsey numbers:*

$$R(k,k) \;\geq\; 2^{\frac{k}{2}}.$$

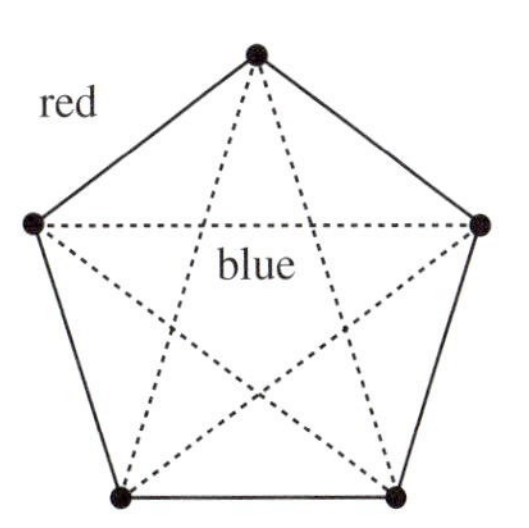

■ **Proof.** We have $R(2,2) = 2$. From (2) we know $R(3,3) \leq 6$, and the pentagon colored as in the figure shows $R(3,3) = 6$.

Now let us assume $k \geq 4$. Suppose $N < 2^{\frac{k}{2}}$, and consider all red-blue colorings, where we color each edge independently red or blue with probability $\frac{1}{2}$. Thus all colorings are equally likely with probability $2^{-\binom{N}{2}}$. Let A be a set of vertices of size k. The probability of the event A_R that the edges in A are all colored red is then $2^{-\binom{k}{2}}$. Hence it follows that the probability p_R for *some* k-set to be colored all red is bounded by

$$p_R \;=\; \text{Prob}\Big(\bigcup_{|A|=k} A_R\Big) \;\leq\; \sum_{|A|=k} \text{Prob}(A_R) \;=\; \binom{N}{k} 2^{-\binom{k}{2}}.$$

Now with $N < 2^{\frac{k}{2}}$ and $k \geq 4$, using $\binom{N}{k} \leq \frac{N^k}{2^{k-1}}$ for $k \geq 2$ (see page 12), we have

$$\binom{N}{k} 2^{-\binom{k}{2}} \;\leq\; \frac{N^k}{2^{k-1}} 2^{-\binom{k}{2}} \;<\; 2^{\frac{k^2}{2}-\binom{k}{2}-k+1} \;=\; 2^{-\frac{k}{2}+1} \leq \frac{1}{2}.$$

Hence $p_R < \frac{1}{2}$, and by symmetry $p_B < \frac{1}{2}$ for the probability of some k vertices with all edges between them colored blue. We conclude that $p_R + p_B < 1$ for $N < 2^{\frac{k}{2}}$, so there *must* be a coloring with no red or blue K_k, which means that K_N does not have property (k,k). □

Of course, there is quite a gap between the lower and the upper bound for $R(k,k)$. Still, as simple as this Book Proof is, no lower bound with a better exponent has been found for general k in the more than 50 years since Erdős' result. In fact, no one has been able to prove a lower bound of the form $R(k,k) > 2^{(\frac{1}{2}+\varepsilon)k}$ nor an upper bound of the form $R(k,k) < 2^{(2-\varepsilon)k}$ for a fixed $\varepsilon > 0$.

Our third result is another beautiful illustration of the probabilistic method. Consider a graph G on n vertices and its chromatic number $\chi(G)$. If $\chi(G)$ is high, that is, if we need many colors, then we might suspect that G contains a large complete subgraph. However, this is far from the truth. Already in the fourties Blanche Descartes constructed graphs with arbitrarily high chromatic number and no triangles, that is, with every cycle having length at least 4, and so did several others (see the box on the next page).

However, in these examples there were many cycles of length 4. Can we do even better? Can we stipulate that there are no cycles of small length and still have arbitrarily high chromatic number? Yes we can! To make matters precise, let us call the length of a shortest cycle in G the *girth* $\gamma(G)$ of G; then we have the following theorem, first proved by Paul Erdős.

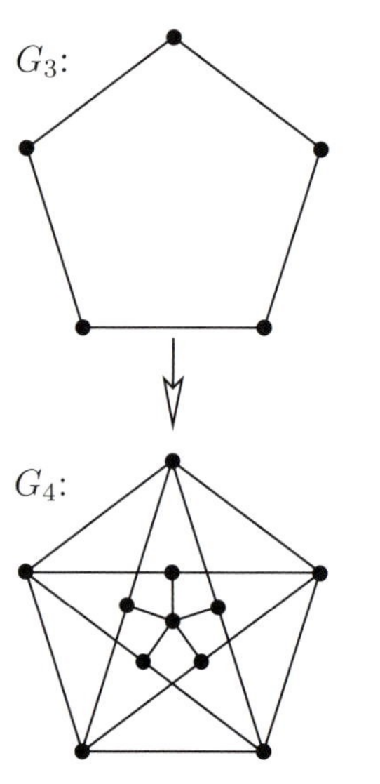

Constructing the Mycielski graph

Triangle-free graphs with high chromatic number

Here is a sequence of triangle-free graphs $G_3, G_4, \ldots$ with

$$\chi(G_n) = n.$$

Start with $G_3 = C_5$, the 5-cycle; thus $\chi(G_3) = 3$. Suppose we have already constructed G_n on the vertex set V. The new graph G_{n+1} has the vertex set $V \cup V' \cup \{z\}$, where the vertices $v' \in V'$ correspond bijectively to $v \in V$, and z is a single other vertex. The edges of G_{n+1} fall into 3 classes: First, we take all edges of G_n; secondly every vertex v' is joined to precisely the neighbors of v in G_n; thirdly z is joined to all $v' \in V'$. Hence from $G_3 = C_5$ we obtain as G_4 the so-called *Mycielski graph*.

Clearly, G_{n+1} is again triangle-free. To prove $\chi(G_{n+1}) = n+1$ we use induction on n. Take any n-coloring of G_n and consider a color class C. There must exist a vertex $v \in C$ which is adjacent to at least one vertex of every other color class; otherwise we could distribute the vertices of C onto the $n-1$ other color classes, resulting in $\chi(G_n) \leq n-1$. But now it is clear that v' (the vertex in V' corresponding to v) must receive the same color as v in this n-coloring. So, all n colors appear in V', and we need a new color for z.

Theorem 3. *For every $k \geq 2$, there exists a graph G with chromatic number $\chi(G) > k$ and girth $\gamma(G) > k$.*

The strategy is similar to that of the previous proofs: We consider a certain probability space on graphs and go on to show that the probability for $\chi(G) \leq k$ is smaller than $\frac{1}{2}$, and similarly the probability for $\gamma(G) \leq k$ is smaller than $\frac{1}{2}$. Consequently, there must exist a graph with the desired properties.

■ **Proof.** Let $V = \{v_1, v_2, \ldots, v_n\}$ be the vertex set, and p a fixed number between 0 and 1, to be carefully chosen later. Our probability space $\mathcal{G}(n,p)$ consists of all graphs on V where the individual edges appear with probability p, independently of each other. In other words, we are talking about a Bernoulli experiment where we throw in each edge with probability p. As an example, the probability $\text{Prob}(K_n)$ for the complete graph is $\text{Prob}(K_n) = p^{\binom{n}{2}}$. In general, we have $\text{Prob}(H) = p^m(1-p)^{\binom{n}{2}-m}$ if the graph H on V has precisely m edges.

Let us first look at the chromatic number $\chi(G)$. By $\alpha = \alpha(G)$ we denote the *independence number*, that is, the size of a largest independent set in G. Since in a coloring with $\chi = \chi(G)$ colors all color classes are independent (and hence of size $\leq \alpha$), we infer $\chi\alpha \geq n$. Therefore if α is small as compared to n, then χ must be large, which is what we want.

Suppose $2 \leq r \leq n$. The probability that a fixed r-set in V is independent is $(1-p)^{\binom{r}{2}}$, and we conclude by the same argument as in Theorem 2

$$\begin{aligned}\text{Prob}(\alpha \geq r) &\leq \binom{n}{r}(1-p)^{\binom{r}{2}} \\ &\leq n^r(1-p)^{\binom{r}{2}} = (n(1-p)^{\frac{r-1}{2}})^r \leq (ne^{-p(r-1)/2})^r,\end{aligned}$$

since $1-p \leq e^{-p}$ for all p.

Given any fixed $k > 0$ we now choose $p := n^{-\frac{k}{k+1}}$, and proceed to show that for n large enough,

$$\text{Prob}\Big(\alpha \geq \frac{n}{2k}\Big) < \frac{1}{2}. \qquad (3)$$

Indeed, since $n^{\frac{1}{k+1}}$ grows faster than $\log n$, we have $n^{\frac{1}{k+1}} \geq 6k \log n$ for large enough n, and thus $p \geq 6k\frac{\log n}{n}$. For $r := \lceil \frac{n}{2k} \rceil$ this gives $pr \geq 3\log n$, and thus

$$ne^{-p(r-1)/2} = ne^{-\frac{pr}{2}}e^{\frac{p}{2}} \leq ne^{-\frac{3}{2}\log n}e^{\frac{1}{2}} = n^{-\frac{1}{2}}e^{\frac{1}{2}} = (\tfrac{e}{n})^{\frac{1}{2}},$$

which converges to 0 as n goes to infinity. Hence (3) holds for all $n \geq n_1$.

Now we look at the second parameter, $\gamma(G)$. For the given k we want to show that there are not too many cycles of length $\leq k$. Let i be between 3 and k, and $A \subseteq V$ a fixed i-set. The number of possible i-cycles on A is clearly the number of cyclic permutations of A divided by 2 (since we may traverse the cycle in either direction), and thus equal to $\frac{(i-1)!}{2}$. The total number of possible i-cycles is therefore $\binom{n}{i}\frac{(i-1)!}{2}$, and every such cycle C appears with probability p^i. Let X be the random variable which counts the number of cycles of length $\leq k$. In order to estimate X we use two simple but beautiful tools. The first is linearity of expectation, and the second is Markov's inequality for nonnegative random variables, which says

$$\text{Prob}(X \geq a) \leq \frac{EX}{a},$$

where EX is the expected value of X. See the appendix to Chapter 14 for both tools.

Let X_C be the indicator random variable of the cycle C of, say, length i. That is, we set $X_C = 1$ or 0 depending on whether C appears in the graph or not; hence $EX_C = p^i$. Since X counts the number of all cycles of length $\leq k$ we have $X = \sum X_C$, and hence by linearity

$$EX = \sum_{i=3}^{k}\binom{n}{i}\frac{(i-1)!}{2}p^i \leq \frac{1}{2}\sum_{i=3}^{k} n^i p^i \leq \frac{1}{2}(k-2)n^k p^k$$

where the last inequality holds because of $np = n^{\frac{1}{k+1}} \geq 1$. Applying now Markov's inequality with $a = \frac{n}{2}$, we obtain

$$\text{Prob}(X \geq \tfrac{n}{2}) \leq \frac{EX}{n/2} \leq (k-2)\frac{(np)^k}{n} = (k-2)n^{-\frac{1}{k+1}}.$$

Since the right-hand side goes to 0 with n going to infinity, we infer that $p(X \geq \frac{n}{2}) < \frac{1}{2}$ for $n \geq n_2$.

Now we are almost home. Our analysis tells us that for $n \geq \max(n_1, n_2)$ there exists a graph H on n vertices with $\alpha(H) < \frac{n}{2k}$ and fewer than $\frac{n}{2}$ cycles of length $\leq k$. Delete one vertex from each of these cycles, and let G be the resulting graph. Then $\gamma(G) > k$ holds at any rate. Since G contains more than $\frac{n}{2}$ vertices and satisfies $\alpha(G) \leq \alpha(H) < \frac{n}{2k}$, we find

$$\chi(G) \geq \frac{n/2}{\alpha(G)} \geq \frac{n}{2\alpha(H)} > \frac{n}{n/k} = k,$$

and the proof is finished. □

Explicit constructions of graphs with high girth and chromatic number (of huge size) are known. (In contrast, one does not know how to construct red/blue colorings with no large monochromatic cliques, whose existence is given by Theorem 2.) What remains striking about the Erdős proof is that it proves the existence of relatively small graphs with high chromatic number and girth.

To end our excursion into the probabilistic world let us discuss an important result in geometric graph theory (which again goes back to Paul Erdős) whose stunning Book Proof is of very recent vintage.

Consider a simple graph $G = G(V, E)$ with n vertices and m edges. We want to embed G into the plane just as we did for planar graphs. Now, we know from Chapter 11 — as a consequence of Euler's formula — that a simple planar graph G has at most $3n - 6$ edges. Hence if m is greater than $3n - 6$, there must be crossings of edges. The *crossing number* $\text{cr}(G)$ is then naturally defined: It is the smallest number of crossings among all drawings of G, where crossings of more than two edges in one point are not allowed. Thus $\text{cr}(G) = 0$ if and only if G is planar.

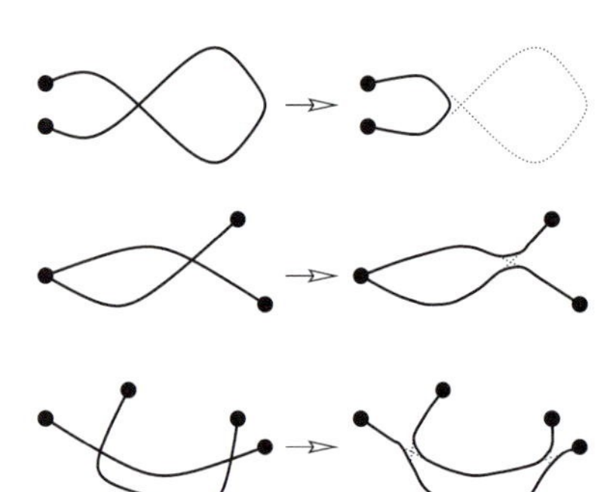

In such a minimal drawing the following three situations are ruled out:

- No edge can cross itself.
- Edges with a common endvertex cannot cross.
- No two edges cross twice.

This is because in either of these cases, we can construct a different drawing of the same graph with fewer crossings, using the operations that are indicated in our figure. So, from now on we assume that any drawing observes these rules.

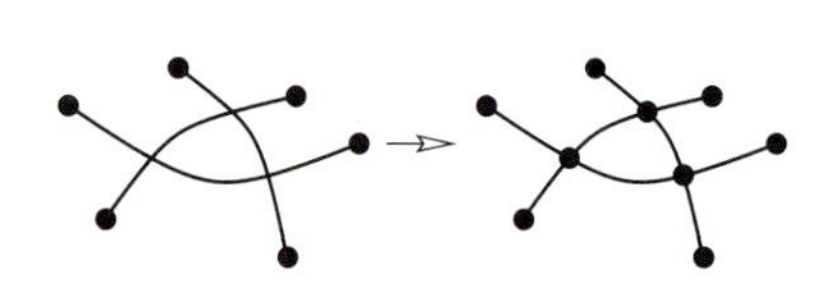

Suppose that G is drawn in the plane with $\text{cr}(G)$ crossings. We can immediately derive a lower bound on the number of crossings. Consider the following graph H: The vertices of H are those of G together with all crossing points, and the edges are all pieces of the original edges as we go along from crossing point to crossing point.

The new graph H is now plane and simple (this follows from our three assumptions!). The number of vertices in H is $n + \text{cr}(G)$ and the number of edges is $m + 2\mathrm{cr}(G)$, since every new vertex has degree 4. Invoking the bound on the number of edges for plane graphs we thus find

$$m + 2\,\mathrm{cr}(G) \;\le\; 3(n + \mathrm{cr}(G)) - 6,$$

that is,

$$\mathrm{cr}(G) \;\ge\; m - 3n + 6. \tag{4}$$

As an example, for the complete graph K_6 we compute

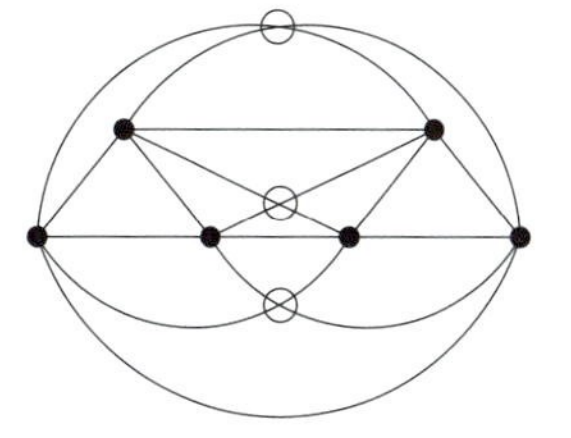

$$\mathrm{cr}(K_6) \ge 15 - 18 + 6 = 3$$

and, in fact, there is an drawing with just 3 crossings.

The bound (4) is good enough when m is linear in n, but when m is larger compared to n, then the picture changes, and this is our theorem.

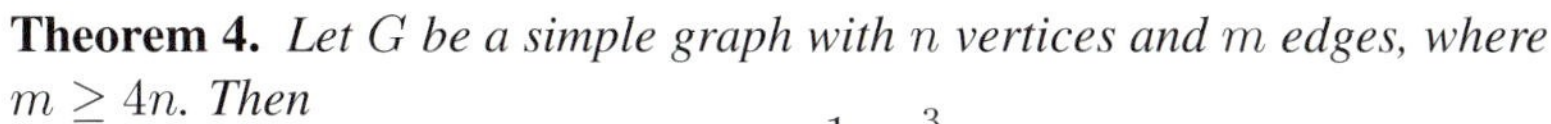

Theorem 4. *Let G be a simple graph with n vertices and m edges, where $m \ge 4n$. Then*

$$\mathrm{cr}(G) \;\ge\; \frac{1}{64}\frac{m^3}{n^2}.$$

The history of this result, called the *crossing lemma*, is quite interesting. It was conjectured by Erdős and Guy in 1973 (with $\frac{1}{64}$ replaced by some constant c). The first proofs were given by Leighton in 1982 (with $\frac{1}{100}$ instead of $\frac{1}{64}$) and independently by Ajtai, Chvátal, Newborn and Szemerédi. The crossing lemma was hardly known (in fact, many people thought of it as a conjecture long after the original proofs), until László Székely demonstrated its usefulness in a beautiful paper, applying it to a variety of hitherto hard geometric extremal problems. The proof which we now present arose from e-mail conversations between Bernard Chazelle, Micha Sharir and Emo Welzl, and it belongs without doubt in The Book.

■ **Proof.** Consider a minimal drawing of G, and let p be a number between 0 and 1 (to be chosen later). Now we generate a subgraph of G, by selecting the vertices of G to lie in the subgraph with probability p, independently from each other. The induced subgraph that we obtain that way will be called G_p.

Let n_p, m_p, X_p be the random variables counting the number of vertices, of edges, and of crossings in G_p. Since $\mathrm{cr}(G) - m + 3n \ge 0$ holds by (4) for *any* graph, we certainly have

$$E(X_p - m_p + 3n_p) \;\ge\; 0.$$

Now we proceed to compute the individual expectations $E(n_p)$, $E(m_p)$ and $E(X_p)$. Clearly, $E(n_p) = pn$ and $E(m_p) = p^2 m$, since an edge appears in G_p if and only if both its endvertices do. And finally, $E(X_p) = p^4\mathrm{cr}(G)$, since a crossing is present in G_p if and only if all four (distinct!) vertices involved are there.

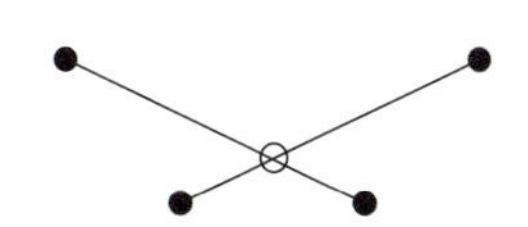

By linearity of expectation we thus find

$$0 \leq E(X_p) - E(m_p) + 3E(n_p) = p^4 \mathrm{cr}(G) - p^2 m + 3pn,$$

which is

$$\mathrm{cr}(G) \geq \frac{p^2 m - 3pn}{p^4} = \frac{m}{p^2} - \frac{3n}{p^3}. \tag{5}$$

Here comes the punch line: Set $p := \frac{4n}{m}$ (which is at most 1 by our assumption), then (5) becomes

$$\mathrm{cr}(G) \geq \frac{1}{64}\left[\frac{4m}{(n/m)^2} - \frac{3n}{(n/m)^3}\right] = \frac{1}{64}\frac{m^3}{n^2},$$

and this is it. □

Paul Erdős would have loved to see this proof.

References

[1] M. AJTAI, V. CHVÁTAL, M. NEWBORN & E. SZEMERÉDI: *Crossing-free subgraphs,* Annals of Discrete Math. **12** (1982), 9-12.

[2] N. ALON & J. SPENCER: *The Probabilistic Method,* Second edition, Wiley-Interscience 2000.

[3] P. ERDŐS: *Some remarks on the theory of graphs,* Bulletin Amer. Math. Soc. **53** (1947), 292-294.

[4] P. ERDŐS: *Graph theory and probability,* Canadian J. Math. **11** (1959), 34-38.

[5] P. ERDŐS: *On a combinatorial problem I,* Nordisk Math. Tidskrift **11** (1963), 5-10.

[6] P. ERDŐS & R. K. GUY: *Crossing number problems,* Amer. Math. Monthly **80** (1973), 52-58.

[7] P. ERDŐS & A. RÉNYI: *On the evolution of random graphs,* Magyar Tud. Akad. Mat. Kut. Int. Közl. **5** (1960), 17-61.

[8] T. LEIGHTON: *Complexity Issues in VLSI,* MIT Press, Cambridge MA 1983.

[9] L. A. SZÉKELY: *Crossing numbers and hard Erdős problems in discrete geometry,* Combinatorics, Probability, and Computing **6** (1997), 353-358.

26.2.98

About the Illustrations

We are happy to have the possibility and privilege to illustrate this volume with wonderful original drawings by Karl Heinrich Hofmann (Darmstadt). Thank you!

The regular polyhedra on page 66 and the fold-out map of a flexible sphere on page 74 are by WAF Ruppert (Vienna).

Jürgen Richter-Gebert provided two illustrations on page 68.

Page 203 features the Weisman Art Museum in Minneapolis designed by Frank Gehry. The photo of its west façade is by Chris Faust. The floorplan is of the Dolly Fiterman Riverview Gallery, which lies behind the west façade.

The portraits of Bertrand, Cantor, Erdős, Euler, Fermat, Herglotz, Hermite, Hilbert, Pólya, Littlewood, and Sylvester are from the photo archive of the Mathematisches Forschungsinstitut Oberwolfach, with permission. (Many thanks to Annette Disch!)

The portrait stamps of Buffon, Chebyshev, Euler, and Ramanujan are from Jeff Miller's mathematical stamps website http://jeff560.tripod.com with his generous permission.

The picture of Hermite is from the first volume of his collected works.

The photo of Claude Shannon was provided by the MIT Museum and is here reproduced with their permission.

The portrait of Cayley is taken from the "Photoalbum für Weierstraß" (edited by Reinhard Bölling, Vieweg 1994), with permission from the Kunstbibliothek, Staatliche Museen zu Berlin, Preussischer Kulturbesitz.

The Cauchy portrait is reproduced with permission from the Collections de l'École Polytechnique, Paris.

The picture of Fermat is reproduced from Stefan Hildebrandt and Anthony Tromba: *The Parsimonious Universe. Shape and Form in the Natural World*, Springer-Verlag, New York 1996.

The portrait of Ernst Witt is from volume 426 (1992) of the Journal für die Reine und Angewandte Mathematik, with permission by Walter de Gruyter Publishers. It was taken around 1941.

The photo of Karol Borsuk was taken in 1967 by Isaac Namioka, and is reproduced with his kind permission.

We thank Dr. Peter Sperner (Braunschweig) for the portrait of his father, and Vera Sós for the photo of Paul Turán.

Thanks to Noga Alon for the portrait of A. Nilli!

Index

Printing and bookbinding: Appl Druck GmbH & Co KG, Wemding